Advances in
Carbohydrate Chemistry and Biochemistry

Volume 67

Advances in Carbohydrate Chemistry and Biochemistry

Editor
DEREK HORTON

Board of Advisors

DAVID C. BAKER
DAVID R. BUNDLE
STEPHEN HANESSIAN
YURIY A. KNIREL
TODD L. LOWARY

SERGE PÉREZ
PETER H. SEEBERGER
J.F.G. VLIEGENTHART
ARNOLD A. STÜTZ

Volume 67

Amsterdam • Boston • Heidelberg • London • New York • Oxford
Paris • San Diego • San Francisco • Singapore • Sydney • Tokyo
Academic Press is an imprint of Elsevier

Academic Press is an imprint of Elsevier
The Boulevard, Langford Lane, Kidlington, Oxford, OX51GB, UK
32, Jamestown Road, London NW1 7BY, UK
Radarweg 29, PO Box 211, 1000 AE Amsterdam, The Netherlands
225 Wyman Street, Waltham, MA 02451, USA
525 B Street, Suite 1900, San Diego, CA 92101-4495, USA

First edition 2012

Copyright © 2012 Elsevier Inc. All rights reserved

No part of this publication may be reproduced, stored in a retrieval system or transmitted in any form or by any means electronic, mechanical, photocopying, recording or otherwise without the prior written permission of the publisher

Permissions may be sought directly from Elsevier's Science & Technology Rights Department in Oxford, UK: phone (+44) (0) 1865 843830; fax (+44) (0) 1865 853333; email: permissions @elsevier.com. Alternatively you can submit your request online by visiting the Elsevier web site at http://elsevier.com/locate/permissions, and selecting *Obtaining permission to use Elsevier material*

Notice
No responsibility is assumed by the publisher for any injury and/or damage to persons or property as a matter of products liability, negligence or otherwise, or from any use or operation of any methods, products, instructions or ideas contained in the material herein. Because of rapid advances in the medical sciences, in particular, independent verification of diagnoses and drug dosages should be made

ISBN: 978-0-12-396527-1
ISSN: 0065-2318

British Library Cataloguing in Publication Data
A catalogue record for this book is available from the British Library

Library of Congress Cataloging-in-Publication Data
A catalog record for this book is available from the Library of Congress

For information on all Academic Press publications
visit our website at store.elsevier.com

Printed and bound in USA

12 13 14 15   10 9 8 7 6 5 4 3 2 1

**Working together to grow
libraries in developing countries**

www.elsevier.com | www.bookaid.org | www.sabre.org

ELSEVIER   BOOK AID International   Sabre Foundation

CONTENTS

Contributors ix

Preface xi

NATHAN SHARON 1925–2011
David Mirelman, Edward A. Bayer, Yair Reisner

Combining Computational Chemistry and Crystallography for a Better Understanding of the Structure of Cellulose
Alfred D. French

I.	Introduction.	20
II.	Information from Crystals of Related Small Molecules	23
	1. Shape of the D-Glucopyranose Ring.	23
	2. Linkage Geometry	28
	3. Conversion to Polymer-Shape Notation	33
	4. Orientation of O-6	35
	5. Crystal Packing and Intermolecular Interactions.	37
	6. Summary of Section on Extrapolation	40
III.	Energy Calculations	41
	1. Results on Individual Isolated Molecules with Empirical Methods	45
	2. Results with Quantum Methods	51
	3. Assessment of ϕ/ψ Mapping	54
	4. Hydroxyl-Group Orientations.	56
IV.	Detection of New Stabilizing Interactions in Cellulose with Atoms-in-Molecules Theory	57
V.	Modeling Crystals of Cellulose	61
VI.	Conclusions.	68
	Appendix. Molecular Structure Drawings for Saccharide Analogues Having β-(1→4) Linkages. . . .	70
	References.	83

Strategies in Synthesis of Heparin/Heparan Sulfate Oligosaccharides: 2000–Present
Steven B. Dulaney and Xuefei Huang

I.	Introduction	96
	1. Background.	96
	2. Challenges in Synthesis of Oligosaccharides of Heparin and HS	98

II.	Linear Synthesis	105
	1. Solution Phase	105
	2. Polymer-Supported Synthesis	107
III.	Active–Latent Glycosylation Strategy	113
IV.	Selective Activation	118
V.	Reactivity-Based Chemoselective Glycosylation	121
VI.	Reactivity-Independent, Pre-Activation-Based, Chemoselective Glycosylation	123
VII.	Chemoenzymatic Synthesis	125
VIII.	Future Outlook	130
	Acknowledgments	130
	References	130

Chemical Synthesis of Glycosylphosphatidylinositol Anchors
BENJAMIN M. SWARTS AND ZHONGWU GUO

I.	Introduction	138
II.	Classic Approaches to GPI Synthesis	142
	1. Synthesis of a *Trypanosoma brucei* GPI Anchor by the Ogawa Group (1991)	142
	2. Synthesis of a Yeast GPI Anchor by the Schmidt Group (1994)	147
	3. Synthesis of a Rat Brain Thy-1 GPI Anchor by the Fraser-Reid Group (1995)	151
	4. Synthesis of a *T. brucei* GPI Anchor by the Ley Group (1998)	157
	5. Synthesis of a Rat Brain Thy-1 GPI Anchor by the Schmidt Group (1999, 2003)	162
	6. Synthesis of a Human Sperm CD52 Antigen GPI Anchor Containing an Acylated Inositol by the Guo Group (2003)	167
	7. Synthesis of a *Plasmodium falciparum* GPI Anchor Containing an Acylated Inositol by the Fraser-Reid Group (2004)	172
	8. Synthesis of a *P. falciparum* GPI Anchor Containing an Acylated Inositol by the Seeberger Group (2005)	177
	9. Synthesis of a *T. cruzi* GPI Anchor by the Vishwakarma Group (2005)	181
	10. Synthesis of a Fully Phosphorylated and Lipidated Human Sperm CD52 Antigen GPI Anchor by the Guo group (2007)	185
III.	Diversity-Oriented Approaches to GPI Synthesis	187
	1. Synthesis of *T. cruzi* GPI Anchors Containing Unsaturated Lipids by the Nikolaev Group (2006)	188
	2. Synthesis of a GPI Anchor Containing Unsaturated Lipid Chains by the Guo Group (2010)	193
	3. Synthesis of a Human Lymphocyte CD52 Antigen GPI Anchor Containing a Polyunsaturated Arachidonoyl Lipid by the Guo Group (2011)	199
	4. Synthesis of "Clickable" GPI Anchors by the Guo Group (2011)	203
	5. General Synthetic Strategy for Branched GPI Anchors by the Seeberger Group (2011)	207
IV.	Conclusions and Outlook	211
	Acknowledgments	212
	References	212

Effect of Protein Dynamics and Solvent in Ligand Recognition by Promiscuous Aminoglycoside-Modifying Enzymes
ENGIN H. SERPERSU AND ADRIANNE L. NORRIS

I.	Introduction	222
II.	Aminoglycoside Antibiotics	222
III.	Aminoglycoside-Modifying Enzymes	224
IV.	Thermodynamic Properties of Enzyme–AG Complexes	225
	1. Enthalpy, Entropy, and Gibbs Energy Changes for AG Binding	225
	2. Proton Linkage	229
	3. Heat-Capacity Changes	231
	4. Solvent Effects	234
V.	Protein Dynamics in Substrate Recognition and Substrate Promiscuity of AGMEs	237
VI.	Conclusions and Future Considerations	241
	Acknowledgments	243
	References	243
	AUTHOR INDEX	249
	SUBJECT INDEX	271

CONTRIBUTORS

Edward A. Bayer, Department of Biological Chemistry, Weizmann Institute of Science, Rehovot 76100, Israel

Steven B. Dulaney, Department of Chemistry, Michigan State University, East Lansing, Michigan, USA

Alfred D. French, Southern Regional Research Center, U.S. Department of Agriculture, New Orleans, Louisiana, USA

Zhongwu Guo, Department of Chemistry, Wayne State University, Detroit, Michigan, USA

Xuefei Huang, Department of Chemistry, Michigan State University, East Lansing, Michigan, USA

David Mirelman, Department of Biological Chemistry, Weizmann Institute of Science, Rehovot 76100, Israel

Adrianne L. Norris, Department of Biochemistry, Cellular and Molecular Biology, The University of Tennessee, Knoxville, Tennessee, USA

Yair Reisner, Department of Immunology, Weizmann Institute of Science, Rehovot 76100, Israel

Engin H. Serpersu, Department of Biochemistry, Cellular and Molecular Biology, The University of Tennessee, Knoxville, Tennessee, USA

Benjamin M. Swarts, Department of Chemistry, Wayne State University, Detroit, Michigan, USA

PREFACE

Crystallographic structure-determination of sugars was pioneered by George Jeffrey in his classic 1939 study of the monosaccharide glucosamine, which showed definitively the $^4C_1$ conformation of the pyranose ring. For polysaccharides, and in particular, the most abundant and widely used carbohydrate, namely, cellulose, a key advance in our knowledge was the 1937 crystallographic study by Meyer and Misch. Their work, building from 1926 proposals by Sponsler and Dore, demonstrated a plausible unit cell for this biopolymer and showed that cellulose is a (1→4)-linked polymer of glucopyranose residues. However, many detailed aspects of the structure of cellulose remain enigmatic even to this day. The present article by French (New Orleans) addresses questions of the fine structure of cellulose by drawing on a combination of crystallographic methodology and computational chemistry, utilizing, in particular, the three-dimensional structure of various cellobiose derivatives. French's contribution complements the recent articles in Volume 64 by Pérez and Samain focusing on structure and engineering of cellulose and that of Mischnik and Momcilovic on chemical derivatization of cellulose, as well as earlier work on cellulose detailed by Marchessault and Sundararajan in Volume 36, and by Jones in Volume 19.

The glycosaminoglycan heparin is a very complex carbohydrate biopolymer that has long been pharmacologically important as an anticoagulant. Its isolation from tissue sources and its properties were first surveyed in this series by Foster and Huggard in Volume 10. Our understanding of its detailed structure and manifold biological functions have presented many challenges, as documented by Casu in Volume 43 and by Casu and Lindahl in Volume 57, with particular emphasis on the pentasaccharide subunit that binds to antithrombin III in a key step of the coagulation process. Major advances in synthetic strategy in recent years have enabled the chemical synthesis of a pentasaccharide drug, fondaparinux (Arixstra®), that is clinically effective for the treatment of deep-vein thrombosis. Dulaney and Huang (East Lansing, Michigan) present in this volume a detailed comparative overview of strategies, reported during the most recent decade, for the chemical synthesis of related pentasaccharide subunits of heparin and heparan sulfate, including both solution-phase and solid-support approaches, the active–latent strategy, and chemoenzymatic methodology.

Another area where synthetic carbohydrate chemists have displayed spectacular virtuosity is in the construction of the glycosylphosphatidylinositol (GPI) glycolipids that anchor cell-surface proteins and glycoproteins to the plasma membrane of eukaryotic cells. Earlier work on the GPI anchor of the protozoan parasite *Trypanosoma cruzi* was described by Lederkremer and Agusti in Volume 62. In this volume,

Swarts and Guo (Detroit, Michigan) survey the intensive current efforts of research groups, including their own and by such noted synthetic investigators as Ogawa, Schmidt, Fraser-Reid, Ley, Seeberger, Vishwakawa, and Nikolaev, to synthesize a range of intact GPI anchors. These anchors play important roles in many biological and pathological events, including cell recognition and adhesion, signal transduction, host defense, and acting as receptors for viruses and toxins. Chemical synthesis of structurally defined GPI anchors and related analogues is a key step toward understanding the properties and functions of these molecules in biological systems and exploring their potential therapeutic applications.

The final article, by Serpersu and Norris (Knoxville, Tennessee), surveys dynamic and thermodynamic aspects of the interactions between various aminoglycoside-modifying enzymes (AGMEs) and their antibiotic ligands. It complements the fundamental articles in Volume 30 by the Umezawa brothers on the structure of the aminoglycoside antibiotics and the mechanisms whereby mutant organisms resist such antibiotics, along with the studies reported in Volume 60 by Willis and Arya on aminoglycoside–nucleic acid recognition. The Tennessee authors stress the key role of two important aminoglycoside-modifying enzymes in their complexation with the aminoglycosides, as studied by the thermodynamics of ligand–protein interactions and the dynamic properties of the protein.

A tribute to the late Nathan Sharon is presented by three of his students and colleagues, David Mirelman, Edward A. Bayer, and Yair Reisner (Rehovot, Israel). Sharon was a major pioneer in carbohydrate biochemistry, and his outstanding contributions to our knowledge of lectins are particularly noteworthy. He was a long-time member of the Board of Advisors of this series.

Stephen J. Angyal, for many years a member of the Board of Advisors and a noted expert on carbohydrate stereochemistry, died on May 14, 2012, and Hassan S. El Khadem, a frequent contributor to this series on nitrogen heterocycles derived from sugars, passed away on May 20, 2012.

DEREK HORTON

Washington, DC
April 2012

NATHAN SHARON

1925–2011

Nathan Sharon, who passed away on June 17, 2011, was one of the world's leading investigators in the field of glycobiology. He was among the first to study and characterize plant lectins, the carbohydrate-binding proteins, with associated hemagglutinating activities, which were later found to have important roles in all forms of life. In addition, he investigated the spatial structures of numerous glycoconjugates. Sharon was a prolific writer, with over 400 publications on the subject, many of which have become citation classics and serve as basic library textbooks. Sharon was President of the International Glycoconjugate Organization from 1981 until 2002 and served for many years on the Editorial Boards of many leading scientific periodicals, including *Advances in Carbohydrate Chemistry and Biochemistry*, *Carbohydrate Research*, *Glycoconjugate Journal*, and *Glycobiology*. At the Weizmann Institute of Science, where Sharon was among the first group of young scientists recruited in the early 1950s, he worked diligently and cared for the Institute for almost 60 years until his demise. Sharon also held several important positions at the Institute: for many years, he was Head of the Department of Biophysics, he then served as Dean of the Faculty of Biochemistry and was an elected President of the Academic Council of the Institute. Sharon was awarded many important prizes and awards, such as The Israel Prize in Biochemical Research and the Rosalind Kornfeld Award for Lifetime Achievement in Glycobiology. Sharon was a member of the Israeli Academy of Sciences and Humanities and a Honorary member of the American Society of Biological Chemists, the American Society for Microbiology, and the Oxford Glycobiology Institute. He held important appointments in many distinguished international centers of research. Among others he served as a Fogarty Scholar in Residence at the National Institutes of Health in Bethesda, Maryland, as a Visiting Professor at the College de France in Paris, and as a Distinguished Visiting Scientist at the Roche Institute for Molecular Biology in Nutley, New Jersey. For many years, Nathan also chaired the Batsheva de Rothschild Fund for the Advancement of Science in Israel.

The Early Years

Nathan Sharon (formerly Shtrikman) was born in Brisk, Poland, on November 4, 1925, and as the Nazis began gaining power in Germany and central Europe, his parents immigrated with him to Palestine in 1934 and settled in Tel Aviv, where he graduated from high school in 1943. He spent several years after high school in active service in the Jewish underground, the Palmach, and following the establishment of the State of Israel in 1948, he joined the Scientific Unit of the Israel Defense Forces. During these years, he also studied chemistry at the Hebrew University in Jerusalem. In 1950, he earned his M.Sc. degree in biochemistry, performing his thesis work on the effects of citric acid in milk products at the laboratories of the Agricultural Research Center in Beit Dagan. He was then accepted by the late Prof. Aharon Katchalski-Katzir to commence his Ph.D. work at the Weizmann Institute and studied under Prof. Katchalski-Katzir's guidance the kinetics of reactions between the reducing groups of sugar moieties and amino groups of proteins and peptides, a subject that Prof. Katchalski-Katzir had begun investigating in the late 1930s. In 1953, Sharon received his Ph.D. degree, which at that time was still officially awarded by the Hebrew University of Jerusalem to graduate students who had successfully completed their thesis work at the Weizmann Institute.

In 1954, he joined the faculty of the Weizmann Institute and, in 1956, he received a Weizmann postdoctoral grant for studies on the mechanisms of protein biosynthesis[1,2] in the laboratory of Nobel laureate Prof. F. Lipman in the Biochemical Research Laboratory of the Massachusetts General Hospital and Harvard Medical School, in Boston, Massachusetts. In 1957, Sharon elected to move to the laboratory of Prof. Roger W. Jeanloz,[3] a pioneer in research on complex carbohydrates, also working at Massachusetts General. In Jeanloz' laboratory, Sharon isolated an unusual diamino sugar from the capsular polysaccharide of *Bacillus subtilis*.[4] Before returning to Israel, Sharon spent an additional year (1958) with Prof. Dan Koshland, Jr., at Brookhaven National Laboratory in uptown New York, working on the mechanism of action of myosin ATPase.[5,6] Upon his return to Israel in 1959, Sharon's first goal was to establish the structure of the diamino sugar that he had isolated in Boston and which he had named bacillosamine.[7] For that purpose, he received an NIH grant, even though nothing was known at the time about the function of the compound. In addition, he also became interested in understanding the mechanism of action of lysozyme, the hen egg-white enzyme, which hydrolyzes the peptidoglycan of bacterial cell wall and also chitin.[8–11]

Love Affair with Soybean Agglutinin and Other Lectins

Sharon's interest in glycoproteins was aroused during the course of investigations of soybean proteins, which he began in the framework of a long-term grant received in

1961 from the U.S. Department of Agriculture, jointly with Prof. Ephraim Katchalski-Katzir, who was the founding Head of the new Department of Biophysics at the Weizmann Institute, where Sharon served as a young faculty member. Prof. Ephraim Katchalski-Katzir was the younger brother of Prof. Aharon Katchalski-Katzir, who was tragically killed in a terrorist attack at Ben Gurion airport in Israel in 1972. The aim of the grant was to conduct fundamental investigations of soy proteins and provide information for their better utilization in human nutrition.

Sharon and his newly recruited research associate, Dr. Halina Lis, who subsequently worked closely with him for over 40 years, chose to focus on the soybean agglutinin protein (SBA) that had been originally isolated and characterized in the 1950s by I. Liener, who had also shown that it contained glucosamine. Since little was known at that time about glycoproteins, the fact that a soy protein was found to contain an amino sugar sparked Sharon's interest, and its study became his main focus.[12] They soon found that SBA contained not only glucosamine but also mannose, and they then isolated from a proteolytic digest of SBA, an asparaginyl-oligosaccharide that contained all the N-acetylglucosamine and mannose residues of the protein.[13] Eventually, they also isolated and characterized the carbohydrate–peptide linking group, namely N-acetylglucosamine–asparagine,[14] which turned out to be identical to that in the original glycopeptides obtained from ovalbumin by Albert Neuberger, another founding father of modern glycoprotein research.[15] The finding of a glycosylated protein from a plant source created great interest, because until that time it was thought that glycoproteins were strictly of animal origin and that any finding of a sugar in a plant protein was most likely due to a non-covalent contamination. Many years later, the complete structure of the carbohydrate of SBA was established by NMR, in joint work with Hans Vliegenthart of the University of Utrecht, as a branched oligomannoside: $Man_9(GlcNAc)_2$,[16] a structure also found in animal glycoproteins. This finding demonstrated that N-glycosylation is a process conserved in both animals and plants. Another unique feature of SBA was that, in contrast to other glycoproteins, all of the SBA molecules carry the same oligosaccharide,[17] and consequently, SBA serves as an excellent source of this glycoconjugate.

One of the most intriguing questions at that time was whether the lectin's specific receptor sites for carbohydrates were, as then suggested, similar to the binding sites of antibodies. Sharon and a young talented graduate student, the late Reuben Lotan, continued to investigate and characterize soy bean agglutinin, which they were able to isolate in quantities by affinity-chromatography methods, many of which were developed at the Department of Biophysics following the examples pioneered by Meir Wilchek and his colleagues.[18] They first demonstrated that SBA is a tetramer, made up of four nearly identical subunits, and that the carbohydrate moieties of SBA were not essential for its biological activity, since they were able to show that the carbohydrate-free SBA, produced in a bacterial expression system,[19] retained its full agglutinating activity for red blood cells.[20,21] Analysis of the amino acid composition of SBA showed that it is similar to that of lectins from other legumes. In particular, it was devoid of sulfur-containing amino acids, in striking contrast to wheat germ agglutinin (WGA), which is rich in such residues. These findings demonstrated that lectins are a diversified group of proteins with respect to size, structural composition, and carbohydrate recognition. Crystals of SBA suitable for X-ray diffraction studies were obtained by Boaz Shaanan and colleagues at Weizmann's Department of Structural Chemistry in 1984,[22] but the high-resolution structure of the lectin, complexed with a ligand, was solved only a decade later by James Sacchettini and coworkers at Albert Einstein College of Medicine, New York.[23]

Sharon's findings on the properties of SBA sparked his interest in additional plant lectins. During the summer of 1971, he spent a few months in the laboratory of Albert Neuberger in London, where he purified wheat germ agglutinin by affinity chromatography and demonstrated that this lectin was not a glycoprotein.[24] He also found that WGA, along with the enzyme lysozyme from egg white, had a very strong affinity for bacterial peptidoglycan, sialylated glycoproteins, and oligosaccharides of chitin. Another lectin, purified and characterized by Reuben Lotan in Sharon's laboratory, was peanut agglutinin (PNA).[25] Its specific carbohydrate-recognition moiety, α-Gal-$(1 \rightarrow 3)$-GalNAc (also known as T-antigen, and a characteristic glycan of O-glycoproteins), was established in collaboration with Miercio Pereira and Elvin Kabat at Columbia University.[26] The carbohydrate specificities of PNA and SBA identified by Sharon and his colleagues became very important features for their subsequent applications, as described in a later section.

Another interesting lectin from a legume was identified in Sharon's laboratory by a young Uruguayan medical student, Jose Iglesias. It came from the seeds of a beautiful tree *Erythrina cristagalli* (ECL) brought by Iglesias from Uruguay in his bag. Iglesias purified the ECL lectin by affinity chromatography and found it to be a glycoprotein containing fucose and xylose, with a recognition specificity for galactose.[27] The carbohydrate structure of the ECL lectin was established in collaboration with the laboratory of Raymond Dwek in Oxford as the branched Asn-linked heptasaccharide α-Man-$(1 \rightarrow 3)$-[α-Man-$(1 \rightarrow 6)$]-[β-Xyl-$(1 \rightarrow 2)$]-β-Man-$(1 \rightarrow 4)$-β-GlcNAc-$(1 \rightarrow 4)$-[α-Fuc-$(1 \rightarrow 3)$]-GlcNAc.[28] It was one of the first reported examples of a plant-specific oligosaccharide that is allergenic to humans. Once the supply of *E. cristagalli* seed-flour that Iglesias had brought from Uruguay was exhausted, Sharon turned his attention to an analogous lectin from *Erythrina corallodendron* (ECorL), a coral tree that grows commonly in Israel; its lectin was originally isolated by Nechama

Gilboa-Garber in Israel.[29] Working on ECorL lectin proved to be highly rewarding because its primary sequence turned out to be homologous to that of other legume lectins.[30] Sharon and his colleagues also investigated how the lectin of ECorL combines with carbohydrates; they performed extensive site-directed mutagenesis in a recombinant form of the lectin.[31,32] Taking also into consideration the three-dimensional structure of the ECorL lectin–ligand complex,[33–35] they concluded that a constellation of three key amino acids, namely aspartic acid, asparagines, and an aromatic residue, are essential for the binding of a terminal galactose moiety.[36,37] Interestingly, they also found that a similar constellation exists in the lectin concanavalin A (ConA) for the binding of mannose. They proposed that homologous lectins having different specificities might bind different saccharides by a similar set of invariable residues that are identically positioned in their tertiary structures. This finding provided further evidence for the hypothesis that Sharon and his colleagues had made in the 1970s,[38] stating that, despite their distinct sugar specificities, legume lectins are members of a single protein family, and the genes coding for them have a common ancestry, and that all of the sequences of legume lectins known (>200) exhibit considerable homology.

Biological Effects of Plant Lectins

Until 1960, when Peter Nowell discovered that phytohemagglutinin (PHA), the lectin from red kidney beans, can act as a mitogen for lymphocytes, it was believed that lymphocytes were end-of-the-line cells that could neither divide nor differentiate.[39] Within a short time, several other lectins were demonstrated to be mitogenic, including the lectin WGA which, as already described, Sharon had purified in Neuberger's laboratory in London and shown it not to be a glycoprotein.[24] Interestingly, as had been previously shown with ConA and low concentrations of mannose, the mitogenic stimulation of lymphocytes by WGA could be inhibited and reversed by low concentrations of chitin oligosaccharides. Accordingly, it was concluded that mitogenic stimulation is the result of the binding of lectins to surface sugars on the lymphocytes, an observation indicating an important biological function.

Another interesting consequence of the mitogenic stimulation of lymphocytes was the finding that they produce a number of lymphokines and cytokines, and that the levels produced by lymphocytes from immunosuppressed patients were much lower. This dramatic difference in the response to PHA-induced lectin stimulation enabled the development of important clinical tests for the assessment of patients' immunocompetence. Studies on the mitogenicity of lectins provided very important information on the cell-surface sugars of lymphocytes and demonstrated differences among various lectins. Sharon and his colleagues found that SBA was mitogenic only in a polymerized form.[21] Furthermore, mitogenicty was evident only after removal of the sialic acid residues, a step that exposed the subterminal galactose and N-acetylgalactosamine residues of the cell-surface glycoproteins and glycolipids, and they demonstrated that the mitogenicity of PNA also required initial removal of the terminal sialic acid residues.[40]

At that time, there were two interesting reports, one by Max Burger from Basel[41] and the second by the group of Leo Sachs at the Weizmann Institute,[42] who found that such lectins as ConA bound to and agglutinated cancer cells or transformed mammalian cells, but not normal parental cells. This report suggested that cancer might be associated with a change in the cell-surface sugars, and Sharon and his colleagues soon discovered that SBA, which has a specificity for galactose and N-acetylgalactosamine, also possessed the remarkable ability to distinguish between normal and malignant cells.[43] However, not all lectins were found to agglutinate malignant cells. Some of them did not because of defects that were found in their N-linked carbohydrate units. PNA was another lectin that agglutinated cells only after their treatment with neuraminidase, a step that exposed galactosyl sites.[40]

Lectins as Tools for the Study of Membranes and Separations of Cells

Interest in lectins and the characterization of their binding affinity to different carbohydrate components intensified in the late 1960s with the realization that they could be extremely valuable reagents for the investigation of different cell-surface sugars. Interest was also growing in understanding the role of different lectins in cell growth and differentiation, as well as in the interactions of cells with their environment and in a variety of pathological processes. Nowell's report[39] that cell-surface carbohydrates can serve as carriers of biological information, as well as Burger's[41] and Sachs'[42] demonstrations that the agglutination of transformed cells by lectins is markedly enhanced as compared to their normal counterparts, raised the hope that a major defect of cancer-cell membranes could be identified.

At that time, around the spring of 1974, one of us (Y. Reisner) joined Sharon's laboratory for Ph.D. studies. This was an exciting era in Nathan's laboratory, during which time he made his great breakthroughs in lectin research and revolutionalized our understanding of the role of these proteins in diverse biological systems. His laboratory flourished under the direction of his chief assistant, Halina Lis, a brilliant intellectual who coauthored several key scientific reviews that have evolved into citation classics. There was also an extraordinary group of bright students, including Mel Schindler, Aya Pruzansky-Jakobovits, and the late Vivian Teichberg and Reuben

Lotan—both of whom passed away prematurely in 2011. All of these colleagues went on to enjoy great success during the course of their scientific careers.

Being Nathan's student was indeed a pleasure for me, Yair Reisner. Nathan's scholarliness, and the encyclopedic knowledge resulting from his phenomenal memory, was truly impressive and indispensable in those pre-internet days. In addition, his great curiosity and pragmatism permitted his students to take initiatives and stray away from rigid, well-accepted dogmas. This was especially true for my Ph.D. studies, when Nathan and I discovered, by chance, an interesting role of carbohydrates in the differentiation of the immune system.

Fascinated by the new hypothesis that sugars of the cancer-cell membrane are defective, I tried to address the lack of data on tumor cells grown *in vivo* because of the absence of appropriate, normal controls. This problem was, and still remains, a major obstacle for such comparative studies. Therefore, I chose to test the agglutinability by lectins of particular thymoma cells that develop spontaneously in aging AKR mice, and for which the normal counterpart was known to be medullar mature thymocytes. To my disappointment, none of the different lectins used, including those from wax beans, *Lotus tetragonolobus*, and soybeans (all purified in Sharon's laboratory), were able to discriminate between the tumor cells and their normal counterparts. Fortunately, PNA had just been purified by the late Reuben Lotan, and he kindly gave me his first batch of the purified lectin. As with the other lectins, PNA did not distinguish between the thymoma cells and their normal counterparts. However, a marked difference was observed between mature and immature normal thymocyte subpopulations, and this difference could be eliminated upon treatment of the mature subpopulation with neuraminidase.[44] This observation, which became a "Citation Classic," served as the basis for our isolation method of the two thymocyte subpopulations, providing good yields and excellent viability of the cells, which are important parameters for studying T-cell maturation routes within the thymus. In addition, this study led to the hypothesis that sialylation of the PNA receptor may be an important step in the maturation of the thymic cells. Indeed, when returning from my postdoctoral studies, my first graduate student, Avishag Toporowicz, demonstrated that the mature thymocytes express higher levels of sialyltransferase.[45]

Years later, this hypothesis was confirmed by evidence presented by Linda Baum and her coworkers from the University of California at Los Angeles and University of California at San Diego, which showed that regulated expression of a single glycosyltransferase, a Gal-(1→3)-GalNAc(2-3)-sialyltransferase, can account for this change in glycosylation.[46] This enzyme sialylates β-Gal-(1→3)-GalNAc, the preferred ligand of PNA, forming the sequence NeuAc-(2→3)-β-Gal-(1→3)-GalNAc, and thus masking the PNA-binding sites. The expression of this enzyme was found to be inversely proportional to that of the PNA receptor. Interestingly, Ellis Reinherz was subsequently able to show that the CD8 glycoprotein, known to be important for T-cell receptor function, is dramatically affected by the sialylation process. This is because of the critical role of the negative charge contributed by the sialic acid component for the CD8 interaction with the MHC–peptide complex on target cells.[47] The PNA receptor continues to serve as a differentiation marker for lymphocytes.[48] It has also been employed for similar purposes in other systems, for example, in embryonal carcinoma cells of mice, as shown by our joint study with Francois Jacob and Gabriel Gachelin of the Pasteur Institute in Paris.[49]

However, the most important outcome of these early studies has perhaps been their exciting clinical application in the field of bone-marrow transplantation. Using the same approach, but with soybean agglutinin, we were able to separate stem cells of mouse bone marrow from the unwanted, life-threatening T cells present in bone-marrow preparations. These cells can cause a lethal complication known as graft-versus-host disease (GVHD).[50] This observation led to my postdoctoral studies with the late Robert Good, one of the fathers of modern immunology and president at the time of Sloan–Kettering Cancer Center, and with Richard O'Reilly who was the head of the bone-marrow transplantation team at the Center. After many hurdles, we were finally able in 1983 to demonstrate[51] that transplantation of bone marrow, depleted of T cells by differential agglutination with SBA, could completely cure three children afflicted with severe combined immune deficiency ("bubble children"), without any GVHD. Following this initial proof-of-concept study, hundreds of infants, for whom matched sibling donors were not available, were successfully treated by this method in many centers throughout the world.

Lectins Involved in Host—Microbial Recognitions

Following the arrival at the Weizmann Institute in 1975 of Itzhak Ofek, a young postdoctoral fellow interested in host–microbial pathogen interactions, Sharon also became interested in lectins of microbial origin. Sharon and Ofek were intrigued by the earlier report[52] by J.P. Duguid from Dundee, who had shown that certain *Escherichia coli* strains could agglutinate erythrocytes and that this phenomenon was inhibited by adding mannose. As a former graduate student of Sharon and a young faculty member in the same department investigating the biosynthesis of bacterial cell-wall components, this writer (D. Mirelman) also became interested in host–pathogen interactions. Together with Ofek and Sharon, we demonstrated that type 1-fimbriated *E. coli* cells (FimH) readily adhered to buccal epithelial cells. This binding could be inhibited by mannose and methyl α-D-mannoside, as well as by pretreatment of the epithelial cells with ConA, but not by other lectins, such as SBA.[53]

Physical shaking of the *E. coli* cells released a proteinous fraction that had an affinity to mannose and was capable of agglutinating red blood cells or yeasts. We concluded that mannose, a common sugar component of mammalian cell-surfaces and of yeasts, acts as a receptor for type 1-fimbriated *E. coli*, which in turn functions as a lectin. Together with N. Firon, we then found that aryl α-glycosides of mannose, such as 2-chloro-4-nitrophenyl α-D-mannopyranoside, were three orders of magnitude more powerful than methyl α-mannoside in inhibiting agglutination of red blood cells or adherence of the fimbriated bacteria to buccal epithelial cells. This observation suggested that the combining site on the FimH is extended with an adjoining hydrophobic region.[54,55] This hypothesis was later confirmed by others by using X-ray crystallography and modeling of FimH–ligand complexes, which revealed interaction of the aglycones with aromatic side chains close to the mannose-binding site.[56]

Since those early studies, there has been much progress in identifying the specificity of surface lectins of different bacterial strains and species. The variety and complexity of carbohydrate specificities found in many of these bacterial lectins appear to serve as key determinants for different animal tropisms and for their susceptibility to infection by a particular bacterium. The idea that inhibition of the *E. coli* mannose-binding FimH lectin by mannosides could be used to prevent a bacterial infection *in vivo* was demonstrated, in conjunction with Moshe Aronson, in a mouse model for bladder infection.[57] Mice bladders that had been infected with fimbriated *E. coli* cells suspended in a solution of methyl α-mannoside were found to have markedly fewer adhered bacteria than those infected with the same bacterial cells suspended in glucose or saline solutions.

The finding that blocking of the lectin, which mediates the initial adhesion stage, could prevent microbial colonization and the subsequent onset of infection has since been demonstrated for many other microorganisms in different animal models. This finding has generated interest in using such an anti-adhesion principle as a therapy for certain human infections. Sharon and Ofek also proposed that lectin recognition of cell-surface sugars plays an important role not only in the infective process but apparently also in innate immunity. This is because bacterial lectins can also bind to carbohydrate receptors on such phagocytic cells as granulocytes and macrophages in the absence of opsonins, a process that often leads to the uptake and killing of the microorganism.[58] In a similar way, lectins on the surfaces of the phagocytic cells can sometimes bind to carbohydrates on the surface of the microorganism, also causing their uptake. Sharon and Ofek named this phenomenon lectino-phagocytosis,[59,60] and it appears to be an early example of an important innate mechanism of defense.

Lectins from legumes were also found to play important roles in plant defenses against soil microorganisms. Together with Sharon, Ezra Galun and the late Reuben

Lotan, we demonstrated that WGA was capable of inhibiting the growth of such fungi as *Trichoderma*, *Penicillia*, and *Aspergilli* by binding specifically to the exposed chitin layer at the tip of the growing hyphae.[61,62]

The experiences I (David Mirelman) shared with Sharon on the role of lectins in bacterial interactions with human cells also led to later investigations on the role of surface lectins in the interactions of the protozoan parasite *Entamoeba histolytica* of the human intestine with the colonic mucosal cells and the intestinal bacterial flora. We found that the amoebic Gal/GalNAc-specific surface lectin, which was discovered by Ravdin and colleagues[63,64] as the adhesin that anchors the parasite to Gal/GalNAc-containing receptors on mucosal cells, also mediates the phagocytic uptake of bacterial cells or fungal cells that possess Gal/GalNAc residues on their surface glycoconjugates.[65] In addition, we found that bacteria having surface lectins that have an affinity toward mannose or glucose bind strongly to the surface of the amoeba cell and are rapidly ingested.[66,67] A remarkable observation was that bacteria lacking surface lectins specific for mannose or glucose, such as *Shigella flexneri* or many gram-positive bacteria that do not have Gal/GalNAc residues on their surfaces, do not interact with the amoeba and are not significantly ingested.[68] On the other hand, we found that the virulence of the parasite toward mammalian cells depends very much on the type of bacteria ingested by the amoeba. The bacteria ingested via the Gal/GalNAc–lectin pathway lead to a decrease in virulence through blocking of the Gal/GalNAc–lectin, whereas most of those ingested because of the mannose-binding lectins increased the virulence of the parasite.[69,70]

Animal Lectins in Cell Recognition

Acceptance of the concept that lectins could also have functions in higher organisms (mammalian, avian, reptilian, and so on) was quite slow and had to await the isolation and characterization of the galactose-specific lectin of the liver by G. Ashwell.[71] Sharon and his two graduate students, the late Vivian Teichberg and Gad Reshef (the latter was killed in the Yom Kippur war in 1973), isolated the first members of the soluble galactose-specific lectins from the electric eel (*Electrophorus electricus*) and from chicken muscle.[72] These lectins were termed galectins, and several of them were later found to be mediators of cellular interactions. One of these, galectin-3, which was found to be involved in the metastatic spread of B16 melanoma cells, was isolated by Sharon's former graduate student, Reuben Lotan.[73]

These early discoveries opened a very active field, and numerous other mammalian lectins have since been discovered, many of which are cell-surface constituents, strategically located to serve for recognition and interaction. It is noteworthy that the molecular arrangement present in the combining sites of legume lectins is also found in certain animal lectins. Comparison of the combining sites of lectins from various sources, including animals, led to the conclusion that structurally different lectins having similar specificities may bind the same saccharide by different sets of combining-site residues, most commonly by hydrogen bonds (some mediated by water) and hydrophobic interactions.[37] All of the foregoing strongly suggests that lectins are products of convergent evolution. Sharon's numerous insightful reviews[37,74–78] on the role of lectins have contributed significantly in convincing the scientific community that mammalian lectins and their specific glyconjugate receptors play a very important role in the modulation of cell–cell and cell–matrix interactions in physiological and pathological conditions.

Teacher, Educator, Collector, Swimmer, and Curious Traveler

In his memoirs, Sharon praises his mentor, Prof. Ephraim Katchalski-Katzir, who for many years was the head of the Biophysics Department, a position Sharon later took on when Prof. Katzir was elected President of the State of Israel. In this context, Sharon wrote: "I am a great believer in serendipity and luck, much more so than in pre-meditated science. Looking back you could build a theory or hypothesize that you came upon a scientific discovery intentionally, and not accidentally, but the truth is that my choice of research area was coincidental in itself. What persuaded me? Curiosity. I was curious. I wanted to know about sugar-containing proteins, glycoproteins, little was known about these in the 1960's and 1970's. I wanted to know the role of sugars in nature. Ephraim Katzir gave me a carte blanche."

Sharon educated generations of younger scientists in the intricacies of complex carbohydrates and lectins. Every year he gave the basic courses in glycobiology at both the Feinberg Graduate School of the Weizmann Institute and Tel Aviv University, and he mentored over 40 graduate students and postgraduate fellows. As already mentioned, Sharon provided all of them with a great deal of freedom and encouraged them to follow their curiosity. Many times, this led to very important discoveries. Sharon then tirelessly probed and critically discussed their results with them and suggested additional experiments and controls to confirm their observations.

Sharon was an excellent and very pedantic writer. In the days before the advent of the computer and word-processing applications, he painstakingly prepared handwritten manuscripts along with his students, and these were then edited and typewritten by his devoted secretaries, Mrs. Dvorah Ochert and later Mrs. Edna Agmon. Afterward, he continued to introduce corrections into the initial draft and rewrote numerous additional revisions of the manuscript until he was eventually satisfied with a final version.

Sharon also gave an annual course on scientific writing to graduate students, and he firmly believed that research scientists should always try to explain and educate lay audiences in novel breakthroughs of science. In this context, he served as a role model through his many years as editor of the Israeli monthly popular-science journal *Mada*, by writing weekly columns on science in the daily newspaper *Haaretz*, by contributing articles to *Scientific American*, and by hosting a weekly hour on *What is New in Science* at the Israeli national radio station.

Nathan Sharon was blessed with an exemplary family life. He is survived by his wife Rachel, who was a veteran schoolteacher, and with whom he was married for over 63 years. They had two daughters, Esther, an economist who has worked for many years in the Ministry of Finance and was Deputy Director of State Revenue, and Osnat, a hematologist in a leading hospital in Israel and the mother of their three grandchildren. Nathan had a number of hobbies. He loved to travel with his wife to exotic places and return with hundreds of photographs, which he was very proud to show. He was an avid stamp collector, always asking for stamps from anyone he met. He was also an ardent lap-pool swimmer, and he always searched for a swimming pool wherever he traveled. On numerous occasions, he participated in the popular annual swim across the Sea of Galilee in Northern Israel.

As is clear from Sharon's very long list of publications, which spans from 1952 to 2011 (see Pubmed), he interacted during his career with hundreds of collaborators and colleagues, with whom he shared his passion for lectins, carbohydrates, and glycoconjugates, and he bonded with them in lifelong friendships. Unfortunately, because of the limited scope of this tribute, we could not list and mention all of the work he did with his colleagues and collaborators. Many of these colleagues, friends, and former students, including the three of us, gathered together from all over the world to celebrate his 80th birthday during a 2-day symposium (November 23–24, 2005) entitled: "Half a Century at the Carbohydrate-Protein Interface," which was held in his honor at the Weizmann Institute in Rehovot. Nathan, as everybody called him, will be remembered by all of his friends and colleagues with great affection and with much respect as a leading pioneer who significantly advanced research in modern glycobiology.

DAVID MIRELMAN, EDWARD A. BAYER, YAIR REISNER

Note on cause of death

In December 2009, Nathan Sharon suffered a massive heart attack. He was operated on and underwent several arterial bypasses which saved his life, but his heart muscles were weakened and his recovery was slow. He passed away on June 17, 2011.

References

1. M. B. Hoagland, P. C. Zamecnik, N. Sharon, F. Lipmann, M. P. Stulberg, and P. D. Boyer, Oxygen transfer to AMP in the enzymatic synthesis of the hydroxamate of tryptophan, *Biochim. Biophys. Acta*, 26 (1957) 215–217.
2. N. Sharon and F. Lipmann, Reactivity of analogs with pancreatic tryptophan-activating enzyme, *Arch. Biochem. Biophys.*, 69 (1957) 219–227.
3. N. Sharon and R. C. Hughes, Roger W. Jeanloz 1917–2007, *Adv. Carbohydr. Chem. Biochem.*, 63 (2010) 2–19.
4. N. Sharon and R. W. Jeanloz, The diaminohexose component of a polysaccharide isolated from *Bacillus subtilis*, *J. Biol. Chem.*, 235 (1960) 1–5.
5. H. M. Levy, N. Sharon, and D. E. Koshland, Purified muscle proteins and the walking rate of ants, *Proc. Natl. Acad. Sci. U.S.A.*, 45 (1959) 785–791.
6. H. M. Levy, N. Sharon, E. Lindemann, and D. E. Koshland, Properties of the active site in myosin hydrolysis of adenosine triphosphate as indicated by the O^{18}-exchange reaction, *J. Biol. Chem.*, 235 (1960) 2628–2632.
7. U. Zehavi and N. Sharon, Structural studies of 4-acetamido-2-amino-2,4,6-trideoxy-D-glucose (*N*-acetylbacillosamine), the *N*-acetyl diaminosugar of *Bacillus licheniformis*, *J. Biol. Chem.*, 248 (1973) 433–438.
8. D. M. Chipman and N. Sharon, Mechanism of lysozyme action, *Science*, 165 (1969) 454–465.
9. D. Mirelman and N. Sharon, Isolation and characterization of the disaccharide *N*-acetyl-glucosaminyl-β(1→4)-*N*-acetylmuramic acid and two tripeptide derivatives of this disaccharide from lysozyme digests of *Bacillus licheniformis* ATCC 9945 cell walls, *J. Biol. Chem.*, 243 (1968) 2279–2287.
10. N. Sharon, T. Osawa, H. M. Flowers, and R. W. Jeanloz, Isolation and study of the chemical structure of a disaccharide from *Micrococcus lysodeikticus* cell walls, *J. Biol. Chem.*, 241 (1966) 223–230.
11. V. I. Teichberg and N. Sharon, A spectrofluorometric study of tryptophan 108 in hen egg-white lysozyme, *FEBS Lett.*, 7 (1970) 171–174.
12. N. Sharon and H. Lis, Lectins: Cell agglutinating and sugar-specific proteins, *Science*, 177 (1972) 949–959.
13. H. Lis, N. Sharon, and E. Katchalski, Soybean hemagglutinin, a plant glycoprotein. I. Isolation of a glycopeptide, *J. Biol. Chem.*, 241 (1966) 684–689.
14. H. Lis, N. Sharon, and E. Katchalski, Identification of the carbohydrate-peptide linking group in soybean agglutinin, *Biochim. Biophys. Acta*, 192 (1969) 364–366.
15. N. Sharon, Albert Neuberger (1908-96): Founder of modern glycoprotein research, *Glycobiology*, 7 (1997) x–xiii.
16. L. Dorland, H. van Halbeek, J. F. G. Vliegenthart, H. Lis, and N. Sharon, Primary structure of the carbohydrate chain of soybean hemagglutinin. A reinvestigation by high resolution ^{1}H NMR spectroscopy, *J. Biol. Chem.*, 256 (1981) 7708–7711.
17. D. A. Ashford, R. A. Dwek, T. W. Rademacher, H. Lis, and N. Sharon, The glycosylation of glycoprotein lectins. Intra- and inter-genus variation in N-linked oligosaccharide expression, *Carbohydr. Res.*, 213 (1991) 215–227.
18. M. Wilchek, T. Miron, and J. Kohn, Affinity chromatography, *Methods Enzymol.*, 104 (1984) 3–56.
19. R. Adar, H. Streicher, S. Rozenblatt, and N. Sharon, Synthesis of soybean agglutinin in bacterial and mammalian cells, *Eur. J. Biochem.*, 249 (1997) 684–689.
20. R. Lotan, H. W. Siegelman, H. Lis, and N. Sharon, Subunit structure of soybean agglutinin, *J. Biol. Chem.*, 249 (1974) 1219–1224.
21. B. Schechter, H. Lis, R. Lotan, A. Novogrodsky, and N. Sharon, Requirement for tetravalency of soybean agglutinin for induction of mitogenic stimulation of lymphocytes, *Eur. J. Immunol.*, 6 (1976) 145–149.

22. B. Shaanan, M. Shoham, A. Yonath, H. Lis, and N. Sharon, Crystallization and preliminary X-ray diffraction studies of soybean agglutinin, *J. Mol. Biol.*, 174 (1984) 723–725.
23. A. Dessen, D. Gupta, S. Sabesan, C. F. Brewer, and J. C. Sacchettini, X-ray crystal analysis of the soybean agglutinin crosslinked with a biantennary analog of the blood group I carbohydrate antigen, *Biochemistry*, 34 (1995) 4933–4942.
24. A. Allen, A. Neuberger, and N. Sharon, The purification and specificity of wheat germ agglutinin, *Biochem. J.*, 131 (1973) 155–162.
25. R. Lotan, E. Skutelsky, D. Danon, and N. Sharon, The purification, composition and specificity of the anti-T lectin from peanut (*Arachis hypogaea*), *J. Biol. Chem.*, 250 (1975) 8518–8523.
26. M. E. A. Pereira, E. A. Kabat, R. Lotan, and N. Sharon, Immunochemical studies on the specificity of peanut (*Arachis hypogaea*) agglutinin, *Carbohydr. Res.*, 51 (1976) 107–118.
27. J. L. Iglesias, H. Lis, and N. Sharon, Purification and properties of a galactose/N-acetyl-D-galactosamine-specific lectin from *Erythrina cristagalli*, *Eur. J. Biochem.*, 123 (1982) 247–252.
28. D. Ashford, R. A. Dwek, J. K. Welply, S. Amatayakul, S. W. Homans, H. Lis, G. N. Taylor, N. Sharon, and T. W. Rademacher, The β(1-2)-D-xylose and α(1-3)-L-fucose substituted N-linked oligosaccharides from *Erythrina cristagalli* lectin. Isolation, characterisation and comparison with other legume lectins, *Eur. J. Biochem.*, 166 (1987) 311–320.
29. N. Gilboa-Garber and L. Mizrachi, A new mitogenic D-galactosephilic lectin isolated from seeds of the coral-tree *Erythrina corallodendron*. Comparison with *Glycine max* (soybean) and *Pseudomonas aeruginosa* lectins, *Can. J. Biochem.*, 59 (1981) 315–322.
30. R. Arango, R. Adar, S. Rozenblatt, and N. Sharon, Expression of *Erythrina corallodendron* lectin in *Escherichia coli*, *Eur. J. Biochem.*, 205 (1992) 575–581.
31. R. Adar, J. Ångström, E. Moreno, K. A. Karlsson, H. Streicher, and N. Sharon, Structural studies of the combining site of *Erythrina collarodendron* lectin. Role of tryptophan, *Protein Sci.*, 7 (1998) 52–63.
32. E. Moreno, S. Teneberg, R. Adar, N. Sharon, K.-A. Karlsson, and J. Ångström, Redefinition of the carbohydrate specificity of *Erythrina corallodendron* lectin based on solid-phase binding assays and molecular modeling of native and recombinant forms obtained by site-directed mutagenesis, *Biochemistry*, 36 (1997) 4429–4437.
33. K. A. Kulkarni, A. Srivastava, N. Mitra, N. Sharon, A. Surolia, M. Vijayan, and K. Suguna, Effect of glycosylation on the structure of *Erythrina corallodendron* lectin, *Proteins*, 56 (2004) 821–827.
34. B. Shaanan, H. Lis, and N. Sharon, Structure of a lectin with ordered carbohydrate, in complex with lactose, *Science*, 253 (1991) 862–866.
35. C. Svensson, S. Teneberg, C. L. Nilsson, F. P. Schwarz, A. Kjellberg, N. Sharon, and U. Krengel, High-resolution crystal structures of *Erythrina cristagalli* lectin in complex with lactose and 2'-α-L-fucosyllactose and their correlation with thermodynamic binding data, *J. Mol. Biol.* (2002) 69–83.
36. R. Lemieux, C. C. Ling, N. Sharon, and H. Streicher, The epitope of the H-type trisaccharide recognized by *Erythrina corallodendron* lectin. Evidence for both attractive polar and strong hydrophobic interactions for complex formation involving a lectin, *Israel J. Chem.*, 40 (2000) 167–176.
37. N. Sharon and H. Lis, The structural basis for carbohydrate recognition by lectins, *Adv. Exp. Med. Biol.*, 491 (2001) 1–16.
38. A. Foriers, C. Wuilmart, N. Sharon, and A. D. Strosberg, Extensive sequence homologies among lectins from leguminous plants, *Biochem. Biophys. Res. Commun.*, 75 (1977) 980–985.
39. P. Nowell, Phytohemagglutinin: An initiator of mitosis in cultures of normal human leukocytes, *Cancer Res.*, 20 (1960) 462–466.
40. A. Novogrodsky, R. Lotan, A. Ravid, and N. Sharon, Peanut agglutinin, a new mitogen that binds to galactosyl sites exposed after neuraminidase treatment, *J. Immunol.*, 115 (1975) 1243–1248.
41. M. Burger and A. R. Goldberg, Identification of a tumor-specific determinant of neoplastic cell surfaces, *Proc. Natl. Acad. Sci. U.S.A.*, 57 (1967) 359–366.

42. M. Inbar and L. Sachs, Interaction of the carbohydrate-binding protein concanavalin A with normal and transformed cells, *Proc. Natl. Acad. Sci. U.S.A.*, 63 (1969) 1418–1425.
43. B. A. Sela, H. Lis, N. Sharon, and L. Sachs, Different locations of carbohydrate-containing sites at the surface membrane of normal and transformed mammalian cells, *J. Membrane Biol.*, 3 (1970) 267–279.
44. Y. Reisner, M. Linker-Israeli, and N. Sharon, Separation of mouse thymocytes into two subpopulations by the use of peanut agglutinin, *Cell. Immunol.*, 25 (1976) 129–134.
45. A. Toporowicz and Y. Reisner, Changes in sialyltransferase activity during murine T cell differentiation, *Cell. Immunol.*, 100 (1986) 10–19.
46. W. Gillespie, J. C. Paulson, S. Kelm, M. Pang, and L. G. Baum, Regulation of α2,3-sialyltransferase expression correlates with conversion of peanut agglutinin (PNA)+ to PNA-phenotype in developing thymocytes, *J. Biol. Chem.*, 268 (1993) 3802–3804.
47. A. M. Moody, D. Chui, P. A. Reche, J. J. Priatel, J. D. Marth, and E. L. Reinherz, Developmentally regulated glycosylation of the CD8$\alpha\beta$ coreceptor stalk modulates ligand binding, *Cell*, 107 (2001) 501–512.
48. M. A. Daniels, K. A. Hogquist, and C. S. Jameson, Sweet 'n' sour: The impact of differential glycosylation on T cell differentiation, *Nat. Immunol.*, 3 (2002) 903–910.
49. Y. Reisner, G. Gachelin, P. Dubois, J. F. Nicolas, N. Sharon, and F. Jacob, Interaction of peanut agglutinin, a lectin specific for terminal D-galactosyl residues, with embryonal carcinoma cells, *Dev. Biol.*, 61 (1977) 20–27.
50. Y. Reisner, L. Itzicovitch, A. Meshorer, and N. Sharon, Hemopoietic stem cell transplantation using mouse bone-marrow and spleen cells fractionated by lectins, *Proc. Natl. Acad. Sci. U.S.A.*, 75 (1978) 2933–2936.
51. Y. Reisner, N. Kapoor, D. Kirkpatrick, M. S. Pollack, S. Cunningham-Rundles, B. Dupont, M. Z. Hodes, R. A. Good, and R. J. O'Reilly, Transplantation for severe combined immunodeficiency with HLA-A,B, D, DR incompatible parental marrow cells fractionated by soybean agglutinin and sheep red blood cells, *Blood*, 61 (1983) 341–348.
52. J. P. Duguid and D. C. Old, Adhesive properties of *Enterobacteriaceae*, in E. H. Beachey, (Ed.), *Bacterial Adherence*, Chapman and Hall, London, 1980, pp. 185–217.
53. I. Ofek, D. Mirelman, and N. Sharon, Adherence of *Escherichia coli* to human mucosal cells mediated by mannose receptors, *Nature*, 265 (1977) 623–625.
54. N. Firon, S. Ashkenazi, D. Mirelman, I. Ofek, and N. Sharon, Aromatic alphaglycosides of mannose are powerful inhibitors of the adherence of type 1 fimbriated *Escherichia coli* to yeast and intestinal epithelial cells, *Infect. Immun.*, 55 (1987) 472–476.
55. N. Firon, I. Ofek, and N. Sharon, Carbohydrate specificity of the surface lectins of *Escherichia coli, Klebsiella pneumoniae,* and *Salmonella typhimurium, Carbohydr. Res.*, 120 (1983) 235–249.
56. J. Bouckaert, M. Berglund, E. Schembri, L. De Genst, M. Cools, M. Wuhrer, C. S. Hung, J. Pinkner, R. Slättegård, A. Zavialov, D. Choudhury, S. Langermann, S. J. Hultgren, L. Wyns, P. Klemm, S. Oscarson, S. D. Knight, and H. De Greve, Receptor binding studies disclose a novel class of high-affinity inhibitors of the *Escherichia coli* FimH adhesin, *Mol. Microbiol.*, 55 (2005) 441–455.
57. M. Aronson, O. Medalia, L. Schori, D. Mirelman, N. Sharon, and I. Ofek, Prevention of colonization of the urinary tract of mice with *Escherichia coli* by blocking of bacterial adherence with methyl β-D-mannopyranoside, *J. Infect. Dis.*, 139 (1979) 329–332.
58. A. Gbarah, D. Mirelman, P. J. Sansonetti, R. Verdon, W. Bernhard, and N. Sharon, *Shigella flexneri* transformants expressing type 1 fimbriae bind to, activate and are killed by phagocytic cells, *Infect. Immun.*, 61 (1993) 1687–1693.
59. I. Ofek and N. Sharon, Lectinophagocytosis: A molecular mechanism of recognition between cell surface sugars and lectins in the phagocytosis of bacteria, *Infect. Immun.*, 56 (1988) 539–547.
60. N. Sharon, Carbohydrates as future anti-adhesion drugs for infectious diseases, *Biochim. Biophys. Acta*, 1760 (2006) 527–537.

61. R. Barkai-Golan, D. Mirelman, and N. Sharon, Studies on growth inhibition by lectins of *Penicillia* and *Aspergilli*, *Arch. Microbiol.*, 116 (1978) 119–124.
62. D. Mirelman, E. Galun, N. Sharon, and R. Lotan, Inhibition of fungal growth by wheat germ agglutinin, *Nature*, 256 (1975) 414–416.
63. J. I. Ravdin and R. L. Guerrant, Role of adherence in cytopathogenic mechanisms of *Entamoeba histolytica*: Studies with mammalian tissue culture cells and human erythrocytes, *J. Clin. Invest.*, 68 (1981) 1305–1313.
64. J. I. Ravdin, C. F. Murphy, R. A. Salata, R. L. Guerrant, and E. L. Hewlett, N-Acetyl-D-galactosamine-inhibitable adherence lectin of *Entamoeba histolytica*. I. Partial purification and relation to amoebic virulence in vitro, *J. Infect. Dis.*, 151 (1985) 804–815.
65. D. Mirelman and J. I. Ravdin, Lectins in *Entamoeba histolytica*, in D. Mirelman, (Ed.), *Microbial Lectins and Agglutinins: Properties and Biological Activity*, John Wiley & Sons, New York, 1986, pp. 319–334.
66. R. Bracha and D. Mirelman, Adherence and ingestion of *Escherichia coli* serotype 055 by trophozoites of *Entamoeba histolytica*, *Infect. Immun.*, 40 (1983) 882–887.
67. R. Bracha, D. Kobiler, and D. Mirelman, Attachment and ingestion of bacteria by trophozoites of *Entamoeba histolytica*, *Infect. Immun.*, 36 (1982) 396–406.
68. R. Verdon, D. Mirelman, and P. J. Sansonetti, A model of interaction between *Entamoeba histolytica* and *Shigella flexneri*, *Res. Microbiol.*, 143 (1992) 67–74.
69. R. Bracha and D. Mirelman, Virulence of *Entamoeba histolytica* trophozoites: Effects of bacteria, microaerobic conditions and metronidazole, *J. Exp. Med.*, 160 (1984) 353–368.
70. F. Padilla-Vaca, S. Ankri, R. Bracha, L. A. Koole, and D. Mirelman, Down regulation of *Entamoeba histolytica* virulence by monoxenic cultivation with *Escherichia coli* 055 is related to a decrease in the expression of the light (35 kilodalton) subunit of the Gal/GalNAc lectin, *Infect. Immun.*, 67 (1999) 2096–2102.
71. G. Ashwell and A. G. Morell, The role of surface carbohydrates in the hepatic recognition and transport of surface glycoproteins, *Adv. Enzymol. Relat. Areas Mol. Biol.*, 41 (1974) 99–128.
72. V. I. Teichberg, I. Silman, D. D. Beitsch, and G. Reshef, A β-D-galactoside binding protein from electric organ tissue of *Electrophorus electricus*, *Proc. Natl. Acad. Sci. U.S.A.*, 72 (1975) 1383–1387.
73. A. Raz and R. Lotan, Lectin-like activities associated with human and murine neoplastic cells, *Cancer Res.*, 41 (1981) 3642–3647.
74. N. Sharon, Lectins: Carbohydrate-specific reagents and biological recognition molecules, *J. Biol. Chem.*, 282 (2007) 2753–2764.
75. N. Sharon, Lectins: Past, present and future, *Biochem. Soc. Trans.*, 36 (2008) 1457–1460.
76. N. Sharon and H. Lis, Lectins: From hemagglutinins to biological recognition molecules. A historical overview, *Glycobiology*, 14 (2004) 53R–62R.
77. N. Sharon and H. Lis, Lectins as cell recognition molecules, *Science*, 246 (1989) 227–234.
78. N. Sharon and H. Lis, Lectins, (1989) Chapman & Hall, , London, .

COMBINING COMPUTATIONAL CHEMISTRY AND CRYSTALLOGRAPHY FOR A BETTER UNDERSTANDING OF THE STRUCTURE OF CELLULOSE

Alfred D. French

Southern Regional Research Center, U.S. Department of Agriculture, New Orleans, Louisiana, USA

I. Introduction	20
II. Information from Crystals of Related Small Molecules	23
1. Shape of the D-Glucopyranose Ring	23
2. Linkage Geometry	28
3. Conversion to Polymer-Shape Notation	33
4. Orientation of O-6	35
5. Crystal Packing and Intermolecular Interactions	37
6. Summary of Section on Extrapolation	40
III. Energy Calculations	41
1. Results on Individual Isolated Molecules with Empirical Methods	45
2. Results with Quantum Methods	51
3. Assessment of ϕ/ψ Mapping	54
4. Hydroxyl-Group Orientations	56
IV. Detection of New Stabilizing Interactions in Cellulose with Atoms-in-Molecules Theory	57
V. Modeling Crystals of Cellulose	61
VI. Conclusions	68
Appendix. Molecular Structure Drawings for Saccharide Analogues Having β-(1→4) Linkages	70
Note Added in Proof	83
References	83

Abbreviations

Å, angstrom, 10^{-10} m; AMBER, assisted model building with energy refinement; B3LYP, Becke, three-parameter, Lee–Yang–Parr exchange–correlation functional, a density functional quantum mechanics method; BCP, bond critical point; BP, bond path; CIF, Crystallographic Information File; CHARMM, Chemistry at HARvard Molecular Mechanics; CSD, Cambridge Structural Database; HF, Hartree–Fock self-consistent field quantum mechanics method; IUPAC, International Union of Pure and Applied Chemistry; MD, molecular dynamics; MDa, megadalton; MM, molecular mechanics; MP2, Moeller–Plesset second-order quantum mechanics method for treating electron correlation; QM, quantum mechanics; REFCODE, the "reference code" denoting a specific crystal structure in the CSD; SMD, solvent model D, where D represents the electronic charge density of the solute; TIP3P, Transferable Intermolecular Potential—Three Point, a potential function for simulating liquid water

I. INTRODUCTION

This article complements two related contributions on cellulose published in 2010 in Volume 65 of *Advances*, one by Pérez and Samain,[1] and the other by Mischnick and Momcilovic.[2] The Pérez and Samain article in particular covers many details of the physical structure of cellulosic materials, including the crystallography and polymorphy of cellulose. Results from other methods, such as magnetic resonance spectroscopy and microscopy, were also surveyed. A strong point of that article is the manner in which the placement of cellulose in the plant cell wall is treated, and characterization of the microfibrils that comprise the cellulosic materials from different sources. The Mischnick and Momcilovic article emphasizes the chemical analysis of both starch and cellulose through chemical substitution, although that work is also inextricably related to physical structure, and some structural concepts are presented there. The present offering treats the structure of cellulose more generally, rather than focusing on polymorphy or on features specific to cellulose from one source or another.

Because of its importance as a major component of plant cell walls, textiles, paper, building materials, and as a fledgling feedstock for biorefineries, cellulose has been studied extensively. Studies continue not only because of new applications but because cellulose has been very reluctant to reveal many of the secrets of its structure. Several continuing controversies exemplify the lack of understanding of cellulose and the structures that Nature creates from it.

One of the longest-running controversies regards the relative orientations of the molecules in cellulose crystals. Are the molecules packed parallel or antiparallel? In parallel packing, all of the reducing ends of the cellulose chains are at one end of the crystal, whereas with antiparallel packing, half of the reducing ends are at one end of the crystal and half at the other. Seventy or eighty years ago, the interpretations were based on philosophy and expectations, and both parallel and antiparallel chain-packing modes were published by the same authors at different times.[3,4] In the 1970s, structures based on antiparallel packing[5] and two types of parallel packing[6–9] were each claimed by different groups to best satisfy fiber-diffraction data.

Over the past decade or so, fiber crystallography has taken giant steps forward with the use of synchrotron and neutron radiation, and with samples that are much more crystalline. These developments resulted in the collection of additional diffraction-intensity data and substantially improved the ratio of data to such variables as the atomic coordinates and temperature factors. The improved resolution from these studies allowed, for example, visualization of the O-6 atoms, allowing a controversy on their orientation to be settled. This work showed that crystallites of native cellulose are composed of parallel chains,[10,11] but that mercerized cellulose and cellulose crystals grown from solution have their structure based on antiparallel chains.[12,13] Mercerization performed in the laboratory involves simply placing such cellulose fibers as cotton or linen in 20% aqueous sodium hydroxide and then rinsing and drying them. This treatment causes a conversion from the native form (cellulose I) into cellulose II, the same form as found for rayon or other forms of cellulose that are regenerated from solution. Cotton fibers, which are composed almost entirely of cellulose, can be rapidly mercerized without apparent major disruptions of the fiber structure. Because of this observation, many workers resisted the idea that such a drastic conversion had occurred. At a symposium on cellulose biogenesis held in 2011 at the 241st National Meeting of the American Chemical Society, a show of hands indicated that roughly 80% of the audience did not believe the conversion of polarity.

Another controversy regards the twisting of cellulose crystals. For the past two decades, it has been a relatively simple matter to make computer models of cellulose crystals, and today's computer systems allow the creation of atomistic models that have the same number of molecules as indicated by experiment. When constructed based on the crystallographically determined atomic coordinates, the models have regular order in three dimensions. However, when force fields are applied to the models by performing energy minimization or molecular dynamics, the crystals twist. Electron micrographs show regions that are periodically twisted,[14] and there has been enthusiasm from modelers[15,16] for the concept of continuously twisted cellulose microfibrils.

To some extent, the concern about crystallite twisting reflects a longer-standing question regarding the presence of a twofold screw axis of symmetry in cellulose molecules. In twofold structures, each glucose residue is rotated 180° about the molecular axis relative to its predecessor. Hydrogen atoms on the C-4 and C-1$'$ atoms are close to each other in models having such symmetry. That distance is so short that such models give lower energy (based on several empirical force fields) if the linkage is twisted away from the twofold conformation. A way to avoid the complicating implications of a continuous twisting of the molecule would be to allow a right-handed twist of the first individual linkage, and a compensating twist to the left for the second linkage, and so on.[17-19] The diffraction studies are not always clear on the symmetry, with intensity sometimes visible in the locations of odd-order meridional reflections that would be missing for ideal twofold structures, but it can be inferred from most modeling studies that twofold structures are indeed not favored.

Other gaps in our knowledge of cellulose are important with respect to potential improvements in the way that cellulose is produced and utilized in practical applications. Many of the reactions of cellulose are thought to occur in noncrystalline regions. The degree of crystallinity is often considered to be important to our understanding, but what does the noncrystalline component of cellulose look like? What does a 70% crystalline rating mean at the atomic level? An important rationale for treating cellulose in the general way adopted for this article is to understand it better in relation to the noncrystalline regions.

Because of the two recent *Advances* articles,[1,2] as well as the availability of the book, *Cellulose Science and Technology*,[20] this article skips over some of the important basic concepts and tools that are covered in those works. Instead, this discussion deals with some specific approaches that can also be applied to other polysaccharides. Included are extrapolations from crystal structures of small molecules determined experimentally, energy mapping by empirical, hybrid, and quantum mechanics methods, and model crystal studies along with calculated diffraction patterns. Atoms-In-Molecules theory,[21] which analyzes the variations in electron density of molecules and their neighbors, has also been enlisted to help identify and understand stabilizing interactions in compounds related to cellulose.

An underlying hypothesis herein is that the components of cellulose will have shapes similar to those of related small molecules. Crystals of small-molecule carbohydrates can often be grown to a size of 0.2 mm needed for X-ray diffraction studies, which can provide high-resolution structural information. Another advantage is that small molecules are more easily studied by computerized molecular modeling at high levels of quantum theory. None of these efforts can aspire to be the final word on their topics; more and more computer power can and will be brought to bear on the questions. The results will lead to a general reduction of assumptions and an increase in the scope of the projects. Also, new crystal structures of small molecules will continue to be solved, providing an increasingly definitive database of information related to plausible atomistic structures of cellulose.

II. INFORMATION FROM CRYSTALS OF RELATED SMALL MOLECULES

The molecular weight of cellulose could be as high as 3.2 MDa (20,000 glucose residues) or more. That size is frequently an important property to consider, but many aspects of cellulose are relatively independent of how many glucose residues are connected within the chains. β-Cellobiose [β-D-glucopyranosyl-(1→4)-β-D-glucopyranose], with its two glucose residues, could be considered to be the shortest chain model, and once the length of chains exceeds six glucose residues or so, the molecule becomes, like cellulose, insoluble in water. Furthermore, the higher cello-oligomers, starting with methyl β-cellotrioside, give powder-diffraction patterns very similar to those of cellulose II from rayon, mercerized linen, or cotton.[22–24] Accordingly, it seems safe to conclude that there is an analogy to a sausage sliced to different lengths—you can get a good taste of the shape properties, even with a very short molecular length.

The ultimate concept of the next six subsections is to build computer models of cellulose by using the coordinates determined precisely from single-crystal studies of small molecules. A question to be answered is thus: "what would a cellulose molecule look like if it were constructed with both the nonreducing residue of crystalline β-cellobiose and the geometry of the glycosidic linkage from a given crystal structure?"[25] After numerous relevant structures have been built, questions can be answered regarding the range of molecular shapes that result, and comparisons can be made with experimental values for cellulose itself. The first step is to make sure that we understand the shape of the primary building block, the glucose monosaccharide residue.

1. Shape of the D-Glucopyranose Ring

The pioneering work on D-glucosamine hydrochloride and hydrobromide by Cox and Jeffrey[26] in 1939 showed that the D-glucopyranose ring adopts the chair shape having carbon 4 high and carbon 1 low, that is, the $^4C_1$ conformation.[27] (The enantiomer, L-glucopyranose, would have the $^1C_4$ conformation. In the remainder of this article, use of the loose term "glucose" implies D-glucopyranose.) This assignment was confirmed for glucose in aqueous solution by Richard Reeves in 1950 by his studies on the complexation of glucose with cuprammonium.[28] Since then, there have been many crystallographic determinations of the ring shape of glucose and its derivatives. Minor variations are seen, but there are no exceptions from the $^4C_1$ shape for β-D-glucopyranose residues. (A number of α-D-glucopyranosyl rings in substituted cyclodextrins have rings with skew or $^1C_4$ shapes.)

A more-detailed description of the ring shape is conveniently assigned in terms of the Cremer–Pople puckering parameters.[29,30] Puckering parameters acknowledge that the main variations in ring shape are differences in the distances of the atoms from a mean plane, and that the bond lengths and covalent bond angles vary to a more limited extent with less effect on the overall shape. The deviation (Q) is routinely expressed in angstrom units, despite the recommended SI units of picometers (1 Å = 100 pm = 0.1 nm). For six-membered rings, there are two other parameters, θ, and Φ. The θ parameter indicates the type of shape, such as chair, half chair, boat, or skew,[27] and the Φ parameter denotes where the deviations from the mean plane are greatest. There are six different boat forms, namely $^{3,O}B$, $B_{1,4}$, $^{2,5}B$, $B_{3,O}$, $^{1,4}B$, and $B_{2,5}$, all with θ values of 90° but with Φ values of 0°, 60°, 120°, 180°, 240°, and 300°, respectively. Only two unique chairs exist, the $^4C_1$ and the $^1C_4$, with θ values of 0° and 180°. (The 1 and 4 atoms are chosen by convention because 1 and 4 are the pair with the lowest numbers; the $^4C_1$ shape is equivalent to, and could otherwise be termed, $^OC_3$ or $^2C_5$. The ring oxygen is the sixth atom in this scheme, but it is the first atom when calculating the puckering parameters described in the paragraph after the next.)

Even though the β-D-glucopyranose rings in gluco-oligosaccharides are firmly in the $^4C_1$ conformation, it is still useful to track the puckering parameters to quantify any deviations from perfection and indicate the degree of strain imposed by the environment and by various substituents. The θ parameter is probably of the greatest interest, as will be seen in the following discussions. That is because the Φ values do not portray substantial variations in ring shape when the θ values are near to zero or to 180°. Furthermore, standard deviations of Φ are likely to be much higher for rings in the chair form.

A convenient way to learn the puckering parameters from diffraction experiments is to search the Cambridge Crystal Structure Database (CSD).[31] Individual structures in the CSD are identified by a unique reference code (REFCODE). Fresh scans were made for the present work, based on several fragments, such as the one shown in Fig. 1.

In Fig. 1, two rings are shown, with all of the oxygen atoms indicated except for O-6 and O-6′. The three-dimensional parameters indicated in the upper right corner denote torsion angles to be determined in the search. TOR1 and TOR2 are the linkage torsion angles ϕ and ψ (see Scheme 1 in Section II.2), and TOR3 (selected) and TOR4 are improper torsion angles used to assure the configuration at C-1′, as shown by the heavy green lines, and C-4 (both are about −31°; a range of −61° to −1° was accepted). However, there were no restrictions on the configuration at other carbon atoms in these searches. Therefore, other aldopyranosyl rings, such as galactopyranose (only for the nonreducing ring) or mannopyranose, can also be included.

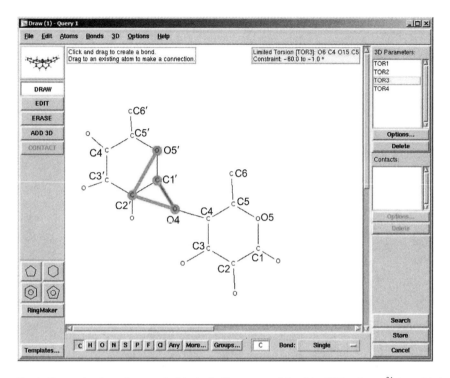

FIG. 1. Screen shot from the query builder in the Conquest module of the CSD software.[31] Atom labels were added. The heavy lines that connect C-1′, O-4, C-2′, and O-5′ define an improper torsion angle used to assure the configuration at C-1′. In the search about to be initiated, the selected structures must have values between −60° and -1° for this improper torsion angle. (See Color Insert.)

After the database was searched, the relevant hits were exported as a .cif file that could subsequently be read into the PLATON software.[32] (The CSD software does not calculate puckering parameters, but the PLATON routine for calculation of geometry does.) PLATON's automatic ring-finding routine usually works well, but for the cellobiose structure assigned the reference code CELLOB04, the nonstandard atomic numbering of the reducing ring caused an out-of-phase result that was readily corrected.

Figure 2 shows histograms of the puckering parameters in rings taken from molecules that have β-(1→4) linkages. Rings having smaller Q values are flatter (less puckered). The smallest amplitudes are from a structure (EYOCUQ01) of lactose determined by powder-diffraction methods.

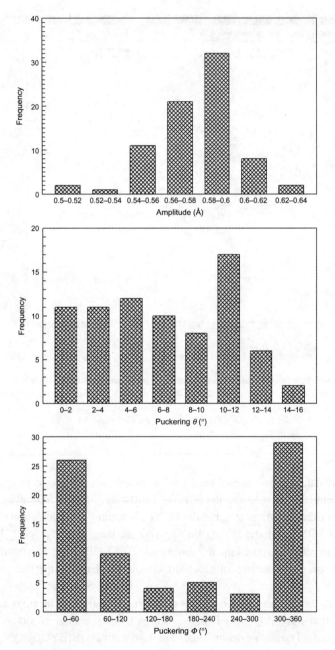

FIG. 2. Histograms of puckering parameters Q, θ, and Φ (top, middle, and bottom) from the search of the CSD described in the text.

A perfect $^4C_1$ ring might be considered to have a θ parameter of 0, but the mean value from the CSD is about 7°. Pyranosyl rings having larger θ values tend toward *E*nvelope or *H*alf-chair forms, which have, respectively, one atom, or two adjacent atoms, out of the plane of the remaining atoms in the ring. This situation is a high-energy condition for rings composed of only single bonds, and so the frequency drops off to zero beyond $\theta = 16°$. Thus, all of the rings in the CSD that met the search criteria had the normal chair form. The most interesting aspect of this search is the frequent incidence of rings having θ values of 10–12°, numbers that are near the observed upper limit. If a greater number of relevant crystal structures were solved and put into the database, or if less-restrictive search criteria were used, the results might be more statistically satisfying. Based on the present fairly flat distribution of different θ values, it seems that the θ parameter can easily change within the limited range observed.

Such variations in structure of the normal chair ring may seem subtle, but it is useful to quantify the geometry for several reasons. For example, if a fiber-diffraction study were to propose a structure that was puckered outside the limits given in Fig. 2,[12] then this deviation becomes at least a noteworthy point in any discussion of the work. Also, puckering parameters provide a simple way to compare the results of theoretical and experimental studies. They can also be used to develop energy surfaces that indicate the high- and low-energy forms.[33–35]

Puckering parameters were thoroughly exploited by Mazeau and Heux in their empirical force-field study of crystalline and amorphous cellulose.[36] Their puckering values for the cellulose Iβ model were in reasonable agreement with the foregoing crystal database scans. They found a mean puckering amplitude of about 0.62 Å, where the maximum in the crystallographic distribution is just slightly lower, about 0.59 Å. Their θ values were less than 20°, and their puckering ϕ values, the most difficult to determine accurately for rings in the chair form, were about 60°, one of the highest histogram bars in the bottom graph of the foregoing Fig. 2. Their models for the amorphous regions, based on randomly oriented molecules, gave quite different results. Some 27% of the glucose residues had θ conformations near 90° (boat or skew forms), and even a number of the structures having nominal chair forms had θ values greater than 30°. Their tangled-chain model may be more applicable to cellulose in regenerated films rather than to the amorphous cellulose that results from mechanical or chemical decrystallization of fairly well-aligned molecules from natural sources. The tangling apparently leads to the distorted rings. However, their predicted glass-transition temperature was higher than any experimental value. Therefore, problems with the force field may exist that also cause it to erroneously predict a large number of distorted residues.

2. Linkage Geometry

Scheme 1 shows the cellobiose molecule with its ϕ and ψ torsion angles, as well as the definitions of the *gt*, *gg*, and *tg* orientations[37,38] of the O-6 atom of the glucopyranose monosaccharide, discussed in Section II.4. Three different scans of the CSD were used to study the geometry of the $(1 \rightarrow 4)$-β-glycosidic linkage. The first scan (Table I), the most inclusive, was based on an analogue having two connected tetrahydropyran rings with unspecified substituents. This is a generous criterion, but it tests the hypothesis that torsion energies of the linkage are the most important factors in determining which linkage geometries would be found in crystals. The second scan was based on the same molecule, but with the addition of a hydroxyl group equivalent to O-3 on the ring at the reducing end of cellobiose. The purpose there was to find the frequency of occurrence of the classic O-3—H...O-5′ hydrogen bond. Finally, an O-6′—H group was added to the ring at the nonreducing end and the scan repeated so that the relationship of the O-6′ orientation to the O-3—H...O-5′ hydrogen bond could be studied.

The two tetrahydropyran rings were constrained: both by the improper dihedrals in the foregoing scan to assure the correct configurations of the atoms at the linkage and two additional torsion angles that were set to $180° \pm 30°$. These torsion angles, C-5′—O-5′—C-1′—O-4 and C-2—C-3—C-4—O-4, were intended to assure that only $^4C_1$ rings would be captured.

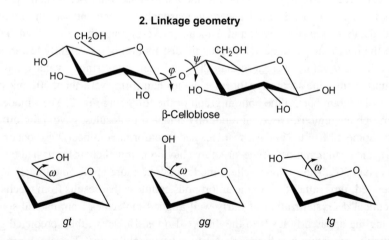

SCHEME 1. β-Cellobiose [4-*O*-β-D-glucopyranosyl-β-D-glucopyranose, β-D-Glc*p*-$(1 \rightarrow 4)$-β-D-Glc*p*], depicting the pyranose rings in the $^4C_1$ conformation and identifying the ϕ,ψ torsion angles along the interglycosidic linkage. The lower structures depict the *gt*, *tg*, and *gg* orientations along the C-5—C-6 bond.

TABLE I
Crystal Structures of Oligosaccharides Having β-(1 → 4) Linkages Used in Fig. 3[a]

REFCODE	Compound	References
ABUCEF	Lewis X trisaccharide methyl glycoside hydrate	39
ACCELL10	β-Cellotriose undecaacetate	40
ACELLO	β-Cellobiose octaacetate	41
ACGLPR	6'-O-Trityl-α-cellobiose heptaacetate	42
ACHITM10	N,N'-Diacetyl-α-chitobiose monohydrate	43
ACLACT	N-Acetyl-lactosamine monohydrate	44
AJUYUZ	[4,6-Di-O-acetyl-2-amino-3-O-(R)-1-carboxyethyl]-2-deoxy-β-D-glucopyranosyl-1',2-lactam)-(1 → 4)-2-acetamido-1,3,6-tri-O-acetyl-2-deoxy-D-glucopyranose chloroform solvate	45
AQOGIW	4-(2-Deoxy-2-azido-3-O-benzyl-6-O-benzoyl-5-O-(naphth-2-ylmethyl)-β-D-glucopyranosyl)-1,6-anhydro-2-O-benzoyl-3-O-benzyl-β-L-idopyranose	46
BAGDIW	(1S)-Spiro[(4-O-(β-D-galactopyranosyl)-1-dehydroxy-D-glucopyranose-1,1'-[(5',7'-bis(p-bromobenzyloxy)-2'-oxaindane)]]	47
BCHITT10	N,N'-Diacetyl-β-chitobiose trihydrate	48
BELJAD	Penta-O-acetyl-avileurekanose C, acetone solvate	49
BERKIS01	Digoxigenin bisdigitoxoside tetrahydrate	50
BIKWOH10	2-Acetamido-1,3,6-tri-O-acetyl-2-deoxy-4-O-(2,3,4,6-tetra-O-acetyl-α-L-idopyranosyl)-α-D-glucopyranose	51
BLACTO02	β-Lactose	52
CAWBIL	Methyl β-lactoside	53
CELLOB04	β-Cellobiose	54
CIVFES10	Digitoxigenin bisdigitoxoside ethyl acetate solvate	50
COFMEP10	Mannotriose trihydrate [β-D-Manp-(1 → 4)-β-D-Manp-(1 → 4)-α-D-Manp trihydrate]	55
DAYFAL	Methyl α-lactoside	56
DETQAV	Tenacigenoside A	57
DIGOXN10	Digoxin	58
DIHTUJ	4-O-β-D-Mannopyranosyl-(1 → 4)-α-D-mannopyranose	59
DUVGIK10	Gitoxigenin bisdigitoxoside ethyl acetate solvate	50
EYOCUQ	4-O-β-D-Galactopyranosyl-α-D-glucopyranose (α-lactose)	60
EYOCUQ01	4-O-β-D-Galactopyranosyl-α-D-glucopyranose (α-lactose)	61
FENSUM	Methyl β-cellobioside heptanitrate	62
FUSXIA	Methyl β-cellobioside 6,6'-dinitrate	63
GACZIT	1-(21-β-Tigloyl-22-α-acetylprotoaescigenin-3-β-O)-2-O-(β-D-glucopyranosyl)-4-O-(β-D-glucopyranosyl)-β-D-glucopyranosiduronic acid (escin 1a) sesquihydrate	64
GITXIN10	Gitoxin	65
GUGKID	1,10-Bis(2,3,6,2',3',4',6'-hepta-O-acetyl-β-D-ureidocellobiosyl)-4,7,13,16-tetraoxa-1,10-diazacyclooctadecane monohydrate	66
GURXPX10	(4-O-Methyl-α-D-glucopyranosyluronic acid)-(1 → 2)-β-D-xylopyranosyl-(1 → 4)-D-xylopyranose trihydrate	67

Continued

TABLE I (Continued)

REFCODE	Compound	References
IDEHEE	Methyl (2,3,4,6-tetra-O-acetyl-β-D-galactopyranosyl)-(1 → 4)-6-O-benzyl-2-deoxy-2-phthalimido-β-D-glucopyranoside	68
IFOVOO01	Methyl 4-O-methyl-β-D-glucopyranosyl-(1 → 4)-β-D-glucopyranoside [methyl 4-O-methyl-β-D-glucopyranosyl-(1 → 4)-β-D-glucopyranoside]	69
IFOVOO03	Methyl 4′-O-methyl-β-cellobioside (methyl 4-O-methyl-β-D-glucopyranosyl-(1 → 4)-β-D-glucopyranoside)	70
KAMHOV	Digoxigenin bisdigitoxoside ethyl acetate solvate	50
KUQGOT	Methyl 4-O-β-D-galactopyranosyl-α-D-mannopyranoside methanol solvate	71
LACTOS11	α-Lactose monohydrate	72
LAKKEO01	α,β-Lactose	73
LOMGOK	Benzyl 2,3-di-O-acetyl-4,6-cyclo-4,6-dideoxy-β-D-galactopyranosyl-(1 → 4)-2,3,6-tri-O-acetyl-β-D-glucopyranoside	74
MCELOB	Methyl β-cellobioside methanol solvate	75
MOVGIN	Disodium β-D-glucopyranosyl-(1 → 4)-α-D-glucopyranose diiodide dihydrate	76
MPYAGL	α-D-Mannopyranosyl-(1 → 3)-β-D-mannopyranosyl-(1 → 4)-2-acetamido-2-deoxy-α-D-glucopyranose	77
NUZRUV	2,3,6,2′,3′,4′,6′-Hepta-O-acetyl-D-lactobionolactone toluene solvate	78
OCATAN	N-β-Lactosylacetamide	79
OLGOSE	Olgose monohydrate	80
QERLAB	11α,12β-Acetonyltenacissoside F acetone solvate	81
QUDRIR	2,3,6,1′,6′-Penta-O-benzoyl-3′,4′-O-isopropylidene-β-lactose	82
RAVNAD	p-Nitrophenyl 4-O-(3,4,6-tri-O-acetyl-2-deoxy-2-phthalimido-β-D-glucopyranosyl)-2,3-di-O-benzoyl-6-O-pivaloyl-β-D-glucopyranoside	83
REMVUA	N-β-Lactosylurea dihydrate	84
RIKMON	(2,3,6-Tri-O-benzoyl-4-deoxy-4-C-ethynyl-β-D-glucopyranosyl)-(1 → 6)-4,5,8-tri-O-acetyl-3,7-anhydro-1,1,2,2-tetradehydro-1,2-dideoxy-1-C-(trimethylsilyl)-D-$glycero$-D-$gulo$-octitol	85
TAQYAL	Methyl β-cellotrioside ethanol solvate monohydrate	22
VEJFAR	(3R,6S,7S)-12-(3-O-Acetyl-4-O-(3-O-acetyl-4-O-(4-O-acetyl-2,6-dideoxy-3-O-methyl-β-D-$arabino$-hexopyranosyl)-2,6-dideoxy-β-D-$arabino$-hexopyranosyl)-2,6-dideoxy-β-D-$arabino$-hexopyranosyl)-6,7,13,15,16-pentaacetoxy-3-n-butyl-8-methoxy-1H,14H-3,4,6,7-tetrahydro-2,9-dioxahexaphene-1,14-dione (FD-594 octaacetate)	86
WEHTEI	β-Cellobiosylnitromethane	87
XOJLUE	4-O-β-D-Galactopyranosyl-1,5-anhydro-D-$gluco$-hexitol (1-deoxylactose)	73
XYLBHA	(1 → 4)- β-D-Xylobiose hexaacetate	88
YONVUT01	Cyclohexyl 4′-O-cyclohexyl-β-cellobioside cyclohexane solvate	89
ZILTUJ	β-Cellotetraose hemihydrate	90
ZZZSWA01	α-Cellobiose octaacetate	91

[a] X-ray structures and corresponding Cambridge Database REFCODEs for each compound in this table are displayed in Appendix.

Of the four atoms required to define a torsion angle at the glycosidic linkage, three of these are C-1', O-4, and C-4 (see Fig. 1). The fourth atom for the ϕ torsion could be H-1', O-5', or C-2'. The hydrogen atom has frequently been chosen by modelers and especially by researchers using NMR. On the other hand, theoretical studies intended to take advantage of the values from X-ray crystallography often do not use hydrogen because it can be poorly located by X-ray work. The O-5' atom is of added significance because it is involved in the exoanomeric effect, and so the angle O-5'—C-1'—O-4—C-4 is the IUPAC standard definition of ϕ.[92] The ψ torsion angle has again often been defined by the relevant proton, H-4, but the standard definition of ψ is the C-3—C-4—O-4—C-1' angle. However, a large body of work has used the definition C-5—C-4—O-4—C-1', and that angle is used in this article. Because C-1' and C-4 both have sp^3 hybridization and tetrahedral geometry, the various definitions are interconvertible by addition or subtraction of 120°. Perhaps the range of values on the axes of ϕ/ψ plots is more important than the choice of atoms. If all workers were to adopt the same effective ranges, equivalent to $-180°$ to $+180°$ for the definitions based on hydrogen atoms, different projects could be more readily compared. Otherwise, four copies of the energy map can be placed side by side and top to bottom, giving 720° along each axis. The desired ranges of the axes can then be enclosed in a new 360° × 360° frame and the remainder cropped (removed) with image manipulation software, as done for Fig. 11 which was based on a figure from the literature.

Figure 3 shows the distribution of the structures found in the scan, again after some manual filtering to remove structures having covalent constraints on the linkage geometries. Some 85 different linkages were available, including the RIKMON structure that has ϕ/ψ values of $-71°$ and 68°. That structure is not shown here because it would expand the range of the map too much, causing a compression of the space in the figure for the other structures. The structures in Fig. 3 range over roughly 4% of the available ϕ/ψ space, and the occupied areas are even smaller. (A grid with 5° × 5° cells overlaid on ϕ/ψ space would only have 50 occupied, out of 5184 grid cells.) Thus, while some flexibility is exhibited, the location of the structures in a small region implies that the probable structures should be predictable in a general sense.

The structures that are listed in Table I are displayed in Appendix. Their linkage geometries, plotted in Fig. 3 have been designated separately, as indicated by the caption. The two major categories are (A) for structures that cannot form inter-residue hydrogen bonds because they do not have either O-3—H or O-2'—H groups (they are both either missing or substituted on 29 disaccharides) and (B) the 42 structures that have both. Of these 42, only 4 do not form the O-3—H...O-5' hydrogen bond. Ten structures had O-3—H groups but no O-2'—H group, and two of them did not form the O-3—H...O-5' hydrogen bond. (The IDEHEE structure, Table I, did not successfully locate the hydrogen

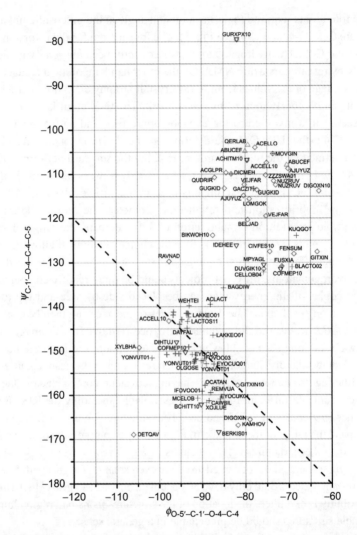

FIG. 3. ϕ/ψ Plot of glycosidic linkage conformations from experimental crystal structures that have β-1→4 linkages and two $^4C_1$ rings. All structures are labeled with their CSD "REFCODE" except for the structures of methyl cellotrioside (TAQYAL) and cellotetraose (ZILTUJ), which are indicated by red-+ symbols. Black + symbols indicate structures having both O-3—H and O-2'—H in equatorial disposition. Inverted blue triangles indicate structures with O-3—H but no O-2'—H, green triangles indicate structures with O-2'—H but no O-3—H group, and magenta diamond symbols indicate structures with either substituted or missing O-2' and O-3—H groups. Orange diamonds indicate structures with O-3—H but which do not form the O-3—H...O-5' hydrogen bond. The dashed diagonal line indicates ϕ/ψ combinations that would correspond to twofold screw-axis symmetry if applied to cellulose. (See Color Insert.)

atom on O-3, but its O-3...O-5' distance is 2.89 Å, suggesting an inter-residue hydrogen bond.) These 10 cover the ψ range of the experimental structures. Three structures had O-2'—H groups but no O-3—H, and they all happen to fall near $\psi = -105°$. There are 12 unique structures that incorporate otherwise unsubstituted lactose but have different substituents at the reducing carbon atom, C-1.

One conclusion is that structures having ψ values greater than $-120°$ are not likely to form O-3—H...O-5' hydrogen bonds. There are 24 structures above the $\psi = -120°$ line in Fig. 3, of which the 5 having the requisite O-3—H group instead participate in other, *intermolecular* hydrogen bonds. A second finding is that many of the hydroxylated structures are near the diagonal twofold line, a number of which are from the ZILTUJ (cellotetraose) or TAQYAL (methyl β-cellotrioside) structures. That observation is significant because both give powder X-ray patterns that are similar to that of cellulose II. (Less understood is the fact that the details of the packing of the tetraose, with two molecules per unit cell, are similar to one of the intermolecular arrangements in the trioside, which has four molecules per unit cell. The other pairing of molecules in the trioside crystal is closer to that found for cellulose II.) A few of the non-hydroxylated structures also have nearly twofold symmetry. Observations of structures near the twofold diagonal line are also significant because di- and oligosaccharide molecules cannot possess actual intramolecular symmetry. That is because the compositions of the reducing and nonreducing residues are nonidentical. Cellulose molecules are sufficiently long that the two end residues are insignificant in terms of affecting the diffraction pattern. Just as the molecules of the cellulose polymer are considered to have a unit cell that is only two glucose residues in length (thus ignoring the ending residues), screw-axis symmetry can be a trait of the long molecules.

3. Conversion to Polymer-Shape Notation

The shape of a polymer is often characterized in terms of the number (n) of residues per turn of the helix and the advance (h) per residue along the helix axis. By convention, left-handed structures are assigned negative values of n. When a polymer has the same geometry for each linkage and monomer, it becomes a helix, at least in the mathematical sense. The equations of Shimanouchi and Mizushima[93] convert geometric data from a helical fragment into the descriptors of polymer shape. For example, if a specific glucose geometry is connected to two copies of itself with identical ϕ and ψ torsion angles and glycosidic bond angles, just three pieces of information are needed, namely the O-1...O-4 distance, the O-1...O-4...O-4' angle, and the O-1...O-4...O-4'...O-4'' torsion angle. A spreadsheet in the appendix of Ref.

25 returns the *n* and *h* values, answering the question, "what would the shape of the cellulose molecule be if it adopted the traits of a particular disaccharide crystal structure."[25] A tool to calculate the *n* and *h* values is especially useful for cellulose chains because of their extended shape and small number of residues per turn (two to three). Such structures are difficult to visualize as helices.

Several structures in Fig. 3 have interesting conversions to *n* and *h* values. The dixylose component in GURXPX10, which has the linkage with the least-negative ψ value, converts to *n/h* values of -3.05 and 4.88 Å, respectively, these constituting the greatest deviation from twofold screw symmetry except for the RIKMON structure (see Table I for name, not shown in Fig. 3). The two residue geometries of RIKMON lead to *n/h* values of 4.43 and 3.07 Å and 4.70 and 3.11 Å.[25] The extreme value of ψ for GURXPX10 is allowed by the absence of C-6, which if present would result in a steric conflict. On the other hand, XYLBHA (xylobiose hexaacetate) has a right-handed linkage conformation closer to twofold screw symmetry, with $\phi = -105°$, $\psi = -149°$ converting to $n = 2.10$, $h = 5.19$ Å. Another structure at the extremes of Fig. 3 is digitalis (DIGOXN10, see Fig. 4). Its trisaccharide fragment has quite different geometries for its two glycosidic linkages. Using the central residue with $\phi/\psi = -79°, -165°$, a structure results that has $n = -2.003$ residues per turn, and $h = 4.99$ Å. In the crystal structure, there is an O-3—H...O-5' hydrogen bond from an axial O-3—H group. The linkage having $\phi/\psi = -63°, -114°$ leads to a helix with $n = -2.88$ and $h = 4.93$ Å, with no inter-residue hydrogen bond. These geometries are similar to those in cellotrioside undecaacetate (ACELLO10, see Fig. 4), which also

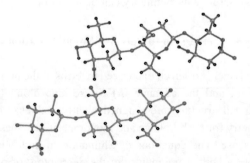

FIG. 4. Similarities between the linkage geometries in DIGOXN10 (upper) and ACCELL10 (lower). The molecular fragments are shifted so that the two linkages giving helical geometries with $n = 2.88$ and 2.91 residues per turn are placed over and under each other, while the residues with nearly twofold linkages are on the left (lower) and right (upper) ends. The acetate groups for ACCELL10 are not shown, nor is the remaining part of the digitalis structure on the reducing end of DIGOXN10. In the twofold linkages, the C-1'—H and C-4—H bonds are parallel to each other. (See Color Insert.)

has two rather different linkages. Their n and h values are 2.10 and 5.23 Å as well as −2.91 and 5.09 Å.

Although hydrogen bonding is widely considered to determine the structures of carbohydrates, there are apparent contradictions. Consider the parallels between ϕ and ψ values for the DIGOXN10 and the peracetylated ACCELL10 trisaccharide molecules (Figs. 3 and 4). Both molecules have a linkage that corresponds to a twofold structure, but only the DIGOXN10 molecule can (and does) make the O-3—H...O-5′ hydrogen bond. Both also have a linkage that corresponds to a nearly threefold structure, as shown in Fig. 4. At those two linkages, neither molecule forms an inter-residue hydrogen bond, although from a chemical standpoint DIGOXN10 could have.

In Ref. 25, many more of these structures were analyzed for the extrapolated n and h values, and the linkages in YONVUT01 (cyclohexyl 4′-O-cyclohexyl-β-cellobioside cyclohexane solvate) were mentioned in another paper.[89] Together with those analyses, the data show that observed β-(1→4) linkages lead to helical geometries that are quite extended, close to the nominal 5.5 Å O-1...O-4 length for a β-glucose residue. The h values are all about 5 Å, and only a few structures correspond to right-handed geometries that deviate significantly from the twofold screw line. Even the most extreme right-handed structure (DETQAV), which has n and h values of 2.35 and 5.10 Å, is rather distant from $n = 3.0$ and thus does not support the proposals for right-handed threefold helices for various related derivatives and complexes of cellulose. This aspect is illustrated in Fig. 5.

Maps such as shown in Fig. 5 have not been produced very often in recent times. As an alternative to the spreadsheet,[25] used to tediously produce Fig. 5, the more-automated procedure of Almond and Sheehan could be used.[95]

4. Orientation of O-6

In addition to information on linkage conformations, and their implications for the helical geometry of cellulose molecules, the crystal structure data also inform on the probable orientations of the hydroxyl groups at position 6, as defined in Fig. 6. The orientation at O-6 is one of the major differences between the structures proposed for native celluloses and those in the literature on small molecules. Populations of the O-6 groups for methyl α- and β-glucoside in solution, according to NMR measurements, are about 45% gg and 45% gt, and 30% gg and 60% gt, respectively, with fewer than 10% in the tg orientation[96] that is observed in crystalline cellulose Iα and Iβ.[10,11] Orientations proposed for cellulose II and III$_I$ are gt.

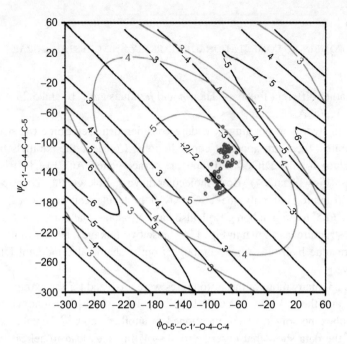

FIG. 5. Conversion of ϕ and ψ values to helical parameters, n and h, based on the geometry of the nonreducing ring in β-cellobiose (CELLOB02)[94] and a glycosidic angle of 115°. The $h=0$ line is not shown, but the contours for $n=5$ and 6 change sign when they cross it. Orange circles indicate the linkage geometries observed experimentally, which are also indicated in Fig. 3. Adapted from Ref. 25. (See Color Insert.)

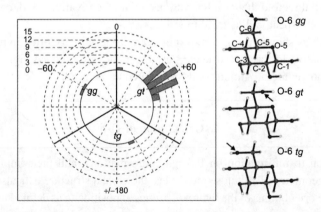

FIG. 6. Distribution of O-6 orientations on rings at the nonreducing ends of cellobiose moieties that have O-3—H groups on the reducing-end ring of the moiety, as found in examples from the CSD. The gt orientations correspond to $+60°$, gg to $-60°$, and tg to $180°$ (the O-6—C-6—C-5—O-5 torsion angle).[37,38] The plot was created with the Vista module of the CCDC software. The O-6 that is oriented to 160.7° (tg) is on a galactose residue in the KUQGOT structure (methyl 4-O-β-D-galactopyranosyl α-D-mannopyranoside methanol solvate). Also shown are glucose rings with the O-6 atoms in the three orientations as indicated by the arrows. Their secondary hydroxyl groups are oriented counterclockwise. (See Color Insert.)

The foregoing collection of crystal structures does offer some support for the proposed structures for cellulose. The existence of *tg* conformations on β-glucose residues was supported by the finding of a *tg* O-6 orientation on a reducing-end ring in the dicyclohexyl cellobioside (YONVUT01) structure (see the later Fig. 9 and related discussion).[89] Further information on the probable orientation of O-6 in cellulose can be gained by examining the orientation of only the nonreducing residues in structures of small molecules having an O-3—H group on the reducing residue. Figure 6 shows that all but five of the O-6'—H groups were *gt* on the 44 structures that had O-3—H groups. This bias toward *gt* structures around β-(1 → 4) linkages may be due to the formation of a (usually weak) bifurcated hydrogen bond so that O-3—H donates to both O-5' and O-6'. That arrangement is observed in the crystal structure (MCELOB) of methyl cellobioside methanol solvate.[75] Other structures have longer O-3—H...O-6' distances, but the O-6' hydroxyl group is usually oriented so that it can be an acceptor.

5. Crystal Packing and Intermolecular Interactions

Besides the patterns of intramolecular hydrogen bonding discussed in the foregoing section, examination of the intermolecular arrangements in different small-molecule crystals can also be informative. One example is pertinent to understanding the mercerization process in the laboratory, treating cellulosic fibers for textiles with sodium hydroxide (NaOH). (This mercerization process is somewhat different from that often used in industry.) The ultimate product of this NaOH treatment is the cellulose II polymorph, but one of the intermediates in this conversion is termed soda cellulose II. That structure gives a fiber-diffraction pattern indicating threefold helical symmetry and a repeat of 14.85 Å ($n=3$, $h=4.95$ Å). The linkage in the MOVGIN structure,[76] a complex of α-cellobiose with NaI and H_2O (Fig. 7), corresponds to a left-handed helix with 2.82 residues per turn, close to a threefold helix. One of the two Na^+ ions is found in a pocket constituted by O-3, O-5', and O-6' as well as an oxygen atom from a water molecule, an I^- ion, and an O-4' atom. Complexation of the Na^+ ion results in alteration of the ψ torsion angle as compared to conformations that have the usual O-3—H...O-5' hydrogen bond. The MOVGIN crystal structure is also interesting because it contains no hydrogen bonds between cellobiose molecules or even between residues in the same molecule. That situation indicates that a major disruption of the native cellulose crystal structure is a likely step in mercerization, as well as indicating a preference for a left-handed threefold helix in the crystal structure of soda cellulose II.

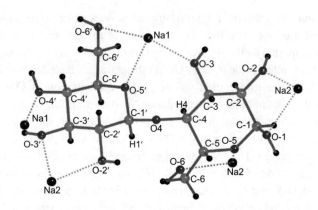

FIG. 7. Structure of the α-cellobiose complex with NaI and H$_2$O (MOVGIN, iodine, and water not shown). The most visible indication of twisting from the twofold screw structure of native cellulose is the difference in orientation of the C-1'—H and C-4—H bonds. In twofold structures, these bonds are parallel to each other. (See Color Insert.)

Another valuable insight into cellulose structure is provided by comparing the IFOVOO01 and ZILTUJ structures in Fig. 8. It has been argued[4] that molecules crystallizing from solution would favor antiparallel packing, with the explanation that the molecules in solution would have random orientations, and coalescence into a crystal would result in equal numbers of molecules having opposite orientations. Because cellulose II in rayon is the product of crystallization from solution, it would be antiparallel, and since cellulose I converts into II without major changes in the fiber, it was expected that cellulose I would also be antiparallel. In this situation, the correct chain orientation in cellulose II was apparently predicted, but on an incorrect basis since there is no requirement for antiparallel packing during crystallization from solution, as illustrated by the IFOVOO01 structure.

In the methyl 4-O-methyl-β-cellobioside structure (IFOVOO01), there is only one molecule per unit cell. The space group is $P1$, that is, only translational symmetry is needed to generate all of the atoms in the crystal. Generation of a second chain by translation along the b-axis created the structure shown in Fig. 8. Parallel packing, with vectors from the nonreducing ends to the reducing ends all pointing in the same direction, is a specific consequence of $P1$ symmetry and having a single molecule per unit cell. In a different crystal form of the same molecule (IFOVOO03), the unit cell has two molecules related by $P2_1$ symmetry. In that case, however, the molecular axes are not parallel to each other.

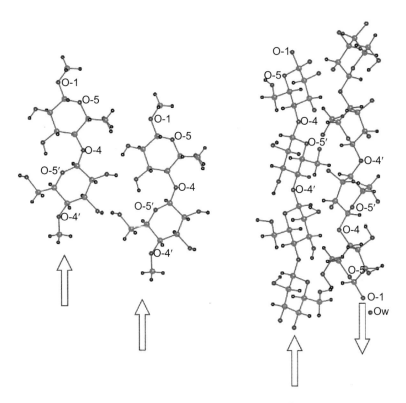

FIG. 8. Molecules from the crystal structures of methyl 4-O-methyl-β-D-glucopyranosyl-(1→4)-β-D-glucopyranoside (methyl 4′-O-methyl-β-cellobioside, IFOVOO01, left) and cellotetraose hemihydrate (ZILTUJ, right). Arrows indicate the direction of the chains in the respective unit cells (up molecules have O-1 higher than O-4). Some hydrogen atoms were not located during the ZILTUJ structure determination. (See Color Insert.)

The structure of cellotetraose hemihydrate also has $P1$ symmetry, but there are two unique molecules per cell, oppositely directed (Fig. 8). With the observation of these and other structures, it is safe to conclude that the philosophical argument used in the early work[4] for an antiparallel structure was wrong. Molecules can indeed crystallize from solution with either parallel or antiparallel packing.

The final example of an instructive crystal structure is cyclohexyl 4′-O-cyclohexyl-β-cellobioside cyclohexane solvate (YONVUT01) (Fig. 9). In its $P1$ unit cell, there are three unique molecules, two directed down in the unit cell and one up.

The tg-oriented O-6 atom in YONVUT01 is indicated in Fig. 9 on the C molecule. It makes a hydrogen bond to O-2′, in the direction opposite to the O-2′—H...O-6

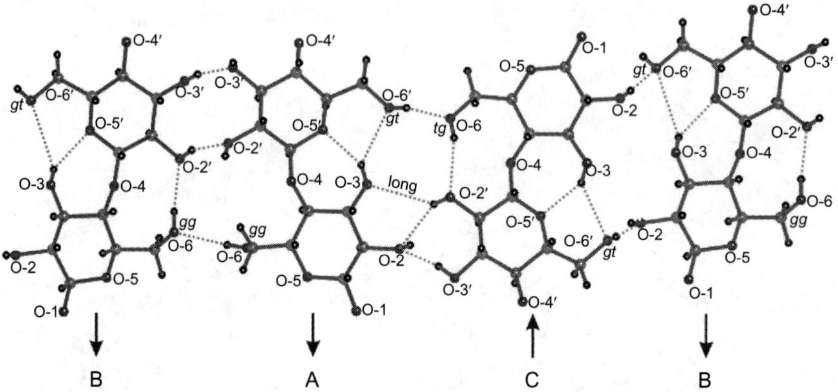

FIG. 9. The carbohydrate component of the crystal structure of cyclohexyl 4′-O-cyclohexyl-β-cellobioside cyclohexane solvate (YONVUT01). The three unique molecules in the unit cell (A, B, and C) are shown, with a copy of B to allow visualization of all hydrogen bonds in the particular crystal plane. The O-6 orientations are labeled gt, gg, and tg, and arrows show the molecular orientations in the unit cell. The cyclohexyl components attached at O-1 and O-4′ are not shown. The figure is from Ref. 89. (See Color Insert.)

hydrogen bond proposed for the majority fraction in native cellulose.[10,11] The B molecule has O-6 in the *gg* position, rotated such that it can also form an O-6—H...O-2′ hydrogen bond. Because the A and B molecules are rotated almost 180° about their O-1...O-4′ axes relative to each other, they are able to form intermolecular hydrogen bonds between like hydroxyl groups, that is, from O-2′ to O-2′, and so on. Finally, the *gt* orientations of all three of the nonreducing O-6 atoms permit the formation of bifurcated O-3—H...O-6′ hydrogen bonds (see sections II.4 and IV).

6. Summary of Section on Extrapolation

In a sense, the foregoing approach constitutes molecular modeling based on strictly experimental results from X-ray diffraction crystallography. (Unlike work with other carbohydrates, neutron diffraction has not been applied to small-molecule relatives of cellulose, although it has been critical to solving the structures of cellulose itself, and has been applied to other small-molecule carbohydrates.[97–99]) The Shimanouchi and Mizushima equations[93] were published more than 50 years ago, when they predated the original crystal structure determinations of cellobiose by more than 5 years. When first introduced, they were used to model cellulose based on standard bond lengths and angles. Then a similar analysis was applied to making a map of a part of ϕ,ψ

space based on an averaged geometry from early crystal structures,[100] similar to the result in Fig. 5. One point of Fig. 5 is to have such a map for all of ϕ,ψ space that uses the current conventions for ϕ and ψ.

A single diffraction experiment is very worthwhile, but it is even more valuable to have a large database of relevant structures. Figure 3 has sufficient data to justify strong suggestions as to the probable shapes of cellulose and its derivatives, and these suggestions indeed agree well with the various structures that have been proposed.[101] According to Fig. 5, which has all of the experimental linkages from Fig. 3 plotted, most cellulose chain shapes are extended and more often than not left handed, with two to three residues per turn. This agrees with the earliest modeling results.[100] At the same time, there is substantial opportunity to add to our knowledge on this subject. Only two of the molecules, namely methyl 4'-O-methyl-β-cellobioside (IFOVOO01 and IFOVOO03) and α-lactose (EYOCUQ and EYOCUQ01), have been found as polymorphs, although many other organic molecules have demonstrated this feature. Also, the number of derivatives of cellobiose that have been subjected to X-ray analysis is surprisingly small in relation to the large numbers of derivatives of cellulose that have been synthesized.

III. Energy Calculations

Although survey and extrapolation methods similar to the foregoing are useful for determining the probable shapes of cellulose and for defining the expected conformational limits, it is often of interest to quantify predictions of shape through energy calculations. The reason for calculating energies is that the probability (P) of finding the ith particular shape is related to that of the shape having the lowest energy (at least for a molecule in the gas phase), by the Boltzmann equation, $P_i = e^{\Delta E/kT}$, where ΔE is the calculated free-energy difference, k is the Boltzmann constant (for individual atoms or the universal gas constant R on a per mole basis), and T is the Kelvin temperature. The shape having the lowest energy is most probable.

In early modeling studies, relative atomic positions within glucose rings were kept fixed (the rigid-residue approximation), and the likelihoods of various proposed shapes were assessed by calculated distances between the atoms on neighboring residues. In fact, such models were initially examined with the help of manual calculators, not computers. Models were described as "fully allowed," "marginally allowed," or "disallowed." In that approach, the energy of an allowed conformation was considered to be zero, and energy of a disallowed model was infinite. A few years later, technology had advanced to the point where the distances were converted into energies based on one of several van der Waals potentials.[102] As with the atoms in the rings, the covalent glycosidic bond-angles were given fixed values in the early programs. The importance of favored values for the torsion angles was recognized from the principles of conformational analysis, but special values for the potential energy resulting from rotations about the glycosidic C-1—O-1 bond were added somewhat later. The Hard Sphere ExoAnomeric (HSEA)[103] and Potential for OligoSaccharide (PFOS)[104] programs included such special terms.

Around the same time, however, software was being developed that treated all of the atoms in the molecule by the principles of classical mechanics.[105] This type of energy calculation became known as molecular mechanics, in contradistinction to quantum mechanics, which treats individual electrons. The small mass of the electron requires a quantum mechanics approach. In molecular mechanics, Hooke's law governing the motion of a spring could be used to approximate a chemical bond, with an ideal length and an increase in energy if the bond is shorter or longer. Typically, the atom positions in a structure are allowed to "relax" or find the positions that give the lowest total energy for the structure through "energy minimization" or "geometry optimization." However, with carbohydrates, the multiple-minimum problem is always present. There will be numerous stable structures, and the energy-minimization routines will only find the lowest energy within a given energy well. The minimizer by itself will not, in general, lead to the structure having the lowest possible energy on a global basis because a simple minimization procedure will not take the structure over energy barriers and seek a lower-energy structure on the other side of the barrier. Therefore, energy minimization is often accompanied by other procedures that permit a search for the global minimum.

The use of "rigid-residue" programs has mostly faded away. The advantage of the relaxed treatment is precisely because the rigid-residue assumption is eliminated. The rigid-residue programs frequently employed the atomic coordinates from a crystal structure as the input for a conformational study. A disqualifying consequence of such an analysis is that the particular crystal structure was most likely to have the lowest energy and other crystallographically observed conformations of the same moiety could have relative energies as high as 100 kcal/mol[106] (1 kcal = 4.18 kJ). Small adjustments in the internal coordinates, such as are permitted by a relaxed-residue, molecular mechanics approach, could have decreased that huge energy difference to as little as 3 or 4 kcal/mol.[107]

It is useful to distinguish the method used for calculating the energy from both the standpoint of the software used to calculate it and the type of calculation to be performed. Thus, if an empirical force field is used to calculate the energy, it is a molecular mechanics (MM) calculation. Both energy minimization and molecular dynamics (MD) can be performed with MM. Both procedures are also performed using electronic structure theory, that is, quantum mechanics (QM), but MD calculations based on QM[108] are used less commonly because of the time required.

Although the MM4 force field has only been implemented in the MM4 program,[109,110] numerous programs have used their own variants of the earlier MM2[111] and MM3[112] force fields, generally annotated with an asterisk or some other mark. Other software programs, such as AMBER[113] or CHARMM,[114] utilize force fields developed for those programs, whereas software such as GROMACS[115,116] and NAMD[117] uses force fields developed for other programs. The development of force fields is an active area that seeks to remedy problems in the earlier versions. An extensive review of the force fields used for carbohydrates is available.[118] Serious modeling work should be based on the most recently developed force field, but consistent with other considerations. A new CHARMM-type force field[119] has already been used several times for modeling projects related to cellulose,[120] and Glycam06[121] is the most recent carbohydrate-aware force field for use with the AMBER program. Programs with collections of different force fields provide a way to show the differences in results that can be expected because of this variety of force fields,[122] but they are not usually the most recent versions.

Generally speaking, the MM3 and MM4 software offers parameters (such as the ideal values for a bond length and the constants for a change in length) for many types of atoms so that many types of molecules can be modeled. Their energy functions are the most complex of the popular force fields and run the most slowly. With the speeds of current computers, that is not necessarily an issue, but the maximum number of atoms (9000 in MM4) could be. The molecular dynamics capabilities of these programs have been ignored, at least by those working on carbohydrates. That is mostly because they are not designed to work with widely used, explicit water models, such as TIP3P.[123] Those explicit water models allow the structure to be efficiently studied in aqueous solution where the conformation is of biological interest. Instead, MM3 studies have been confined to energy minimization with various elevated dielectric constants. However, water is one of the many molecules that these programs can model.

The AMBER and CHARMM programs depend on streamlined potential energy functions, but their authors feel that the answers are usually as good as those from more complex functions, considering that the unknown aspects of the torsional potentials and electrostatics are more of a problem.[124] Other tradeoffs in various programs include the use of single precision for greater speed, or double precision for greater accuracy.

A classic goal of disaccharide modeling is the determination of the range of likely interglycosidic conformations in ϕ/ψ space. This can be accomplished with several modeling protocols, including energy minimizations at each point on a grid, unconstrained minimization, and various forms of MD simulation. All of these protocols have a fundamental problem, namely "sampling." If full consideration is not given to all of the likely structures, then the sampling is inadequate and the results are likely to be incorrect. The computed energy of an isolated (in vacuum) α-glucose molecule in the usual $^4C_1$ shape varies over a range of 13 kcal/mol depending on the combinations of individual orientations of the exocyclic substituents. Some 150 combinations are stable at the B3LYP/6-31+G(d,p) level of quantum mechanics theory.[125] That number of stable structures is a significant decrease from the 729 structures that are generated by putting each of the five O—H groups and the C-6—O group into each of the three staggered orientations ($729 = 3^6$). Cellobiose has 10 rotatable groups besides the ϕ and ψ torsion angles, giving 59,049 potential combinations with an even larger range of energies for stable structures. Some strategy must be used to try to find the global minimum, or the results will not be completely representative of the particular method of calculating the energy.

With minimization, the obvious approach is to start the calculations with as many different combinations of exocyclic group orientations as can be afforded, perhaps using a separate calculation to identify rapidly those combinations most likely to have low energy. For example, the O-6 atoms can be placed in the gg, gt, and tg orientations, and the hydroxyl groups can be put into various combinations of clockwise or counterclockwise (Fig. 6) orientations. When the exocyclic groups are included, energy surfaces for the disaccharide are actually hypersurfaces, but the usual practice is to display only the energy contours versus the ϕ and ψ torsion angles.[126] However, a variety of additional information, such as exocyclic orientation,[127] ring puckering,[128] or hydrogen bonding,[100] can be overlaid on the energy contours.

In MD studies, the counterpart approach to diminishing problems of sampling is to use sufficiently long simulations. However, much of the conformational surface would never be visited during even lengthy simulations, and barriers between important conformers would not be crossed sufficiently often at room temperatures to obtain valid relative frequencies that could be used in energy calculations. Useful techniques include assisting the structure to take the different shapes, for example, metadynamics,[129] or umbrella sampling.[130] Another approach utilizes the consequence of higher temperatures to visit higher-energy conformations during the simulation, combined with interchanges with structures simulated at cooler temperatures to obtain the dynamics at room temperature (a method called Replica Exchange MD).[131]

1. Results on Individual Isolated Molecules with Empirical Methods

It is fairly rare for a modeling researcher to utilize more than one software system. Instead, the literature on the same or similar molecules, such as cellobiose or lactose, describes work performed in various ways by different groups, employing different software. In each of the calculations in Fig. 10, the same set of 155 starting geometries was used at each ϕ/ψ point. A further procedure was to expand the glycosidic bond angle to 150° before starting each minimization. That step decreases the interpenetration of the adjacent glucose rings in the regions of high-energy structures.[132] Interpenetration can result in an insurmountable energy barrier for the minimization process and thus generate high but meaningless final energies. In our work, interpenetration can occur because, before minimization for each starting geometry, the ϕ and ψ torsion angles were adjusted to the ϕ/ψ point in question with the internal coordinates of the glucose residues otherwise being held rigid. That approach is different[133] from the technology that is sometimes incorporated in modeling software for "scanning" a potential energy surface wherein which each calculation is started with the energy-minimized structure from the previous ϕ/ψ point. When carbohydrate structures are minimized, inelastic structural deformations can occur, such as changes in ring shape or rotations about exocyclic bonds. At subsequent minimizations, the molecular structure typically does not revert to the starting geometry initially intended from those alterations, another manifestation of the multiple-minimum problem. Therefore, the calculated energies will depend on what ϕ/ψ points preceded them. Starting each minimization with a fresh copy of the starting geometry avoids those complications.

Figure 10 shows a series of ϕ/ψ surfaces made with MM4,[109,110] AMBER/Glycam06,[121] and CHARMM/CSFF.[134] The energy maps in the left column all used the dielectric constant recommended for vacuum calculations, namely 1.5 for MM4 and 1.0 for the others. In the case of the Glycam06 and CSFF maps, there is a caution that their force-field parameters may not be optimal for vacuum calculations. Instead, the parameters have been adjusted to yield the best match with the experimentally observed properties in solution when using a specific water model during MD studies.

In energy-minimization studies, it is impractical to incorporate enough explicit water molecules in the model to constitute a solution. A simple way to mimic the effect of a condensed phase is to increase the dielectric constant (ε),[135] and maps in the right column were produced with dielectric constants selected to give the best visual fit with the distribution of conformations observed in crystal structures. Similarities are found for all three vacuum maps. The lowest energy on each map is found at the left or right edge ($\phi = -300°$ or $+60°$, $\psi = -120°$), and not in the center near

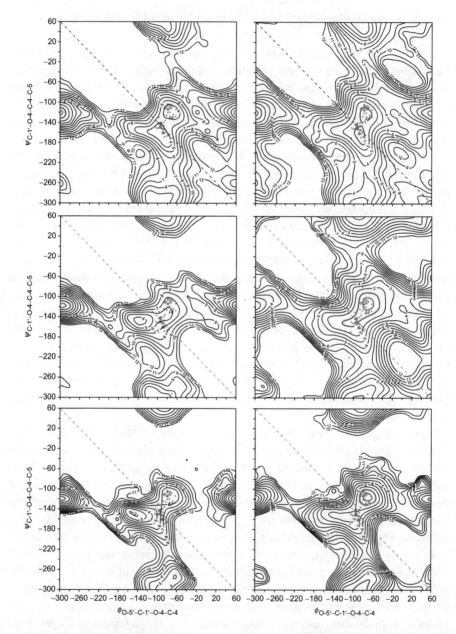

FIG. 10. Energy maps for MM4 (top left: $\varepsilon = 1.5$, top right $\varepsilon = 7.5$), AMBER/Glycam (middle left $\varepsilon = 1.0$, middle right $\varepsilon = 8.0$), and CHARMM/CSFF (bottom left 1.0, bottom right 4.0). Magenta contours are for 0.25 kcal/mol and the blue for 1.0 kcal/mol. Orange dots represent the crystal structures described in Table I and Fig. 3. Calculations by G. P. Johnson and A. D. French. (See Color Insert.)

the crystal structures from Fig. 3 and Table I. That edge-of-the-map location is, however, in agreement with the experimental gas-phase studies of Simon's group,[136] as well as the QM studies of Strati et al.[137] (see following discussion). The minimum at the top or bottom of each map is the highest in energy, and the GURXPX10 crystal structure (see Table I, $\phi = -82°$, $\psi = -80°$) has a high (5–10 kcal/mol) energy, due in part to its xylobiose fragment not having C-6 groups. The MM4 maps are the most predictive in the sense that the crystal structures are well distributed in the central area of two nearly equal secondary minima; the twofold screw-axis conformations are on a ridge that is not very much higher in energy.

The main effect of the elevated dielectric constant is to decrease the strength of the intramolecular hydrogen bonds by a factor of the default dielectric constant in a vacuum divided by the dielectric constant. Thus, for example, the purely electrostatic CHARMM hydrogen bonding is decreased by a factor of 1/4. The maps from calculations at elevated values of the dielectric constant all decrease the energy of the minima near the crystal structures, relative to the edge-of-the-map gas-phase minima. The AMBER/Glycam and CHARMM/CSFF maps show the central minimum to be the global minimum, favoring it by more than 2 kcal/mol and about 0.5 kcal/-mol, respectively. The MM4 $\varepsilon = 7.5$ map shows the $\phi = -300°$ or $+60°$, $\psi = -120°$ minimum to be still favored, but by less than 0.5 kcal/mol, whereas it was favored by more than 3 kcal/mol in the vacuum calculation. The corresponding energy of the dixylose moiety in the GlcA-Xyl-Xyl trisaccharide structure (GURXPX10) was substantially decreased by all calculations at elevated values of the dielectric constant. The MM4 calculations incorporate the external anomeric torsional effect,[110] which lowers the energy barrier at ψ torsion intervals of $-120°$ by about 1 kcal/mol. Despite that fact, the twofold screw-axis diagonal line coincides with a small barrier between two adjacent minima.

Another approach involved Replica Exchange Molecular Dynamics studies of cellobiose, tetraose, and hexaose, all in aqueous solution using TIP3P (a potential function for simulating liquid water), calculated with AMBER/Glycam.[138] This was a very expensive calculation in terms of computer time. The resulting frequency plot for all three temperature ranges of the disaccharide was quite similar to the 1- and 2-kcal/-mol contour lines from the $\varepsilon = 8$ Glycam map in Fig. 10. There was little indication of end effects from the three different lengths of the molecule, with very similar ϕ and ψ values regardless of position in the hexaose. A finding in that work that there was a slightly shorter per-residue extension of the longer molecules may have failed to account for the deviations from twofold screw symmetry.

Both methyl α- and methyl β-cellobioside were studied with vacuum-state and solvated molecular dynamics and the new carbohydrate-aware CHARMM force field.[139] These were also expensive analyses, with 15 simulations of the vacuum state, each for a microsecond, and 5 simulations of the solution state, each for 50 ns, for both cellobiosides. The CHARMM parameterization used extensive MP2/cc-pVTZ quantum mechanics studies of analogues of the disaccharides that were based on tetrahydropyran for the nonreducing ring and cyclohexane for the reducing ring. Because the simulations were so lengthy, sufficient data were collected to justify contouring the energy surface out to 6 kcal/mol above the minimum (Fig. 11), based on the frequencies of each ϕ/ψ conformation. On the solvated maps in that work, the lowest energy is found near the center of the map and the crystal structures. The map for methyl β-cellobioside in a vacuum shows a preference for the alternative $\phi = -300°$ or $+60°$, $\psi = -120°$ minimum by more than 1 kcal/mol, but the vacuum map for methyl α-cellobioside indicates a preference of about 0.2 kcal/mol for the central region. The southern minimum for both compounds in solution is the second lowest in energy; the eastern minimum is higher by about 2 kcal/mol than the southern one. Except for the relative energies of the secondary minima, the β-cellobioside map is remarkably similar to one in Ref. 25 that was based on a hybrid calculation. The authors also studied methyl β-maltoside in the same paper. Their vacuum map showed a low-energy minimum (<2 kcal/mol) at $\phi = 170$ that is a high-energy region (>10 kcal/mol) in most other maltose modeling studies. Their solvated maltoside map did not show the alternative minimum and was more similar to other work.

Despite that problem with the vacuum maltoside map, these new CHARMM MD simulations in vacuum seem to provide a better fit of the cellobiose-type crystal structures than the foregoing energy-minimization results in Fig. 10 with the AMBER/Glycam or CHARMM/CSFF force fields. The rationalization of the experimental crystal structures is better with the CHARMM-solvated calculations than with the vacuum studies, but these CHARMM-solvated studies are different from, rather than superior to, the preceding Glycam results at high dielectric constant or in TIP3P water.[138] In all cases, the new CHARMM study with explicit solvent, and the elevated dielectric CHARMM/CSFF and AMBER/Glycam-solvated studies, indicates a higher energy to one side of the twofold screw conformation line, whereas the MM4 calculations herein show two nearly equal minima on either side of the twofold line.

Both cellobiose and 4-thiocellobiose (the O4 of the glycosidic linkage is replaced by sulfur) were studied with a nonintegral QM::MM hybrid method.[140] Although the MM3 force field seems to work reasonably well for cellobiose,[127,141] the torsional parameters for the C—S bonds in the thio compound were not part of the original MM3 parameterization. Instead of deriving O—C—S—C and C—S—C—C parameters, conformational energy surfaces were made for the "stripped" analogues of the disaccharides with both MM3 (using MM3's automatic trial parameter generator) and HF/6-31G(d) quantum mechanics. The differences between the QM and MM3

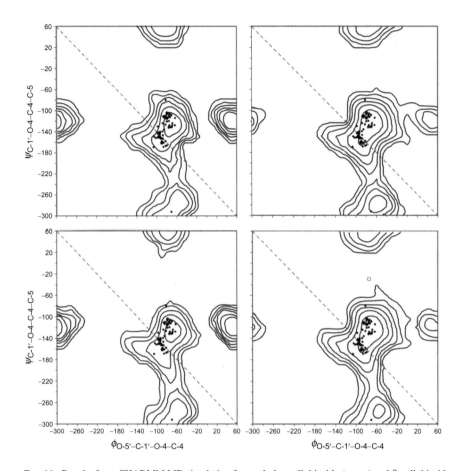

FIG. 11. Results from CHARMM MD simulation for methyl α-cellobioside (upper) and β-cellobioside, lower. Vacuum simulations are on the left and simulations in TIP3P water are on the right. Crystal conformations are indicated by black dots, and ϕ and ψ combinations that would correspond to structures having a twofold screw axis are indicated by the diagonal line. Contours are in relative kcal/mol values, with the outer contours at 6 kcal/mol. Contour lines were taken from Ref. 139.

surfaces were added to the full MM3 surfaces for the disaccharides. That step corrects for the lack of well-developed parameters for the torsion angles in the linkage. This strategy was quite effective for modeling sucrose,[107] and a capability to add such a correction map has been built into the current CHARMM package.[119]

Figure 12 shows the difference between the HF/6-31G(d) QM surface and the MM3 (96) surface for the thiocellobiose analogue. There is enough difference between the two maps to warrant a hybrid approach for modeling 4-thiocellobiose with MM3. The difference between the final, hybrid maps for $^4C_1$ cellobiose and thiocellobiose was

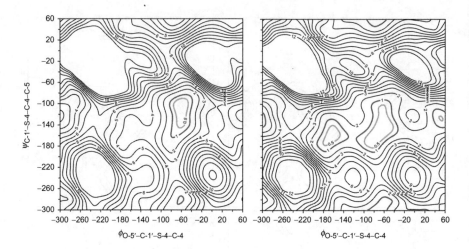

FIG. 12. Energy maps using HF/6-31G(d) (left) and MM3(96) (right) for the analogue of 4-thiocellobiose (tetrahydropyran rings, linked through a sulfur atom). The MM3 calculation utilized parameters generated automatically for the ϕ and ψ torsional energies because parameters for C—O—C—S and O—C—S—C angles had not been part of the official development of MM3. Green contour lines indicate a 0.5-kcal/mol contour. The substantial differences in energies led to the use of a hybrid method for mapping the disaccharide.[140] In that simple, nonintegral hybrid method,[107] the difference between these two maps at each ϕ/ψ point was added to the MM3 map for the disaccharide. Calculations by G. P. Johnson, L. Peterson, A. D. French, and P. Reilly. (See Color Insert.)

remarkably small.[140] The main differences were in the heights of the barriers between the minima, with a slightly larger area within the 1 kcal/mol contour of the map for thiocellobiose. One distinguishing feature of that work was the large number of starting geometries. With inclusion of $^4C_1$, $^1S_3$, and ^2S_O rings, 8219 cellobiose structures were used, with 2277 for 4-thiocellobiose in the $^4C_1$ conformation.

The specific purpose of those energy maps was to assess the distortion of glycosidic linkages when bound at the active sites of glycosidase enzymes. The conformations of monosaccharide rings have long been known to undergo distortion during the process of cleavage of the glycosidic linkage. If the ring is not in the normal chair, it is considered to be distorted and to have increased potential energy. However, the degree of linkage distortion is not so easily assessed by a visual examination of the structure, but it can be considerably aided by knowledge of an energy surface. Do the ϕ and ψ values correspond to a high energy? If so, the linkage is "distorted." Examination of the crystal structures of the complexes of glycosidases with their carbohydrate substrates indicated that the distortion effectively aided in making the linkage easier to attack in some cases.

2. Results with Quantum Methods

Fairly modest QM methods take roughly five orders of magnitude more time to compute than MM methods, and so the types of problems that can be studied are restricted in comparison. Early efforts were focused on understanding the anomeric effect,[142] and subsequently on providing reasonable standard values for parameterization of torsion angles, especially those associated with glycosidic linkages.[143] In the year 2000, it was shown that QM maps for disaccharide analogues lacking hydroxyl groups were remarkably predictive for the crystallographically observed conformations of the disaccharides themselves.[144] That predictive ability was improved by including methyl groups as stand-ins for the CH_2OH groups. As mentioned previously, the work of Strati et al.[137] predicted that the gas-phase structure of cellobiose is very different from the conformations observed in condensed phases, based on minimization studies of 26 conformers with the B3LYP/6-311++G(d,p) level of theory.

Subsequently, energy surfaces using HF/6-31G(d) and HF/6-311++G(d,p) were constructed for cellobiose.[145] Those maps covered only the quarter of the total ϕ/ψ space that is near the crystal structures but were supplemented by freely minimized structures with both HF and B3LYP calculations. Those free minimizations confirmed the results of Strati et al., while the maps located the conformations determined by crystallography in a secondary minimum that is otherwise (similar to the $\varepsilon=8$ AMBER/Glycam map in Fig. 10). Follow-on work identified the 23 combinations of exocyclic group orientations that gave lowest energy for the 81 points on the small ϕ/ψ plot.[146]

More recently, conformational space was mapped for cellobiose by Schnupf and Momany.[126] They used B3LYP theory and a mixed basis set, 4-31G, for the carbon atoms and 6-31+G(d) for the oxygen atoms to decrease the computation time. Also, they used the COSMO continuum-solvation model. This model accounts for dielectric screening that occurs when polar molecules are dissolved in polar solvents. More recent continuum methods also include cavitation, dispersion, and contributions of structural change in the solvent to the full free energy of solvation.[147] Still, their use of COSMO showed an effect on the energy surface that was along the lines of the effects of solvent modeling in the empirical force-field results shown in Figs. 10 and 11. Namely, it decreased the energy around the crystal structures relative to the gas-phase minimum at the edge of the map. However, COSMO solvation did not diminish it enough to favor the conformations observed in the condensed phase.

An unusual proposal in the Schnupf and Momany work favored the construction of energy maps for each local minimum for each starting geometry so that the reader could "digest" the conformational behavior of the molecule, instead of viewing adiabatic maps. That concept was rebutted[133] with the consideration that some 23 different starting geometries had been needed to yield the lowest energy at each of 81 ϕ/ψ map points in a quarter of ϕ/ψ space,[146] and there are typically four minima for the entire ϕ/ψ space. Thus, a minimum of 92 maps would be required to consider all minima, prior to "digestion" of the information. That would cover the energies in a vacuum, but then the solvated energies would be needed. According to their proposal, 184 different maps would therefore have to be published. Despite their complaints about the adiabatic maps, however, they published both vacuum-state and solvated adiabatic maps. Each showed minima on both sides of the line for twofold screw-axis structures.

Issue was also taken[133] with their mapping strategy. That strategy used the structure from the previous minimization to start the next one. As discussed in the previous section, such scanning of the conformations has the disadvantage that inelastic structural conversions occur on proceeding from one conformation to the next. When it comes time to calculate the energy for the structure that is 360° from the original, and thus in principle identical, other details of the structure, such as hydroxyl-group orientations or ring shapes, would have changed, so the calculated energy would not be the same.[133]

Another problem with that Schnupf and Momany work was the limited number of starting geometries. Of course, limits to the number are a consequence of time-consuming QM studies, but low-energy structures had been previously identified[146] and were not used. New calculations[148] were undertaken that used those previous structures, as well as the starting geometries from the Schnupf and Momany work.[126] The new project was designed to allow maximum comparability of the two studies, while taking advantage of different mapping strategies[132,133] and a more complete, SMD[147] solvation model.

Figure 13 shows the vacuum-state and solvated adiabatic maps from the new calculations.[148] Both favor twofold screw-axis structures. In the case of the vacuum-state map, where the twofold structure is only in a secondary minimum, the favored twofold conformation came from a starting geometry that was not included in the Schnupf and Momany work. Interestingly, the favored structures in vacuum at the B3LYP level of theory were generally the same as the exocyclic arrangements favored in the earlier HF study.[146] Regarding the solvated map, its preferred starting structures were generally from the Schnupf and Momany set. Because of that, the different result is attributed to the improved SMD solvation model.

One feature of the solvated map is that it is "flatter" than the vacuum-state map. Thus, the paths between minima are lower in energy, and more of the ϕ/ψ space is covered at energies of less than 12 kcal/mol. This is also true among the individual starting geometries; there is less variation in calculated energy at each ϕ/ψ map point. Figure 14 bears this out, showing the distributions and ranges of energies at four different ϕ/ψ points for both the vacuum and solvated maps.

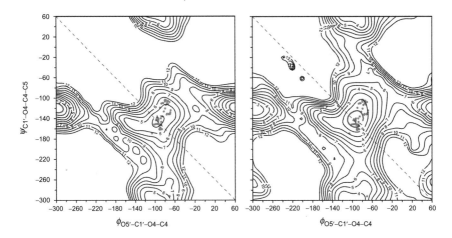

FIG. 13. Energy surfaces for β-cellobiose calculated with geometries from mixed-basis-set energy minimization (see text). The left map is for a vacuum calculation, and the right incorporates SMD solvation. Used with permission from Ref. 148. (See Color Insert.)

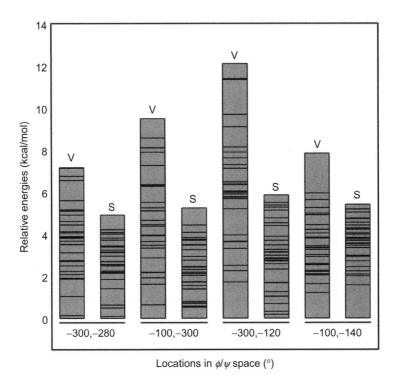

FIG. 14. Relative energy distributions at the indicated ϕ/ψ map points for the 38 starting geometries of β-cellobiose from Ref. 148. Each bar has horizontal lines (including the top and bottom, some overlap or nearly overlap, making wider lines) for the 38 different calculated energies at each ϕ/ψ point. The S and V designations above each bar indicate whether the energies were calculated in vacuum (V) or SMD continuum solvent (S).

3. Assessment of ϕ/ψ Mapping

Taken together, the foregoing energy maps reveal some important points, especially in relationship to the linkage geometries that have been plotted from crystal structures. The first point, raised in the foregoing Section I.2, is that the experimentally observed geometries are found in only a small region of conformational space. This is so, despite the diversity of chemical structures among the disaccharides and derivatives (Table I and Appendix). Therefore, carbohydrate structures in the solid state should be predictable, regardless of many specific details of the molecule's environment and even its substituents.

In crystals, it is almost certain that the cellobiose molecule does not exist in its intrinsic minimum-energy shape. For example, an isolated molecule will make intramolecular hydrogen bonds for the simple reason that there are no other molecules to which it can make hydrogen bonds of lower energy. While the molecule in the crystal will have the lowest free energy, that energy will have both intermolecular and intramolecular contributions. Orientations of exocyclic groups will change to allow the stronger intermolecular interactions to replace the weaker interactions of the isolated molecule. Nevertheless, the situation can be viewed as one in which the molecule generally takes a minimum-energy form, but the interactions in each specific crystal constitute perturbations from that ideal form. When numerous experimentally determined structures are available, as in the case for cellobiose and lactose, the quality of prediction can be monitored by two measures. Qualitatively, observed structures should have a low corresponding energy, and the areas of low energy should be occupied by observed structures. A quantitative examination can be obtained by fitting an exponential, Boltzmann-like curve to the corresponding energies, resulting in a "temperature."[132] The higher the temperature, the greater range of energies that would be observed.

Efforts with MM3 similar to those in Fig. 10 also showed that better qualitative fits of crystal structures to energy surfaces were obtained when an elevated dielectric constant was employed. The same was true for qualitative and quantitative fits when a hybrid mapping technique was used.[132] When dielectric constants of 3.5 or 7.5 were used, the "temperature" was about 450 K, but it was considerably higher (924 K) when the MM3 vacuum dielectric of 1.5 was employed.[132] Those temperature values were based on corresponding relative energies for crystal structure conformations on energy maps for eight disaccharides, all calculated with the same method.

To some workers, this need to elevate the dielectric constant to successfully predict the conformations in crystals with isolated-molecule calculations indicated an imbalance in the force field. Other workers have maintained that it is not appropriate to try to predict conformations in the crystal, and that changing the dielectric constant is an assault on the careful work done to parameterize the force field. The Glycam and CSFF examples in Fig. 10 indicate that those force fields also become more predictive of the crystal structures when the dielectric constant is elevated, but the MM4 maps in that same figure do not vary as much. That observation would appear to result from the functional forms of the equations used to calculate hydrogen bonding. In the Glycam and CSFF force fields, hydrogen bonding is purely electrostatic, but the MM4 equation has both an electrostatic contribution and a modified Buckingham-potential component, the latter being immune to changes in the dielectric constant. The 1992 and newer versions of MM3 also use a Buckingham term but do incorporate the dielectric constant. In any case, the maps produced with elevated dielectric constants have lower energy for the region of the crystal structures relative to the minimum at the right and left edges of the maps than do the vacuum maps. Also, their contouring in the region of the crystal structures for the CSFF and Glycam force fields in Fig. 10 is much more compatible with the distribution of crystal structures.

From the preceding paragraph, it seems that there is reason to believe that the crystal structures can be predicted with isolated models, although the use of the elevated dielectric constant is obviously a crude tool. A likely explanation was already given, namely, that the elevated dielectric constant decreases the strength of the intramolecular hydrogen bonds that create steep-sided energy wells in vacuum-state calculations. That decrease in strength is analogous to the situation in condensed phases, where stronger intermolecular hydrogen bonds can form and intramolecular bonds are less relevant. Its applicability, if not validity, was confirmed by an excellent comparison (not shown) of the AMBER/Glycam energy map of Fig. 10 computed with dielectric 8.0 and the Replica Exchange MD run with explicit TIP3P water.[138]

Another question that can be explored is whether the conformations in solution resemble the conformations in crystals. In the aforementioned review, two more or less standard approaches were taken. In the case of the CHARMM calculations in Fig. 11,[139] free-energy contour surfaces were determined, both with and without explicit solvent water. Instead of umbrella sampling or Replica Exchange MD, the authors simply ran very lengthy simulations. There was a difference between the vacuum-state and solvated runs for the methyl β-cellobioside that was analogous to the difference for the CHARMM/CSFF and AMBER/Glycam energy-minimization mapping studies presented in Fig. 10. That is, the minimum on the right- and left-hand edges of the map was favored in the vacuum-state calculations, whereas the central minimum near the crystal structures was favored in the solution studies.

Comparison of the B3LYP results for β-cellobiose in Fig. 13 with the CHARMM calculations for methyl β-cellobioside in Fig. 11 reveals several differences. Whereas the twofold line passes through the center of the region bounded by the 1 kcal/mol contour line of the QM plots, the CHARMM minima are somewhat removed from the line. Another difference is that the CHARMM surfaces are somewhat flatter than the QM surfaces, and the transition barriers between minima calculated by CHARMM are lower. Also, the relative energies of the three main minima are different.

Differences are inevitable at the present state of the art. For one thing, the molecules are different. The dispositions of the methyl group made only a small difference between the maps for the α- and β-cellobiosides but, nevertheless, could influence whether clockwise or reverse-clockwise orientations of the hydroxyl groups are favored. The fact that the individual atomic charges vary somewhat with conformation in the QM calculations and are fixed for the CHARMM calculations could also affect the results. In favor of the MD calculations, the lengthy simulations should have overcome the sampling issues, whereas the QM calculations were limited to just 38 unique arrangements of the exocyclic groups. Starting geometries as yet untested could alter the adiabatic energy surface. Also, because the minimizations are constrained to grid points on a map, free-energy calculations are not used in adiabatic mapping studies.

4. Hydroxyl-Group Orientations

As mentioned in the preceding paragraph and shown in Fig. 14, the orientations of the exocyclic groups are major factors in the calculated energy. Equatorial secondary hydroxyl groups form apparent cooperative, clockwise, or counterclockwise rings of hydrogen bonds in the lowest-energy vacuum models.[149] This subject has been explored in greater depth by using 1,2-dihydroxycyclohexane as a model compound.[150] In that project, one of the hydroxyl groups was held fixed as an acceptor, and the other hydroxyl group was rotated in increments, with energies calculated by quantum mechanics. That energy curve was compared with the one from the rotation of a single hydroxyl group attached to cyclohexane. A net variation of 4 kcal/mol was observed in vacuum, with the minimum given by an arrangement that appears to resemble a hydrogen bond between the two hydroxyl groups. However, the putative hydrogen bond would have a small O—H...O angle of 105° and a long H...O distance of about 2.42 Å. This is typical of one component of one of the clockwise or counterclockwise rings. As described in the following, analysis of the electron density around the hydroxyl groups does not indicate that a hydrogen bond has actually formed, at least in the sense of the Atoms-In-Molecules approach[151] (see the following section). Despite not finding a bond critical point, some sign of interaction can be visualized with a very weak, lens-shaped second derivative (the Laplacian) of the electron density.[152,153] Repeating the 1,2-dihydroxycyclohexane calculations with the SMD continuum solvent model (as used in Fig. 13) diminished the net attraction between the two hydroxyl groups to about 0.5 kcal/mol.

Besides the computations in that work, the CSD[31] was scanned for sequences of three secondary hydroxyl groups that are adjacent and equatorially oriented, analogous to those in β-glucopyranose. Clockwise or counterclockwise orientations of hydroxyl groups similar to those described in the preceding paragraph were found in the experimental crystal structures. However, arrangements corresponding to two sequential "bonds" were rare, fewer than would be expected based on there being no energy preference for such structures. When intraresidue O—H...O interactions were found on adjacent hydroxyl groups of pyranose rings, invariably a stronger hydrogen bond was also involved. The donated hydrogen atom in these intraresidue attractions is also linked, with shorter H...O distances and larger O—H...O angles, either to an adjacent ring on the same molecule or to a different molecule.

IV. Detection of New Stabilizing Interactions in Cellulose with Atoms-in-Molecules Theory

Richard Bader's quantum theory of Atoms-in-Molecules (AIM)[21] provides a consistent and relatively convenient method to determine atom–atom interactions and other properties. The electron (or charge) density throughout a molecule or larger structure can be obtained, either from quantum mechanics calculations or from X-ray diffraction experiments, and then analyzed with AIM theory. A complementary experimental approach is called deformation density analysis.[154] In both experimental analyses, diffraction-intensity data are collected with greater redundancy to achieve the highest accuracy. After a conventional refinement of atom positions against the observed diffraction-structure factors that assumes spherical atoms, a multipole refinement is used. After such refinements, distortions of the electron cloud around an atom can be visualized by subtracting a spherical model of the electron density. The remaining electron density indicates, for example, lone pairs of electrons or areas of negative and positive electron density that are indicative of hydrogen bonding.[155]

Figure 15 shows the key element of AIM analyses. In the molecular graphs produced by this theory, the lines between atoms that look like the bonds from our conventional cartoons of molecular structures are actually the bond paths that are

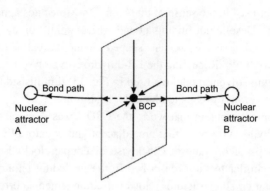

FIG. 15. Definitions of bond path and bond critical point (BCP) for a bond between two atoms (nuclear attractors) A and B. Arrows indicate directions of increasing electron density. The BCP is at the minimum along a line (the bond path) that corresponds to a maximum in the electron density in a plane that is perpendicular to the line.

determined by the analysis of the electron density. A bond critical point is identified where the electron density is at a minimum but increases toward two nuclei in one dimension while being at a maximum in a plane perpendicular to the paths of increasing electron density as the two nuclei are approached. These bond paths need not be straight lines (Fig. 15).

Purely theoretical AIM studies were carried out by Chen and Naidoo, who analyzed hydrogen bonds in model disaccharides.[156] Extensive surveys of non-carbohydrate structures for hydrogen bonding have been carried out with relatively small basis sets and are reported in Ref. 157. Other workers have used AIM theory to study the stacking of sugars with aromatic rings through π-electron interactions.[158]

Figure 16 shows the results from an AIM analysis of the lowest energy structure for the minimum at the bottom of the vacuum QM map in Fig. 13. As is typical for computed carbohydrate structures at energy minima, the secondary hydroxyl groups are arranged in counter- (reverse-) clockwise orientations that would appear to form weak hydrogen bonds. As previously noted,[159,160] however, the AIM analysis does not consider these atoms (or those in similar clockwise orientations) to be engaged in "bonding," and no bond paths or bond critical points are shown. Even so, there is definitely a lower energy for isolated (vacuum) structures with such arrangements of the secondary hydroxyl groups.[150] Other interactions are shown, such as those between the hydrogen atoms on C-3 and C-1'. Several empirical relationships have been used to estimate the energy of such interactions, based on the values of the electron density, especially λ_3, the positive eigenvalue of the Hessian matrix of second derivatives of the electron density.[161,162]

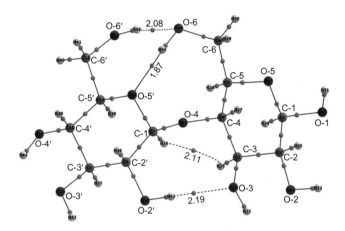

FIG. 16. Molecular graph for β-cellobiose, created with the AIMStudio module of the AIMAll package. The structure having $\phi = -100°$ and $\psi = -300°$ corresponds to the third-lowest minimum, at the bottom of the vacuum map (Fig. 13). With O-6 on the reducing residue in the *tg* orientation and O-6 on the nonreducing residue in the *gt* orientation, it has a relative energy of 4.56 kcal/mol. Distances (Å) for atom–atom noncovalent interactions have been added. The short hydrogen bond (1.87 Å) between O-6 and O-5′ merits a solid bond path according to the defaults in the software, whereas weaker interactions are shown with dotted lines. Green dots represent the bond critical points. (See Color Insert.)

The first simultaneous application of AIM and difference-density theories to a carbohydrate molecule was to crystalline sucrose.[163] The structure of trehalose dihydrate was the next to be studied with these methods.[164] In the dihydrate trehalose crystal, weak intermolecular interactions such as C—H...O hydrogen bonds to its water molecules were identified, as well as conventional O—H...O hydrogen bonds. More recently, the same methods were applied to methyl β-cellobioside.[165] That compound is especially interesting to students of cellulose structure because of its many similarities to the cellulose III$_I$ polymorph. One finding in that work was the confirmation of the bifurcated intramolecular O-3—H...O-5′ and O-3—H...O-6′ hydrogen bonds, with the requisite bond critical points for both interactions. Although this particular crystal structure furnishes the best known example of those bifurcated interactions with O-5′ and O-6′, many other examples might be found, as discussed in the foregoing section on O-6 orientation.

Additional insight into the stability of this crystal structure, and by analogy, to cellulose III$_I$, was furnished by applying AIM analysis to the stacks of cellobioside molecules in the *a*-axis direction, as shown in Fig. 17. Therein, the crystallographic coordinates were used with a "single-point" QM calculation to obtain the wave

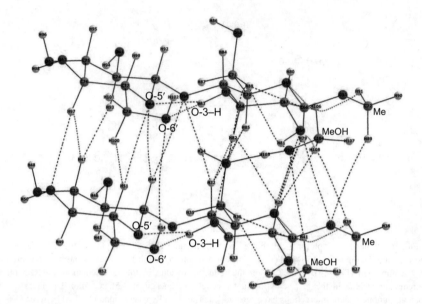

FIG. 17. Atoms-In-Molecules molecular graphs for the methyl β-cellobioside–methanol adduct, showing bond paths (BPs) and the atoms, based on the MCELOB crystal structure.[75] Two complete cellobioside molecules are shown, one above the other, with the reducing-end methyl groups at the right side (Me). The methanol molecules (MeOH) are also shown. The white spheres are hydrogen atoms, the large red spheres are oxygen atoms, and the gray spheres are carbon. Covalent BPs are solid black lines, and weak interactions, including O—H...O and C—H...O hydrogen bonds, are shown as dashed lines. A strong conventional hydrogen bond from the upper MeOH to a lower O-6 is shown as a solid line, and the slightly weaker bifurcated O-3-H...O-5 and ...O-6 bonds are shown as dashed lines. This drawing was produced by AIMStudio from the AIMAll[166] software, based on B3LYP/6-311++G(3df,3pd) quantum calculations. (See Color Insert.)

function (description of the distribution of electron density), but a similar drawing (with fewer interactions) can be created based on direct analysis of the electron density from the X-ray experiment. In Fig. 17, there are many dashed lines between the upper and lower molecules. Most of these are C—H...O interactions,[151] but there are also some H...H interactions. The O-3—H...O-5' and ...O-6' bifurcated hydrogen bonds can be seen for both molecules. Also of note is the O-6'—H...O-6' hydrogen bond between the two molecules, which has an interatomic H...O distance of about 3.6 Å. That bond indicated by the software could be an artifact of the incomplete surroundings of the two isolated molecules in the computer calculation. In the crystal, O-6'—H is a donor, forming a strong hydrogen bond to O-3' on an adjacent molecule that was not included in these AIM calculations. That and other molecules were omitted because the AIM calculations are based on output from a quantum mechanics calculation that uses a large basis set [6-311++G(3df,3pd)] which is time consuming, even when just two molecules are included.

V. Modeling Crystals of Cellulose

The study of cellulose crystals is important because cellulose is often encountered in a crystalline state. Its biosynthetic formation results in almost simultaneous crystallization, the extent of which depends on the plant (or animal) source. Cellulose materials regenerated from solution display a range of crystallinity. This crystallinity is attributable to the existence in the fibers of small crystallites that can be isolated through removal of noncrystalline material by acid hydrolysis. These small particles were known and characterized many years ago[167,168] but are now termed nanoparticles. Cellulose nanocrystals are currently undergoing substantial research and development.[169]

The crystallinity of cellulose is a major factor in "biomass recalcitrance."[170] This factor is responsible for the heterogeneous nature of many chemical reactions of cellulose unless special effort is taken to bring the cellulose sample into solution; the crystallinity inhibits dissolution of cellulose in most solvents. Substantial effort has been dedicated to learning the extent and character of crystallinity of cellulosic materials, especially the orientation of the crystallites relative to the fiber axes. Correlations have often been drawn between the orientation angle of the crystallite and their ability to elongate.[171] Conversely, the noncrystalline parts of the cellulosic materials are also of substantial interest because those regions are where many of the interactions occur between cellulosic materials and reagents or probes. Unless special efforts are taken to bring the cellulose sample into solution, most derivatizing reagents do not penetrate the crystals and are confined to the crystallite surfaces and regions where the cellulose chains are not incorporated into the crystals.[172] This feature leaves many of the cellulose molecules mostly unreacted and the diffraction pattern of the sample is mostly unchanged.[173] Textile fabrics are often given a treatment ("industrial mercerization") with warm sodium hydroxide that partly decrystallizes the cellulose but does not cause substantial conversion into cellulose II. That partial decrystallization improves the uptake of dyes. The nature of noncrystalline cellulose is more mysterious than that of the crystalline material. Although a particular cellulose sample might be said to be 70% crystalline, the atomistic representation of such a state of the material has not been provided. (A model has been developed for completely amorphous material such as might be found in films of regenerated cellulose.[36])

There are several good reasons to make models of cellulose crystals. Initially, the idea was to test the proposed crystal structures for the various polymorphs.[174] The software that is often used for solving and refining fiber-diffraction patterns minimizes the discrepancy between the observed and the calculated diffraction intensities, as well as the total potential energy of the system.[175,176] However, the potential energy values calculated in those programs are based on rather simple equations and parameters that have not been developed to the same level as force fields for stand-alone modeling programs. Therefore, modeling with a better-developed force field was a valuable test for the structures proposed in the early 1990s that were based on fiber-diffraction work.

Another use for models of cellulose crystals was to actually determine the molecular structures and packing arrangements for the various cellulose polymorphs, making use of unit-cell dimensions that were in the literature. This was an attractive proposition for cellulose Iα[177] and more so for cellulose III$_I$.[178] Cellulose Iα has only one cellobiose moiety per unit cell, and cellulose III$_I$ has only one glucopyranose residue per asymmetric unit. In the case of III$_I$, the only variables that required systematic variation were the orientations of the hydroxyl groups on O-2, O-3, and O-6, and whether the chains were "up" or "down" in the unit cell. The O-6 orientation was already known to be gt from NMR results. With three staggered orientations for the three hydroxyl groups, there were 27 possible "up" models and 27 "down" models for evaluation. Although it has been typical to base models on X-ray data that provide at the minimum the unit-cell dimensions or even the reported coordinates, at least one extensive study was based on only the $P2_1$ space group, followed by various modeling methods.[179] The results from that work were in reasonably good agreement with the experimental values.

As a first approximation, the suitability of the proposed models (or, as always, the validity of the calculation) could be assessed by the value of the calculated energy and the dimensional changes that occurred in the model after its energy had been minimized or the model had undergone molecular dynamics simulation. One of the long-standing targets has been to ascertain whether the calculated energy of cellulose II is lower than that of cellulose I. There is some experimental evidence for this premise,[180] and it is assumed to be true because of the irreversibility of the mercerization reaction that converts I into II. Calculations do not, however, always give the anticipated result.[181]

Modern determinations of crystal structure[10,11,13] with synchrotron X-ray and neutron diffraction data allow more confidence in the structural details of the various polymorphs. Also, modern computational facilities allow models of realistic size to be studied, and so concerns about long-range forces are lessened. The models take two forms. In one, the so-called "mini-crystal" approach, cellulose chains having lengths as small as eight or as long as 40 or more glucose residues are simply placed on a lattice that is based on repetitions of the crystallographic unit cell. This is the more "natural" model, with the details of the model dependent on the original coordinates and the potential energy function employed. Alternatively, the structure can be composed of a repeated unit with periodic boundaries. The periodically repeated entity is some multiple of the original unit cell and provides a way to introduce long-range effects without explicitly including all of the atoms. It also controls to some degree the changes that can occur during minimization or MD simulation.[36,182] (Almost all such studies have been carried out with empirical force fields. A QM study is found in Ref. 183.)

With better-established modern crystal structures as a basis for building model crystals of cellulose, the models can be used in speculations as to the mechanisms of conversions induced by temperature. Recent work found a reasonable agreement in predicted properties when either the AMBER/Glycam06 or the new CHARMM force fields were employed[184] to model a transformation that occurs at about 500 K. Transitions of metastable structures to more stable ones have also been studied. One pioneering example has been the not-entirely-successful effort to understand the mechanism of reversion of cellulose III$_I$ back to cellulose Iβ, a process that occurs experimentally in high-moisture environments.[185]

A particularly intriguing use of modern crystal modeling has been the study of enzyme interactions with crystals of cellulose. Those efforts aim to understand how an enzyme might disassemble the cellulose to produce glucose for subsequent fermentation to alcohol.[186,187] A simpler calculation involves the energetics of removal of a chain from the surface of the crystal.[188]

One feature of the current, large mini-crystal models is a twist (see Figs. 18 and 19).[15,16] The crystallites are twisted about their long axes, on the order of a degree per glucose unit of length. This seems to occur with more than one force field but is dependent on the number of chains in the crystal model. Twisting and other alterations in the structure clearly depend on the particular force field and the length of simulation, with convergence not achieved after 800 ns of simulation.[189] It would be easy to ascribe the twist to a failure of the modeling except for the observations of periodic twisting in electron micrographs.[14] Other work, involving atomic force microscopy of numerous microfibrils, indicates that many are not twisted, and that the twisted ones are the exception.[190] Therefore, scrutiny of the models is in order. In the work of Matthews et al.,[15] dimensions of the unit cell changed substantially and the O-6 groups on half of the chains assumed the *gg* orientation, in contradiction to the results of high-quality experimental fiber crystallography and to the findings on O-6 orientation in the foregoing section.

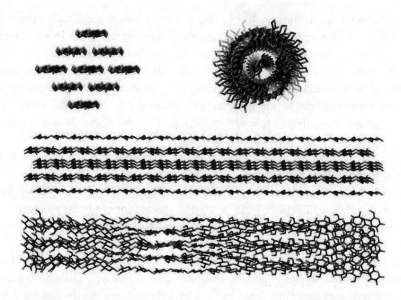

FIG. 18. Models of cellulose crystals with nine chains (3 × 3) before (upper left and center) and after (upper right and bottom) energy minimization with AMBER/Glycam06. Compare the extent of twisting with the lesser amount in the 100-chain model in Fig. 19. (See Color Insert.)

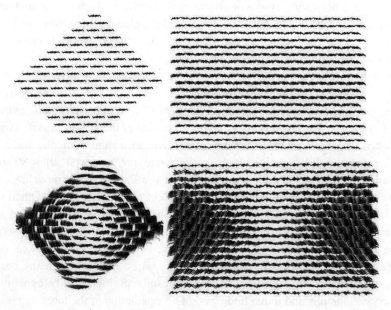

FIG. 19. Models of cellulose crystals with 100 chains (10 × 10) before (upper) and after (lower) energy minimization with AMBER/Glycam06. (See Color Insert.)

The most obvious test would be to calculate the diffraction patterns from the models for comparisons with the experimental patterns. What would be the effect of twist on the diffraction pattern?

The problem was that there was no software for calculating the diffraction pattern from such nonperiodic structures as twisted crystals. Whereas the facility to calculate powder-diffraction patterns from such periodic structures as normal crystals is readily available, the twisted models deviate from periodicity. A normal crystal can be created by repetition of the unit cell in each of its three directions, and so its diffraction pattern is calculated for just those atoms in the unit cell. In a nonperiodic structure, however, the positions of all of the atoms must be included explicitly in the calculation.

One solution is to use the Debye scattering equation that is used for small-angle scattering experiments.[191] Software for that purpose became available in a timely way to address questions about the large cellulose models.[192] A limitation, however, is that the results of a Debye scattering equation are presented as a powder pattern. Powder patterns are very useful for many purposes in the study of cellulose because they can usually be obtained rapidly, especially when the samples are in the form of powders or textiles. However, all of the diffraction information obtained is spherically averaged, which results in overlap of the intensity data and decreases the information content of the diffraction pattern. This is a severe disadvantage in complex studies, such as structure determination, when the largest amount of independent data is needed.

In many cases, cellulose diffraction data are available as a fiber pattern that contains considerably more information than powder patterns. Special software was created by Yoshiharu Nishiyama that can calculate fiber-diffraction patterns from the nonperiodic computer models.[193] Figure 21 shows calculated fiber-diffraction patterns from models based on both the original coordinates and the twisted models of crystals with 16 and 36 chains. The model diffraction patterns are positioned so that there is a progression of order. Thus, the half-pattern on the far left is from a molecular dynamics model with only 16 chains. It is followed by the half-pattern from the model with 16 chains in their original positions. Next on the right is the half-pattern from the molecular dynamics of a twisted model with 36 chains, followed by the half-pattern from the model with its 36 chains, all in the original crystallographic positions. All of these models were surrounded by a single shell of water molecules when the diffraction patterns were calculated.

The presence or absence of water in the calculations made only a small difference in the patterns, except that the small-angle scattering was decreased. When water was not present, that scattering was strong enough to shift the position of the 1 − 1 0 peak. That is an artifact of the calculations being carried out on an isolated model crystal instead of the matrix of crystallites in an actual fiber.

Although a subtle difference in the patterns of twisted and untwisted crystal models was discussed,[193] there is certainly no gross difference between the patterns from twisted and untwisted models. These patterns do offer comment on another issue, however. Cellulose crystals are often considered to be composed of a crystalline core surrounded by a surface of amorphous molecules (Fig. 20). Such a sharp distinction seems improbable, however. Instead, there is likely to be a gradual loss of order going from the center of the crystal out to the surface, allowing additional freedom on the surface for the exocyclic groups to rotate to different orientations. Little variation in such crude diffraction patterns as these can be expected from variation in the orientation of hydrogen atoms of hydroxyl groups because hydrogen has very low ability to scatter X-rays. The 16-chain model with original coordinates could be considered to be a crystalline core, and the 36-chain model based on MD coordinates can be considered as representing a crystalline core of 16 chains, surrounded by 20 more chains, all molecules at the surface.

Because each of the half-patterns in Fig. 21 shows progressively greater crystallinity (progressively narrower spot widths), it can be said that the surface chains contribute to the overall crystallinity of the sample, despite the idea that they are not held in place by being completely surrounded by neighbors like the inner molecules. This idea contradicts the two-phase conceptual model in Fig. 20 but supports the

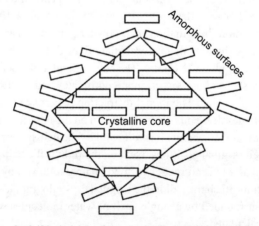

6×6 (36-chain) model crystal with 4×4 (16-chain) core

FIG. 20. Diagrammatic model typical of proposed elementary fibrils of cellulose, following biosynthesis. This is a two-phase model, with a crystalline core and an amorphous surface. This picture contrasts with the foregoing models that have a gradual departure from perfect order at greater distances from the center of the crystal, for example, Figs. 18 and 19. This model elementary fibril is also smaller than that indicated by diffraction patterns from cotton cellulose.

concept of a gradual loss of order as the surface is approached from the inside of the crystal. Of course, this is X-ray crystallinity, which takes little account of the positions of hydrogen atoms. Surface hydroxyl groups are much more likely than interior ones to adopt alternative orientations.

Unfortunately, the changes in the models that result from MD or energy minimization are not limited to twisting of the models. The length expands by 0.2 Å per cellobiose unit with the Glycam06 force field, whereas the width in the a-axis direction contracts by 0.3 Å.[183] These altered values are still a reasonable approximation to the experimental results, but they do affect the calculated diffraction patterns in measurable ways. For example, the 004 peaks are closer to each other in Fig. 21 for the patterns based on the MD models than they are for the 004 peaks for the patterns based on the original coordinates.

The 36-chain model is often considered to be the size of the biosynthetic unit, but its patterns show broader peaks than the experimental patterns from ramie

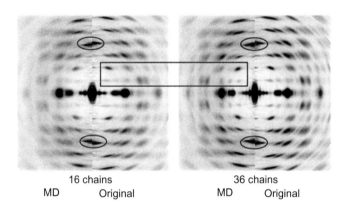

16 chains   36 chains
MD   Original   MD   Original

FIG. 21. Juxtaposed half-diffraction patterns computed from models that have 4×4 and 6×6 arrays of cellulose chains. Each model is surrounded by a solvation shell of water. The chains in the "Original" models were arranged according to the coordinates of Nishiyama et al. for cellulose Iβ[10] with slight deviations arising from the restrained MD equilibration process, and the MD models resulted from unrestrained 100 ps simulations. The spots along the vertical lines through the centers of the combined patterns are "meridional reflections" and have Miller indices of the form 00X, where X is the layer-line number in either direction from the horizontal center. The most intense of the meridionals is 004, in the ovals. Their spacings are different for the Original and MD models because the models expand under the influence of the AMBER/Glycam06 force field. Other discrepancies also occur because of changes due to the force field. When the crystal expands, the distance shrinks between the upper and lower 004 spots because the diffraction pattern represents reciprocal space. In the rectangle are spots on the first and second layer lines for the 4×4 original and 6×6 MD models. The larger MD model has increased resolution of the spots, indicating higher crystallinity, despite the twisting and otherwise decreased order. Adapted from information in Ref. 193.

cellulose.[183] That broadening indicates that the 36-chain model is too small for ramie (or cotton). A recent determination of crystallite size by neutron diffraction of deuterated spruce wood[194] showed a considerably smaller crystallite size than the ramie crystallites indicated by the experimental pattern in Ref. 193. The conclusion of those workers is that there are about 24 chains in the (possibly twisted) crystals, with considerable disorder on the crystallite surfaces.

The sizes of small crystals such as those in cellulose are often assessed with the Scherrer equation, which depends on the peak width at half height.[195] That equation employs a constant that depends on the shape of the crystal. With the ability to calculate the diffraction patterns from detailed models, values of the constant can be tested for specific model shapes.[193]

Ultimately, as hinted in the opening paragraphs of this section, the most important use of model crystals could be a development of an understanding of the boundaries between crystalline material and the noncrystalline material that is susceptible to normal heterogeneous reactions. Just as proposed fibrillar structures can be tested with calculated diffraction patterns, various ideas about decrystallized material can also be examined, following models with decreasing order to the point where there is no indication of order in the calculated pattern.

VI. Conclusions

This article has focused on some fundamental aspects of cellulose structure. In the second section, the numerous crystal structures of related small molecules were exploited to summarize the ranges of conformational variables, such as the degree and type of ring puckering, the torsion angles of the glycosidic linkages, and the orientations of the primary and secondary hydroxyl groups. These analyses serve to define the expected ranges for the much larger molecules of cellulose, for which such information is both more difficult to obtain and less well founded. Much more could be done with such small-molecule crystals, however. Cellulose has been derivatized in many different ways, but the number of similar reactions carried out on cellobiose is surprisingly low. If crystals of such derivatives and of complexes with solvent molecules were available, they would provide improved statistics on the ranges of conformational descriptors that are expected for cellulose itself, as well as on specific interactions in solvent complexes to explain the mechanisms of solvation.

The third section reviewed energy calculations while taking into account the results from the crystallographic experiments in the preceding section. There is still a large variability in the approaches and methods for calculating energies, and it is not surprising that the results between the different methods used do not always agree completely. However, in the energy-minimization studies with elevated dielectric constants, the explicit-solvent molecular dynamics studies, and the continuum-solvation studies, hope for a consensus is emerging. Somewhat surprisingly, solution-state calculations on a single parent disaccharide molecule seem to allow prediction of conformations in crystal structures for a range of related molecules, despite their having a wide range of substitution types. This correlation is also despite the absence of the neighboring molecules in the crystal, indicating a minimal effect of crystal packing on conformation. The finding that an alternative conformation is preferred in vacuum, both by computation-aided experiment and earlier by pure computation, provides a worthwhile benchmark for modeling methodology.

The quantum theory of Atoms-In-Molecules is a relative newcomer to the field of computational methodology applied to cellulose, or all carbohydrates for that matter. In calling attention to its capabilities for detecting and quantifying interatomic interactions, it is hoped to stimulate others to try using this approach. A wider experience and compilation of results under varying conditions should allow a more-thorough assessment of the utility of the method. However, even with the present limited experience, the method has shown interactions that are contributing to the stabilization of a given molecular structure because the space between the involved atoms is experiencing at least a small increase in the time-averaged electron density compared to noninteracting atoms.

The final section covers crystal models. A number of groups are using them for several purposes. Whether to use periodic boundary conditions (the infinite crystal model) or the mini-crystal depends on the goal of the calculation, but it is a more complete test of the force field when the finite, mini-crystal models are used. Another argument in favor of the mini-crystals is that the actual structures being modeled are finite, and it is now routine to make models of a representative size, at least in cross section. The new ability to calculate diffraction patterns from the nonperiodic models should be useful on several fronts. After all, when a crystal model yields a diffraction pattern similar to the one observed by experiment, that model constitutes a plausible atomistic representation of the averaged structure. A limitation in the case of such small crystals is that hydrogen atoms do not impact the calculated intensities, and the hydrogen atoms could be very disordered, even when the diffraction pattern indicates fairly high crystallinity. This problem is somewhat familiar, as the different methods of assessing cellulose crystallinity, such as chemical reactions, NMR spectroscopy, liquid or vapor adsorption, or infrared spectroscopy, all measure different phenomena.[196] Even among X-ray methods, there is considerable controversy.[197]

APPENDIX. MOLECULAR STRUCTURE DRAWINGS FOR SACCHARIDE ANALOGUES HAVING β-(1→4) LINKAGES

Drawings of the molecules listed in Table I are displayed on the following pages. They are provided to aid in understanding the chemical structure as well as to illustrate the diversity of structures that have, with one exception, a similar linkage conformation. The one exception is for the RIKMON structure that exists by itself at the bottom of the plots of the linkage geometries in ϕ/ψ space. A quick assessment of the linkage conformation can be had by examining the C-1—H and C-4—H bonds. If those bonds are parallel, the linkage would correspond to a twofold screw axis when applied to cellulose.

The structures are based on the coordinates from the Cambridge Crystal Structure Database, except that in the case of multiple molecules per asymmetric unit only one example of a given molecule is shown. All solvent molecules have been removed. The molecules are positioned with the nonreducing residue on the left, with the left-most nonreducing ring that makes a β-(1→4) linkage drawn in the standard position with the C-5—O-5 bond in the plane of the paper and above the C-2—C-3 bond. The GUGKID structure has a twofold rotation axis of chemical symmetry, so there is also a β-(1→4)-linked moiety on the right. Hydrogen atoms are black, carbons are light gray (green in online version), oxygens are dark gray (red in online version), nitrogen is black-large (blue in online version), and sulfur in RIKMON is white (yellow in online version). Only the major-fraction disordered atoms were plotted. Some hydrogen atoms were not located in the crystal structure determination.

ABUCEF

COMBINING COMPUTATIONAL CHEMISTRY AND CRYSTALLOGRAPHY 71

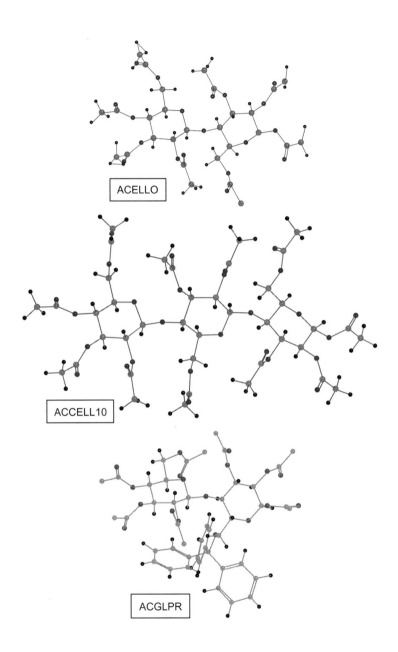

ACHITM10

ACLACT

AJUYUZ

AQOGIW

COMBINING COMPUTATIONAL CHEMISTRY AND CRYSTALLOGRAPHY 73

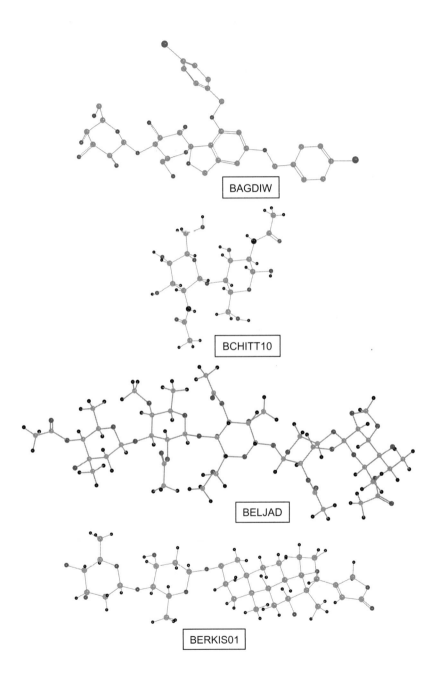

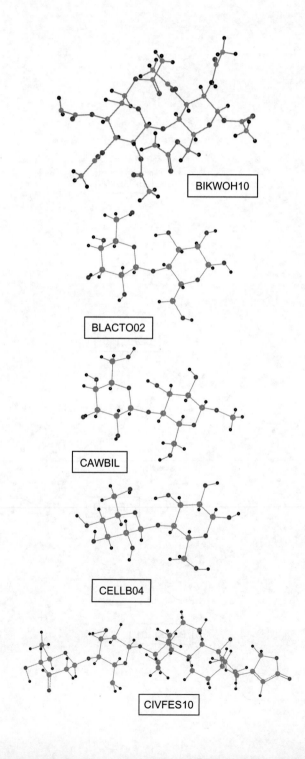

COMBINING COMPUTATIONAL CHEMISTRY AND CRYSTALLOGRAPHY

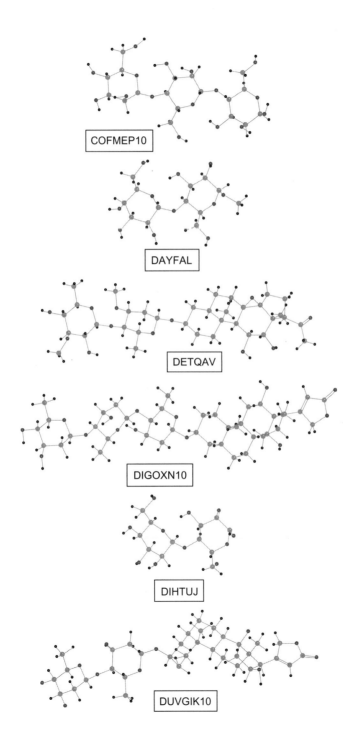

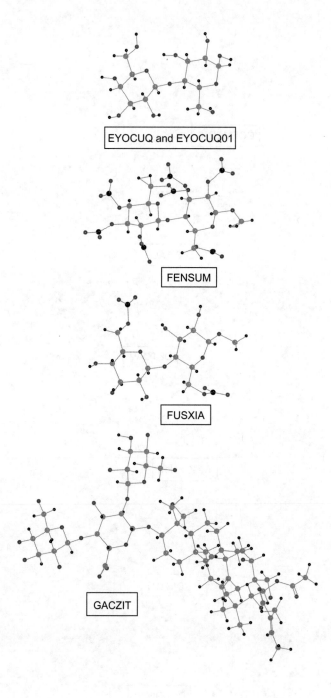

GITXIN10

GUGKID

GURXPX10

IDEHEE

IFOVOO01 and IFOVOO03

KAMHOV

KUQGOT

LACTOS11

LOMGOK

COMBINING COMPUTATIONAL CHEMISTRY AND CRYSTALLOGRAPHY 79

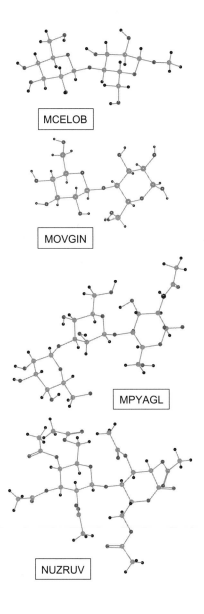

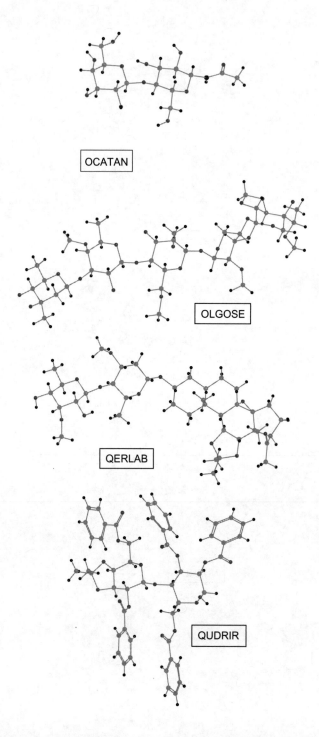

COMBINING COMPUTATIONAL CHEMISTRY AND CRYSTALLOGRAPHY 81

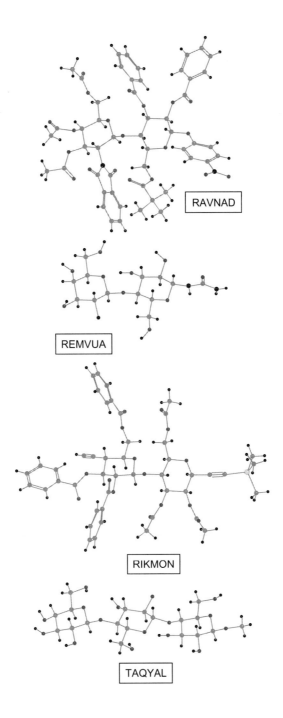

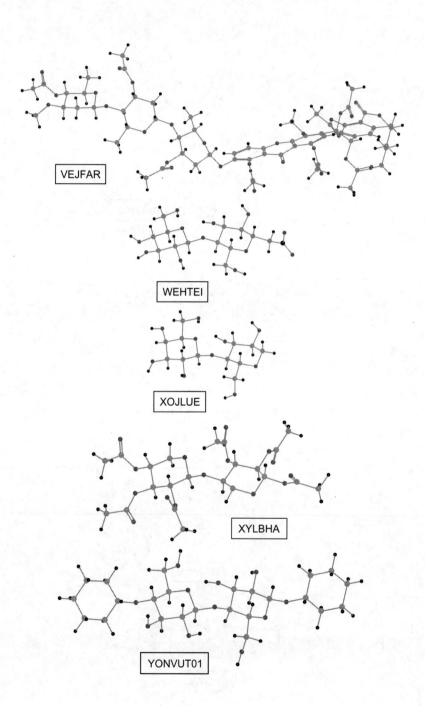

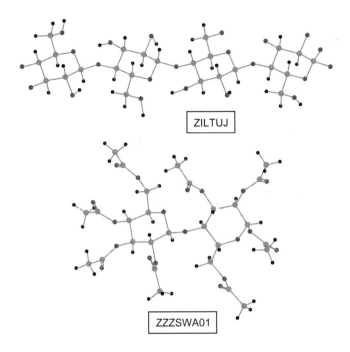

NOTE ADDED IN PROOF

C. Burden, W. Mackie, and B. Sheldrick, 4-O-[β]-D-Galactopyranosyl-[α]-D-mannopyranose hemiethanol dihydrate, *Acta Crystallogr*, C. 42 (1986) 177–179.

REFERENCES

1. S. Pérez and D. Samain, Structure and engineering of celluloses, *Adv. Carbohydr. Chem. Biochem.*, 65 (2010) 25–116.
2. P. Mischnick and D. Momcilovic, Chemical structure analysis of starch and cellulose derivatives, *Adv. Carbohydr. Chem. Biochem.*, 65 (2010) 117–210.
3. K. H. Meyer and H. Mark, Über den Bau des krystallisierten Anteils der Cellulose, *Ber. Dtsch. Chem. Ges. (A and B Series)*, 61 (1928) 593–614.
4. K. H. Meyer and L. Misch, Positions des atomes dans le nouveau modèle spacial de la cellulose, *Helv. Chim. Acta*, 20 (1937) 232–244.
5. A. D. French, The crystal structure of native ramie cellulose, *Carbohydr. Res.*, 61 (1978) 67–80.
6. A. Sarko and R. Muggli, Packing analysis of carbohydrates and polysaccharides. III. Valonia cellulose and cellulose II, *Macromolecules*, 7 (1974) 486–494.

7. K. H. Gardner and J. Blackwell, The structure of native cellulose, *Biopolymers*, 13 (1974) 1975–2001.
8. C. Woodcock and A. Sarko, Packing analysis of carbohydrates and polysaccharides. 11. Molecular and crystal structure of native ramie cellulose, *Macromolecules*, 13 (1980) 1183–1187.
9. A. D. French and P. S. Howley, Comparisons of structures proposed for cellulose, in C. Schuerch, (Ed.), *Cellulose and Wood—Chemistry and Technology*, Wiley, New York, 1989, pp. 159–167.
10. Y. Nishiyama, H. Chanzy, and P. J. Langan, Crystal structure and hydrogen-bonding system in cellulose Iβ from synchrotron X-ray and neutron fiber diffraction, *J. Am. Chem. Soc.*, 124 (2002) 9074–9082.
11. Y. Nishiyama, J. Sugiyama, H. Chanzy, and P. Langan, Crystal structure and hydrogen bonding system in cellulose Iα, from synchrotron X-ray and neutron fiber diffraction, *J. Am. Chem. Soc.*, 125 (2003) 14300–14306.
12. P. Langan, Y. Nishiyama, and H. Chanzy, A revised structure and hydrogen-bonding system in cellulose II from a neutron fiber diffraction analysis, *J. Am. Chem. Soc.*, 121 (1999) 9940–9946.
13. P. Langan, Y. Nishiyama, and H. Chanzy, X-ray structure of mercerized cellulose II at 1 Ångstrom resolution, *Biomacromolecules*, 2 (2001) 410–416.
14. S. J. Hanley, J.-F. Revol, L. Godbout, and D. G. Gray, Atomic force microscopy and transmission electron microscopy of cellulose from *Micrasterias denticulata*; evidence for a chiral helical microfibril twist, *Cellulose*, 4 (1997) 209–220.
15. J. F. Matthews, C. E. Skopec, P. E. Mason, P. Zuccato, R. W. Torget, J. Sugiyama, M. E. Himmel, and J. W. Brady, Computer simulation studies of microcrystalline cellulose Iβ, *Carbohydr. Res.*, 341 (2006) 138–152.
16. T. Yui, S. Nishimura, S. Akiba, and S. Hayashi, Swelling behavior of the cellulose Iβ crystal models by molecular dynamics, *Carbohydr. Res.*, 341 (2006) 2521–2530.
17. R. H. Atalla, Conformational effects in the hydrolysis of cellulose, *Hydrolysis of Cellulose: Mechanisms of Enzymatic and Acid Catalysis, Advances in Chemistry*, Vol. 181, American Chemical Society, pp. 55–69.
18. A. D. French, Computer models of cellulose, in C. Schuerch, (Ed.), *Cellulose and Wood—Chemistry and Technology*, Wiley, New York, 1989, pp. 103–118.
19. R. J. Viëtor, R. H. Newman, M.-A. Ha, D. C. Apperley, and M. C. Jarvis, Conformational features of crystal-surface cellulose from higher plants, *Plant J.*, 30 (2002) 721–731.
20. J.-L. Wertz, O. Bédué, and J. P. Mercier, Cellulose Science and Technology, (2010) EPFL Press, Lausanne.
21. R. F. W. Bader, Atoms In Molecules: A Quantum Theory, (1990) Oxford University Press Inc., New York, International Series of Monographs on Chemistry.
22. S. Raymond, B. Henrissat, D. T. Qui, A. Kvick, and H. Chanzy, The crystal structure of methyl β-cellotrioside monohydrate 0.25 ethanolate and its relationship to cellulose II, *Carbohydr. Res.*, 277 (1995) 209–229.
23. B. Henrissat, S. Pérez, I. Tvaroška, and W. T. Winter, Multidisciplinary approaches to the structures of model compounds for cellulose II, in R. H. Atalla, (Ed.), *The Structures of Cellulose: Characterization of the Solid States, ACS Symposium Series*, Vol. 340, pp. 38–67.
24. B. J. Poppleton and A. Mcl. Mathieson, Crystal structure of β-D-cellotetraose and its relationship to cellulose, *Nature*, 219 (1968) 1046–1048.
25. A. D. French and G. P. Johnson, What crystals of small analogs are trying to tell us about cellulose structure, *Cellulose*, 11 (2004) 5–22.
26. E. G. Cox and G. A. Jeffrey, Crystal structure of glucosamine hydrobromide, *Nature*, 143 (1939) 894–895.
27. IUPAC-IUBMB Nomenclature of carbohydrates, *Adv. Carbohydr. Chem. Biochem.*, 52(1997), (1996) 43–177http://www.chem.qmul.ac.uk/iupac/2carb/06n07.html#07.

28. R. E. Reeves, The shape of pyranoside rings, *J. Am. Chem. Soc.*, 72 (1950) 1499–1506.
29. G. A. Jeffrey and J. H. Yates, Stereographic representation of the Cremer—Pople ring-puckering parameters for pyranoid rings, *Carbohydr. Res.*, 74 (1979) 319–322.
30. D. Cremer and K. J. Szabó, Ab initio studies of six-membered rings, present status and future developments, in E. Juaristi, (Ed.), *Methods in Stereochemical Analysis, Conformational Behavior of Six-Membered Rings, Analysis, Dynamics, and Stereoelectronic Effects*, VCH Publishers, New York, 1995, p. 59.
31. I. J. Bruno, J. C. Cole, P. R. Edgington, M. Kessler, C. F. Macrae, P. McCabe, J. Pearson, and R. Taylor, New software for searching the Cambridge Structural Database and visualising crystal structures, *Acta Crystallogr. B*, 58 (2002) 389–397.
32. A. L. Spek, *PLATON, A Multipurpose Crystallographic Tool*, Utrecht University, Utrecht, The Netherlands, 2008, http://www.cryst.chem.uu.nl/platon/pl000000.html.
33. M. K. Dowd, A. D. French, and P. J. Reilly, Modeling of aldopyranosyl ring puckering with MM3(92), *Carbohydr. Res.*, 264 (1994) 1–19.
34. X. Biarnés, A. Ardèvol, A. Planas, C. Rovira, A. Laio, and M. Parrinello, The conformational free energy landscape of β-D-glucopyranose. Implications for substrate preactivation in β-glucoside hydrolases, *J. Am. Chem. Soc.*, 129 (2007) 10686–10693.
35. V. Spiwok, B. Králová, and I. Tvaroska, Modelling of β-D-glucopyranose ring distortion in different force fields: A metadynamics study, *Carbohydr. Res.*, 345 (2010) 530–537.
36. K. Mazeau and L. Heux, Molecular dynamics simulations of bulk native crystalline and amorphous structures of cellulose, *J. Phys. Chem. B*, 107 (2003) 2394–2403.
37. A. Sarko and R. H. Marchessault, Supermolecular structure of polysaccharides, *J. Polym. Sci. C*, 28 (1969) 317–331.
38. R. Sundararajan and R. H. Marchessault, Bibliography of crystal structures of polysaccharides 1976, *Adv. Carbohydr. Chem. Biochem.*, 36 (1979) 315–332.
39. S. Pérez, N. Mouhous-Riou, N. E. Nifant'ev, Y. E. Tsvetkov, B. Bachet, and A. Imberty, Crystal and molecular structure of a histo-blood group antigen involved in cell adhesion: The Lewis x trisaccharide, *Glycobiology*, 6 (1996) 537–542.
40. S. Pérez and F. Brisse, The crystal and molecular structure of a trisaccharide, β-cellotriose undecaacetate: 1,2,3,6-Tetra-O-acetyl-4-O-[2,3,6-tri-O-acetyl-4-O-(2,3,4,6-tetra-O-acetyl-β-D-glucopyranosyl)-β-D-glucopyranosyl]-β-D-glucopyranose, *Acta Crystallogr. B*, 33 (1977) 2578–2584.
41. F. Leung, H. D. Chanzy, S. Pérez, and R. H. Marchessault, Crystal structure of β-D-acetyl cellobiose, $C_{28}H_{38}O_{19}$, *Can. J. Chem.*, 54 (1976) 1365–1371.
42. T. Taga, S. Sumiya, K. Osaki, T. Utamura, and K. Koizumi, Structure of 1,2,3-tri-O-acetyl-4-O-(2,3,4,6-tetra-O-acetyl-β-D-glucopyranosyl)-6-O-triphenylmethyl-α-D-glucopyranose (6'-O-trityl-α-cellobiose heptaacetate), *Acta Crystallogr. B*, 37 (1981) 963–966.
43. F. Mo and L. H. Jensen, The crystal structure of a β-(1→4) linked disaccharide, α-N,N'-diacetylchitobiose monohydrate, *Acta Crystallogr. B*, 34 (1978) 1562–1569.
44. F. Longchambon, J. Ohanessian, H. Gillier-Pandraud, D. Duchet, J.-C. Jacquinet, and P. Sinaÿ, Structure of N-acetyllactosamine (2-acetamido-2-desoxy-4-O-β-D-galactopyranosyl-α-D-glucopyranose), *Acta Crystallogr. B*, 37 (1981) 601–607.
45. D. Keglevic, B. Kojic-Prodic, and Z. B. Tomisic, Synthesis and conformational analysis of the repeating units of bacterial spore peptidoglycan, *Carbohydr. Res.*, 338 (2003) 1299–1308.
46. J.-C. Lee, X.-A. Lu, S. S. Kulkarni, Y.-S. Wen, and S.-C. Hung, Synthesis of heparin oligosaccharides, *J. Am. Chem. Soc.*, 126 (2004) 476–477.
47. G. Rihs and P. Traxler, X-ray analysis of a novel spirocyclic diglycoside, a degradation product of the antibiotics papulacandin A, B and C [Rontgenstrukturanalyse eines Neuartigen Spirocyclischen Diglycosides, eines Abbauproduktes der Antibiotika Papulacandin A, B und C], *Helv. Chim. Acta*, 64 (1981) 1533–1539.

48. F. Mo, On the conformational variability of the N-acetylglucosamine β-(1→4) linked dimer, crystal and molecular structure of β-N,N'-diacetylchitobiose trihydrate, *Acta Chem. Scand. A*, 33 (1979) 207–218.
49. E. Kupfer, K. Neupert-Laves, M. Dobler, and W. Keller-Schierlein, Metabolic products of microorganisms. 210. Avileurekanoses A and C and further degradation products of avilamycins A and C, *Helv. Chim. Acta*, 65 (1982) 3–12.
50. K. Go and K. K. Bhandary, Structural studies on the biosides of *Digitalis lanata*: Bisdigitoxosides of digitoxigenin, gitoxigenin and digoxigenin, *Acta Crystallogr. B*, 45 (1989) 306–312.
51. A. Neuman, J. Becquart, D. Avenel, H. Gillier-Pandraud, and P. Sinaÿ, Structure crystalline du 2-acétamido-1,3,6-tri-*O*-acétyl-2-désoxy-4-*O*-(2,3,4,6-tétra-*O*-acétyl-α-L-idopyranosyl)-α-D-glucopyranose, *Carbohydr. Res.*, 139 (1985) 23–34.
52. S. Garnier, S. Petit, and G. Coquerel, Dehydration mechanism and crystallisation behaviour of lactose, *J. Ther. Anal. Calorim.*, 68 (2002) 489–502.
53. R. Stenutz, M. Shang, and A. S. Serianni, Methyl β-lactoside (methyl 4-O-β-D-galactopyranosyl-β-D-glucopyranoside) methanol solvate, *Acta Crystallogr. C*, 55 (1999) 1719–1721.
54. E. Kalenius, T. Kekalainen, R. Neitola, K. Beyeh, K. Rissanen, and P. Vainiotalo, Size- and structure-selective noncovalent recognition of saccharides by tetraethyl and tetraphenyl resorcinarenes in the gas phase, *Chem. Eur. J.*, 14 (2008) 5220–5228.
55. W. Mackie, B. Sheldrick, D. Akrigg, and S. Pérez, Crystal and molecular structure of mannotriose and its relationship to the conformations and packing of mannan and glucomannan chains and mannobiose, *Int. J. Biol. Macromol.*, 8 (1986) 43–51.
56. Q. Pan, B. C. Noll, and A. S. Serianni, Methyl 4-O-β-D-galactopyranosyl α-D-glucopyranoside (methyl α-lactoside), *Acta Crystallogr. C*, 61 (2005) o674–o677.
57. X.-L. Wang, Q.-F. Li, K.-B. Yu, S.-L. Peng, Y. Zhou, and L.-S. Ding, Four new pregnane glycosides from the stems of *Marsdenia tenacissima*, *Helv. Chim. Acta*, 89 (2006) 2738–2744.
58. K. Go, G. Kartha, and J. P. Chen, Structure of digoxin, *Acta Crystallogr. B*, 36 (1980) 1811–1819.
59. B. Sheldrick, W. Mackie, and D. Akrigg, The crystal and molecular structure of *O*-β-D-mannopyranosyl-(1→4)-α-D-mannopyranose (mannobiose), *Carbohydr. Res.*, 132 (1984) 1–6.
60. C. Platteau, J. Lefebvre, F. Affouard, and P. Derollez, Ab initio structure determination of the hygroscopic anhydrous form of α-lactose by powder X-ray diffraction, *Acta Crystallogr. B*, 60 (2004) 453–460.
61. C. Platteau, F. Affouard, J.-F. Willart, P. Derollez, and F. Mallet, Structure determination of the stable anhydrous phase of α-lactose from X-ray powder diffraction, *Acta Crystallogr. B*, 61 (2005) 185–191.
62. A. V. Nikitin, V. I. Andrianov, R. M. Myasnikova, S. I. Firgang, A. I. Usov, V. F. Sopin, and A. I. Pertsin, Crystal and molecular structure of β-methylcellobioside heptanitrate, *Kristallografiya (Russ.) (Crystallogr.Rep.)*, 31 (1986) 676–681.
63. A. V. Nikitin, T. A. Shibanova, S. I. Firgang, A. I. Usov, V. F. Sopin, and R. M. Myasnikova, Crystal and molecular structure of methyl-β-cellobioside 6,6'-dinitrate, *Kristallografiya (Russ.) (Crystallogr. Rep.)*, 32 (1987) 896–900.
64. X.-W. Yang, J. Zhao, Y.-X. Cui, X.-H. Liu, C.-M. Ma, M. Hattori, and L.-H. Zhang, Anti-HIV-1 protease triterpenoid saponins from the seeds of *Aesculus chinensis*, *J. Nat. Prod.*, 62 (1999) 1510–1513.
65. K. Go and G. Kartha, Structure of gitoxin, *Acta Crystallogr. Sect. B Struct. Crystallogr. Cryst. Chem.*, 36 (1980) 3034–3040.
66. S. Porwanski, F. Dumarcay-Charbonnier, S. Menuel, J.-P. Joly, V. Bulach, and A. Marsura, Bis-β-cyclodextrinyl- and bis-cellobiosyl-diazacrowns: Synthesis and molecular complexation behaviors toward Busulfan anticancer agent and two basic aminoacids, *Tetrahedron*, 65 (2009) 6196–6203.

67. R. A. Moran and G. F. Richards, The crystal and molecular structure of an aldotriouronic acid trihydrate: O-(4-O-Methyl-α-D-glucopyranosyluronic acid)-(1-2)-O-β-D-xylopyranosyl-(1-4)-D-xylopyranose trihydrate, *Acta Crystallogr. B*, 29 (1973) 2770–2783.
68. J. Xia, T. Srikrishnan, J. L. Alderfer, R. K. Jain, C. F. Piskorz, and K. L. Matta, Chemical synthesis of sulfated oligosaccharides with a β-D-Gal-(1→3)-[β-D-Gal-(1→4)-(α-L-Fuc-(1→3)-β-D-GlcNAc-(1→6)]-α-D-GalNAc sequence, *Carbohydr. Res.*, 329 (2000) 561–577.
69. A. Rencurosi, J. Rohrling, J. Pauli, A. Potthast, C. Jager, S. Pérez, P. Kosma, and A. Imberty, Polymorphism in the crystal structure of the cellulose fragment analogue methyl 4-O-methyl-β-D-glucopyranosyl-(1-4)-β-D-glucopyranoside, *Angew. Chem. Int. Ed.*, 41 (2002) 4277–4281.
70. F. Malz, Y. Yoneda, T. Kawada, K. Mereiter, P. Kosma, T. Rosenau, and C. Jager, Synthesis of methyl 4′-O-methyl-β-D-cellobioside-$^{13}C_{12}$ from D-glucose-$^{13}C_6$. Part 2: Solid-state NMR studies, *Carbohydr. Res.*, 342 (2007) 65–70.
71. X. Hu, Q. Pan, B. C. Noll, A. G. Oliver, and A. S. Serianni, Methyl 4-O-β-D galactopyranosyl α-D-mannopyranoside methanol 0.375-solvate, *Acta Crystallogr. C*, 66 (2010) o67–o70.
72. J. H. Smith, S. E. Dann, M. R. J. Elsegood, S. H. Dale, and C. G. Blatchford, α-Lactose monohydrate: A redetermination at 150 K, *Acta Crystallogr. Sect. E Struct. Rep. Online*, 61 (2005) o2499–o2501.
73. K. P. Guiry, S. J. Coles, H. A. Moynihan, and S. E. Lawrence, Effect of 1-deoxy-D-lactose upon the crystallization of D-lactose, *Cryst. Growth Des.*, 8 (2008) 3927–3934.
74. W. Frey, R. Sardzik, and V. Jager, Crystal structure of benzyl 2,3-di-O-acetyl-4,6-cyclo-4,6-dideoxy-β-D-galactopyranosyl-(1-4)-2,3,6-tri-O-acetyl-β-D-glucopyranoside, $C_{29}H_{36}O_{14}$, *Z. Kristallogr. New Cryst. Struct.*, 223 (2008) 259–261.
75. J. T. Ham and D. G. Williams, The crystal and molecular structure of methyl β-cellobioside-methanol, *Acta Crystallogr. B*, 26 (1970) 1373–1383.
76. Z. Peralta-Inga, G. P. Johnson, M. K. Dowd, J. A. Rendleman, E. D. Stevens, and A. D. French, The crystal structure of the α-cellobiose·2NaI·2 H_2O complex in the context of related structures and conformational analysis, *Carbohydr. Res.*, 337 (2002) 851–861.
77. V. Warin, F. Baert, R. Fouret, G. Strecker, G. Spik, B. Fournet, and J. Montreuil, The crystal and molecular structure of O-α-D-mannopyranosyl-(1→3)-O-β-D-mannopyranosyl-(1→4)-2-acetamido-2-deoxy-α-D-glucopyranose, *Carbohydr. Res.*, 76 (1979) 11–22.
78. D. Schollmeyer, R. Stadler, and V. von Braunmühl, Crystal structure of 2,3,6,2′,3′,4′,6′,-heptaacetyl-D-lactobionolactone toluene solvate (2/1), $C_{59}H_{76}O_{36}$, *Z. Kristallogr. N. Cryst. Struct.*, 213 (1998) 410–412.
79. T. Lakshmanan, D. Sriram, and D. Loganathan, β-1-Acetamido-4-O-β-D-galactopyranosyl-D-glucopyranose dihydrate, *Acta Crystallogr. C*, 57 (2001) 825–826.
80. A. K. Ganguly, O. Z. Sarre, A. T. McPhail, and R. W. Miller, X-ray crystal and molecular structure of olgose, a major degradation product of the oligosaccharide antibiotic everninomicin D, *Chem. Commun.* (1979) 22–24.
81. Q.-F. Li, X.-L. Wang, K.-B. Yu, S.-L. Peng, and L.-S. Ding, 11α,12β-Acetonyltenacissoside F acetone solvate: A new polyoxypregnane glycoside from the stems of *Marsdenia tenacissima*, *Acta Crystallogr. Sect. E Struct. Rep. Online*, 62 (2006) o5255–o5256.
82. G.-W. Xing, L. Chen, and F.-F. Liang, Facile synthesis of tumor-associated carbohydrate antigen ganglioside GM3 from sialic acid, lactose, and serine, *Eur. J. Org. Chem.*, 34 (2009) 5963–5970.
83. K. A. Abboud, S. S. Toporek, and B. A. Horenstein, 2,3-Di-O-benzoyl-1-O-p-nitrophenyl-6-O-pivaloyl-4-O-(3′,4′, 6′-tri-O-acetyl-2′-deoxy-2′-N-phthalimido-β-D-1′-glucopyranosyl)-β-D-glucopyranose, *Acta Crystallogr. C*, 53 (1997) 742–744.
84. M. M. Olmstead, M. Hu, M. J. Kurth, J. M. Krochta, and Y.-L. Hsieh, Lactosylurea dihydrate, *Acta Crystallogr. C*, 53 (1997) 915–916.
85. A. Ernst and A. Vasella, Oligosaccharide analogues of polysaccharides: Part 8. Orthogonally protected cellobiose-derived dialkynes. A convenient method for the regioselective bromo- and

protodegermylation of trimethylgermyl- and trimethylsilyl-protected dialkynes, *Helv. Chim. Acta*, 79 (1996) 1279–1304.
86. T. Eguchi, K. Kondo, K. Kakinuma, H. Uekusa, Y. Ohashi, K. Mizoue, and Y.-F. Qiao, Unique solvent-dependent atropisomerism of a novel cytotoxic naphthoxanthene antibiotic FD-594, *J. Org. Chem.*, 64 (1999) 5371–5376.
87. P. Koll, M. Petrusova, L. Petrus, B. Zimmer, M. Morf, and J. Kopf, Crystal and molecular structures of β-cellobiosylnitromethane and of β-maltosylnitromethane heptaacetate, *Carbohydr. Res.*, 248 (1993) 37–43.
88. F. Leung and R. H. Marchessault, Crystal structure of β-D, 1→4 xylobiose hexaacetate, *Can. J. Chem.*, 51 (1973) 1215–1222.
89. Y. Yoneda, K. Mereiter, C. Jaeger, L. Brecker, P. Kosma, T. Rosenau, and A. French, Van der Waals versus hydrogen-bonding forces in a crystalline analog of cellotetraose: Cyclohexyl 4'-O-cyclohexyl β-D-cellobioside cyclohexane solvate, *J. Am. Chem. Soc.*, 130 (2008) 16678–16690.
90. K. Gessler, N. Krauss, T. Steiner, C. Betzel, A. Sarko, and W. Saenger, β-D-Cellotetraose hemihydrate as a structural model for cellulose II. An X-ray diffraction study, *J. Am. Chem. Soc.*, 117 (1995) 11397–11406.
91. C. Bruhn, Y. Arendt, and D. Steinborn, Crystal structure of O1,O2,O3,O6-tetraacetyl-O4-[tetra-O-acetyl-β-D-glucopyranosyl]-α-D-glucopyranose (octa-O-acetyl-α-D-cellobiose), C28H38O19, *Z. Kristallogr. N. Cryst. Struct.*, 216 (2001) 587–588.
92. IUPAC-IUB Joint Commission on Biochemical Nomenclature (JCBN), Symbols for specifying the conformation of polysaccharide chains: Recommendations 1981, *Eur. J. Biochem.*, 131 (1983) 5–7. http://www.chem.qmul.ac.uk/iupac/misc/psac.html#220.
93. T. Shimanouchi and S.-I. Mizushima, On the helical configuration of a polymer chain, *J. Chem. Phys.*, 33 (1955) 707–711.
94. S. S. C. Chu and G. A. Jeffrey, The refinement of the crystal structures of β-D-glucose and cellobiose, *Acta Crystallogr. B*, 24 (1968) 830–838.
95. A. Almond and J. K. Sheehan, Glycosaminoglycan conformation: Do aqueous molecular dynamics simulations agree with X-ray fiber diffraction? *Glycobiology*, 10 (2000) 329–338.
96. C. Thibaudeau, R. Stenutz, B. Hertz, T. Klepach, S. Zhao, Q. Wu, I. Carmichael, and A. S. Serianni, Correlated C−C and C−O bond conformations in saccharide hydroxymethyl groups: Parametrization and application of redundant $^1H-^1H$, $^{13}C-^1H$, and $^{13}C-^{13}C$ NMR J-couplings, *J. Am. Chem. Soc.*, 126 (2004) 15668–15685.
97. G. M. Brown and H. A. Levy, α-D-Glucose: Further refinement based on neutron-diffraction data, *Acta Crystallogr. B*, 35 (1979) 656–659.
98. G. M. Brown and H. A. Levy, Further refinement of the structure of sucrose based on neutron-diffraction data, *Acta Crystallogr. B*, 29 (1973) 790–797.
99. M. E. Gress and G. A. Jeffrey, A neutron diffraction refinement of the crystal structure of β-maltose monohydrate, *Acta Crystallogr. B*, 33 (1977) 2490–2495.
100. G. N. Ramachandran, C. Ramakrishnan, and V. Sasisekaharan, Stereochemistry of polypeptide and polysaccharide chain conformations, in G. N. Ramachandran, (Ed.), *Aspects of Protein Structure*, Academic Press, London, 1963, pp. 121–135.
101. P. Zugenmaier, Crystalline Cellulose and Cellulose Derivatives: Characterization and Structure, (2008) Springer, Berlin, 286pp.
102. D. A. Rees and R. J. Skerrett, Conformational analysis of cellobiose, cellulose, and xylan, *Carbohydr. Res.*, 7 (1968) 334–348.
103. R. U. Lemieux, K. Bock, L. T. J. Delbaere, S. Koto, and V. S. Rao, The conformations of oligosaccharides related to the ABH and Lewis human blood group determinants, *Can. J. Chem.*, 58 (1980) 631–653.
104. I. Tvaroška and S. Pérez, Conformational-energy calculations for oligosaccharides: A comparison of methods and a strategy of calculation, *Carbohydr. Res.*, 149 (1986) 389–410.

105. G. A. Jeffrey and R. Taylor, The application of molecular mechanics to the structures of carbohydrates, *J. Comput. Chem.*, 1 (1980) 99–109.
106. V. Ferretti, V. Bertolasi, G. Gilli, and C. A. Accorsi, Structure of 6-kestose monohydrate, $C_{18}H_{31}O_{16} \cdot H_2O$, *Acta Crystallogr. C*, 40 (1984) 531–535.
107. A. D. French, A.-M. Kelterer, C. J. Cramer, G. P. Johnson, and M. K. Dowd, A QM/MM analysis of the conformations of crystalline sucrose moieties, *Carbohydr. Res.*, 326 (2000) 305–322.
108. R. Car and M. Parrinello, Unified approach for molecular dynamics and density-functional theory, *Phys. Rev. Lett.*, 55 (1985) 2471–2474.
109. J.-H. Lii, K.-H. Chen, and N. L. Allinger, Alcohols, ethers, carbohydrates, and related compounds. IV. Carbohydrates, *J. Comput. Chem.*, 24 (2003) 1504–1513.
110. J.-H. Lii, K.-H. Chen, G. P. Johnson, A. D. French, and N. L. Allinger, The external anomeric torsional effect, *Carbohydr. Res.*, 340 (2005) 853–862.
111. L. Nørskov-Lauritsen and N. L. Allinger, A molecular mechanics treatment of the anomeric effect, *J. Comput. Chem.*, 5 (1984) 326–335.
112. N. L. Allinger, M. Rahman, and J.-H. Lii, A molecular mechanics force field (MM3) for alcohols and ethers, *J. Am. Chem. Soc.*, 112 (1990) 8293–8307.
113. D. A. Case, T. E. Cheatham, III, T. Darden, H. Gohlke, R. Luo, K. M. Merz, Jr., A. Onufriev, C. Simmerling, B. Wang, and R. Woods, The Amber biomolecular simulation programs, *J. Comput. Chem.*, 26 (2005) 1668–1688.
114. B. R. Brooks, C. L. Brooks, III, A. D. Mackerell, L. Nilsson, R. J. Petrella, B. Roux, Y. Won, G. Archontis, C. Bartels, S. Boresch, A. Caflisch, L. Caves, Q. Cui, A. R. Dinner, M. Feig, S. Fischer, J. Gao, M. Hodoscek, W. Im, K. Kuczera, T. Lazaridis, J. Ma, V. Ovchinnikov, E. Paci, R. W. Pastor, C. B. Post, J. Z. Pu, M. Schaefer, B. Tidor, R. M. Venable, H. L. Woodcock, X. Wu, W. Yang, D. M. York, and M. Karplus, CHARMM: The biomolecular simulation program, *J. Comput. Chem.*, 30 (2009) 1545–1615.
115. B. Hess, C. Kutzner, D. van der Spoel, and E. Lindahl, GROMACS 4: Algorithms for highly efficient, load-balanced, and scalable molecular simulation, *J. Chem. Theory Comput.*, 4 (2008) 435–447.
116. R. Schulz, B. Lindner, L. Petridis, and J. C. Smith, Scaling of multimillion-atom biological molecular dynamics simulation on a petascale supercomputer, *J. Chem. Theory Comput.*, 5 (2009) 2798–2808.
117. E. Humeres, C. Mascayano, G. Riadi, and F. González-Nilo, Molecular dynamics simulation of the aqueous solvation shell of cellulose and xanthate ester derivatives, *J. Phys. Org. Chem.*, 19 (2006) 896–901.
118. B. L. Foley, M. B. Tessier, and R. J. Woods, Carbohydrate force fields, *WIREs Comput. Mol. Sci.* (2011) doi:10.1002/wcms.89.
119. O. Guvench, E. Hatcher, R. M. Venable, R. W. Pastor, and A. D. MacKerell, Jr., CHARMM Additive all-atom force field for glycosidic linkages between hexopyranoses, *J. Chem. Theory Comput.*, 5 (2009) 2353–2370.
120. A. S. Gross and J.-W. Chu, On the molecular origins of biomass recalcitrance: The interaction network and solvation structures of cellulose microfibrils, *J. Phys. Chem. B*, 114 (2010) 13333–13341.
121. K. N. Kirschner, A. B. Yongye, S. M. Tschampel, J. González-Outeiriño, C. R. Daniels, B. L. Foley, and R. J. Woods, GLYCAM06: A generalizable biomolecular force field. Carbohydrates, *J. Comput. Chem.*, 29 (2008) 622–655.
122. C. A. Stortz, G. P. Johnson, A. D. French, and G. I. Csonka, Comparison of different force fields for the study of disaccharides, *Carbohydr. Res.*, 344 (2009) 2217–2228.
123. W. L. Jorgensen, J. Chandrasekhar, J. D. Madura, R. W. Impey, and M. L. Klein, Comparison of simple potential functions for simulating liquid water, *J. Chem. Phys.*, 79 (1983) 926–935.
124. P. A. Kollman, J. W. Caldwell, W. S. Ross, D. A. Pearlman, D. A. Case, S. DeBolt, T. E. Cheatham, III, D. Ferguson, and G. Seible, AMBER: A program for simulation of biological and organic

molecules, in P. von Ragué Schleyer, (Ed.), *Encyclopedia of Computational Chemistry*, Vol. 1, John Wiley, Chichester, 2008, pp. 11–13.
125. A. D. French and G. P. Johnson, Computerized molecular modeling of carbohydrates, in Z. A. Popper, (Ed.), *The Plant Cell Wall: Methods and Protocols,* Springer, New York, 2011, pp. 21–42.
126. U. Schnupf and F. A. Momany, Rapidly calculated DFT relaxed iso-potential maps: β-Cellobiose, *Cellulose,* 18 (2011) 859–887.
127. M. K. Dowd, A. D. French, and P. J. Reilly, Conformational analysis of the anomeric forms of sophorose, laminarabiose, and cellobiose using MM3, *Carbohydr. Res.,* 233 (1992) 15–34.
128. C. A. Stortz and A. D. French, Disaccharide conformational maps: Adiabaticity in analogues with variable ring shapes, *Mol. Simulat.,* 34 (2008) 373–389.
129. V. Spiwok and I. Tvaroška, Metadynamics modelling of the solvent effect on primary hydroxyl rotamer equilibria in hexopyranosides, *Carbohydr. Res.,* 344 (2009) 1575–1581.
130. L. Perić-Hassler, H. S. Hansen, R. Baron, and P. H. Hünenberger, Conformational properties of glucose-based disaccharides investigated using molecular dynamics simulations with local elevation umbrella sampling, *Carbohydr. Res.,* 345 (2010) 1781–1801.
131. R. K. Campen, A. V. Verde, and J. D. Kubicki, Influence of glycosidic linkage neighbors on disaccharide conformation in vacuum, *J. Phys. Chem. B,* 111 (2007) 13775–13785.
132. A. D. French, A.-M. Kelterer, G. P. Johnson, M. K. Dowd, and C. J. Cramer, Constructing and evaluating energy surfaces of crystalline disaccharides, *J. Mol. Graph. Model.,* 18 (2000) 95–107.
133. A. D. French, In defense of adiabatic φ/ψ mapping for cellobiose and other disaccharides, *Cellulose,* 18 (2011) 889–896.
134. M. Kuttel, J. W. Brady, and K. J. Naidoo, Carbohydrate solution simulations: Producing a force field with experimentally consistent primary alcohol rotational frequencies and populations, *J. Comput. Chem.,* 23 (2002) 1236–1243.
135. M. Kuttel, Conformational free energy maps for globobiose (α-D-Galp-(1→4)-β-D-Galp) in implicit and explicit aqueous solution, *Carbohydr. Res.,* 343 (2008) 1091–1098.
136. E. J. Cocinero, D. P. Gamblin, B. G. Davis, and J. P. Simons, The building blocks of cellulose: The intrinsic conformational structures of cellobiose, its epimer, lactose, and their singly hydrated complexes, *J. Am. Chem. Soc.,* 131 (2009) 11117–11123.
137. G. L. Strati, J. L. Willett, and F. A. Momany, Ab initio computational study of β-cellobiose conformers using B3LYP/6-311++G**, *Carbohydr. Res.,* 337 (2002) 1833–1849.
138. T. Shen, P. Langan, A. D. French, G. P. Johnson, and S. Gnanakaran, Conformational flexibility of soluble cellulose oligomers: Chain length and temperature dependence, *J. Am. Chem. Soc.,* 131 (2009) 14786–14794.
139. E. Hatcher, E. Säwén, G. Widmalm, and A. D. MacKerell, Jr., Conformational properties of methyl β-maltoside and methyl α- and β-cellobioside disaccharides, *J. Phys. Chem. B,* 115 (2011) 597–608.
140. G. Johnson, L. Peterson, A. French, and P. Reilly, Twisting of glycosidic bonds by hydrolases, *Carbohydr. Res.,* 344 (2009) 2157–2166.
141. S. Mendonca, G. P. Johnson, A. D. French, and R. A. Laine, Conformational analyses of native and permethylated disaccharides, *J. Phys. Chem. A,* 106 (2002) 4115–4124.
142. G. A. Jeffrey, J. A. Pople, and L. Radom, The application of ab initio molecular orbital theory to the anomeric effect. A comparison of theoretical predictions and experimental data on conformations and bond lengths in some pyranoses and methyl pyranosides, *Carbohydr. Res.,* 25 (1972) 117–131.
143. I. Tvaroška and J. P. Carver, Ab initio molecular orbital calculation of carbohydrate model compounds. 2. Conformational analysis of axial and equatorial 2-methoxytetrahydropyrans, *J. Phys. Chem.,* 98 (1994) 9477–9485.
144. A. D. French, A.-M. Kelterer, G. P. Johnson, M. K. Dowd, and C. J. Cramer, HF/6-31G* energy surfaces for disaccharide analogs, *J. Comput. Chem.,* 22 (2001) 65–78.

145. A. D. French and G. P. Johnson, Quantum mechanics studies of cellobiose conformations, *Can. J. Chem.*, 84 (2006) 603–612.
146. A. D. French and G. P. Johnson, Roles of starting geometries in quantum mechanics studies of cellobiose, *Mol. Simulat.*, 34 (2008) 365–372.
147. A. V. Marenich, C. J. Cramer, and D. G. Truhlar, Universal solvation model based on solute electron density and on a continuum model of the solvent defined by the bulk dielectric constant and atomic surface tensions, *J. Phys. Chem. B*, 113 (2009) 6378–6396.
148. A. D. French, G. P. Johnson, C. J. Cramer, and G. I. Csonka, Conformational analysis of cellobiose by electronic structure theories, *Carbohydr. Res.*, 350 (2012) 68–76.
149. S. N. Ha, L. J. Madsen, and J. W. Brady, Conformational analysis and molecular dynamics simulations of maltose, *Biopolymers*, 27 (1988) 1927–1952.
150. A. D. French and G. I. Csonka, Hydroxyl orientations in cellobiose and other polyhydroxyl compounds: Modeling versus experiment, *Cellulose*, 18 (2011) 897–909.
151. U. Koch and P. L. A. Popelier, Characterization of C–H–O hydrogen bonds on the basis of the charge density, *J. Phys. Chem.*, 99 (1995) 9747–9754.
152. G. I. Csonka, N. Anh, J. G. Ángyán, and I. G. Csizmadia, Investigation of intramolecular hydrogen bonding in 1, 2-ethanediol using density functional methods, *Chem. Phys. Lett.*, 245 (1995) 129–135.
153. G. I. Csonka and I. G. Csizmadia, Density functional conformational analysis of 1, 2-ethanediol, *Chem. Phys. Lett.*, 243 (1995) 419–428.
154. T. S. Koritsanszky and P. Coppens, Chemical applications of X-ray charge-density analysis, *Chem. Rev.*, 101 (2001) 1583–1627.
155. F. Longchambon, H. Gillier-Pandraud, R. Wiest, B. Rees, A. Mitschler, R. Feld, M. Lehmann, and P. Becker, Etude structurale et densité de déformation électronique X-N à 75 K dans la région anomère du β-DL-arabinose, *Acta Crystallogr. B*, 41 (1985) 47–56.
156. J. Y.-J. Chen and K. J. Naidoo, Evaluating intramolecular hydrogen bond strengths in (1–4) linked disaccharides from electron density relationships, *J. Phys. Chem. B*, 107 (2003) 9558–9566.
157. S. J. Grabowski, (Ed.), *Hydrogen Bonding—New Insights*, (2006) Springer, Dordrecht, 519pp.
158. M. D. Díaz, M. del Carmen Fernández-Alonso, G. Cuevas, F. J. Cañada, and J. Jiménez-Barbero, On the role of aromatic–sugar interactions in the molecular recognition of carbohydrates: A 3D view by using NMR, *Pure Appl. Chem.*, 80 (2008) 1827–1835.
159. G. I. Csonka, I. Kolossvary, P. Császár, K. Eliás, and I. G. Csizmadia, The conformational space of selected aldo-pyranoses, *J. Mol. Struct. (THEOCHEM)*, 395–396 (1997) 29–40.
160. R. A. Klein, Electron density topological analysis of hydrogen bonding in glucopyranose and hydrated glucopyranose, *J. Am. Chem. Soc.*, 124 (2002) 13931–13937.
161. E. Espinosa, M. Souhassou, H. Lachekar, and C. Lecomte, Topological analysis of the electron density in hydrogen bonds, *Acta Crystallogr. B*, 55 (1999) 563–572.
162. E. Espinosa, E. Molins, and C. Lecomte, Hydrogen bond strengths revealed by topological analyses of experimentally observed electron densities, *Chem. Phys. Lett.*, 258 (1998) 170–173.
163. D. M. M. Jaradat, S. Mebs, L. Chęińska, and P. Luger, Experimental charge density of sucrose at 20 K: Bond topological, atomic, and intermolecular quantitative properties, *Carbohydr. Res.*, 342 (2007) 1480–1489.
164. E. D. Stevens, M. K. Dowd, G. P. Johnson, and A. D. French, Experimental and theoretical electron density distribution of α,α-trehalose dihydrate, *Carbohydr. Res.*, 345 (2010) 1469–1481.
165. E. D. Stevens, M. K. Dowd, G. P. Johnson, and A. D. French, Electron density studies of methyl cellobioside, Proceedings, 2011 Beltwide Cotton Conferences, Atlanta, Georgia, January 4–7, 2011, pp. 1416–1418.
166. T. Keith, AIMAll, Version 10.05.04, aim.tkgristmill.com, 2010.
167. B. G. Rånby, The mercerization of cellulose. II. A phase transition study with X-ray diffraction, *Acta Chem. Scand.*, 6 (1952) 116–127.

168. R. H. Marchessault, All things cellulose: A personal account of some historic events, *Cellulose*, 18 (2011) 1377–1379.
169. Y. Habibi, L. Lucia, and O. J. Rojas, Cellulose nanocrystals: Chemistry, self-assembly and applications, *Chem. Rev.*, 110 (2010) 3479–3500.
170. M. E. Himmel, S. Y. Ding, D. K. Johnson, W. S. Adney, M. R. Nimlos, J. W. Brady, and T. D. Foust, Biomass recalcitrance: Engineering plants and enzymes for biofuels production, *Science*, 315 (2007) 804–807.
171. K. Ward, Jr., Crystallinity of cellulose and its significance for the fiber properties, *Text. Res. J.*, 20 (1950) 363–372.
172. H. Steinmeier and P. Zugenmaier, "Homogeneous" and "heterogeneous" cellulose tri-esters and a cellulose triurethane: Synthesis and structural investigations of the crystalline state, *Carbohydr. Res.*, 164 (1987) 97–105.
173. D. V. Parikh, D. P. Thibodeaux, and B. Condon, X-ray crystallinity of bleached and crosslinked cottons, *Text. Res. J.*, 77 (2007) 612–616.
174. A. D. French, D. P. Miller, and A. Aabloo, Miniature crystal models of cellulose polymorphs and other carbohydrates, *Int. J. Biol. Macromol.*, 15 (1993) 30–36.
175. P. J. C. Smith and S. Arnott, LALS: A linked-atom least-squares reciprocal-space refinement system incorporating stereochemical restraints to supplement sparse diffraction data, *Acta Crystallogr. A*, 34 (1978) 3–11.
176. P. Zugenmaier and A. Sarko, The variable virtual bond modeling technique for solving polymer structures, in A. D. French and K. C. H. Gardner, (Eds.) *Fiber Diffraction Methods, ACS Symposium Series*, Vol. 141, pp. 225–237.
177. A. Aabloo and A. D. French, Preliminary potential energy calculations of cellulose Iα crystal structure, *Macromol. Theory Simul.*, 3 (1994) 185–191.
178. Z. M. Ford, E. D. Stevens, G. P. Johnson, and A. D. French, Determining the crystal structure of cellulose III$_I$ by modeling, *Carbohydr. Res.*, 340 (2005) 827–833.
179. R. J. Viëtor, K. Mazeau, M. Lakin, and S. Pérez, A priori crystal structure prediction of native celluloses, *Biopolymers*, 54 (2000) 342–354.
180. B. G. Rånby, The mercerization of cellulose. I. A thermodynamic discussion, *Acta Chem. Scand.*, 6 (1952) 101–115.
181. L. M. J. Kroon-Batenburg and J. Kroon, The crystal and molecular structures of cellulose I and II, *Glycoconj. J.*, 14 (1997) 677–690.
182. M. Bergenstråhle, L. A. Berglund, and K. Mazeau, Thermal response in crystalline Iβ cellulose: A molecular dynamics study, *J. Phys. Chem. B*, 111 (2007) 9138–9145.
183. Y. Nishiyama, G. P. Johnson, A. D. French, V. T. Forsyth, and P. Langan, Neutron crystallography, molecular dynamics, and quantum mechanics studies of the nature of hydrogen bonding in cellulose Iβ, *Biomacromolecules*, 9 (2008) 3133–3140.
184. J. F. Matthews, M. Bergenstråhle, G. T. Beckham, M. E. Himmel, M. R. Nimlos, J. W. Brady, and M. F. Crowley, High-temperature behavior of cellulose I, *J. Phys. Chem. B*, 115 (2011) 2155–2166.
185. T. Yui, N. Okayama, and S. Hayashi, Structure conversions of cellulose III$_I$ crystal models in solution state: A molecular dynamics study, *Cellulose*, 17 (2010) 679–691.
186. G. Bellesia, A. Asztalos, T. Shen, P. Langan, A. Redondoe, and S. Gnanakaran, In silico studies of crystalline cellulose and its degradation by enzymes, *Acta Crystallogr. D Biol. Crystallogr.*, 66 (2010) 1184–1188.
187. G. T. Beckham, Y. J. Bomble, E. A. Bayer, M. E. Himmel, and M. F. Crowley, Applications of computational science for understanding enzymatic deconstruction of cellulose, *Curr. Opin. Biotechnol.*, 22 (2011) 231–238.
188. G. T. Beckham, J. F. Matthews, B. Peters, Y. J. Bomble, M. E. Himmel, and M. F. Crowley, Molecular-level origins of biomass recalcitrance: Decrystallization free energies for four common cellulose polymorphs, *J. Phys. Chem. B*, 115 (2011) 4118–4127.

189. J. F. Matthews, G. T. Beckham, M. Bergenstraå hle-Wohlert, J. W. Brady, M. E. Himmel, and M. F. Crowley, Comparison of cellulose Iβ simulations with three carbohydrate force fields, *J. Chem. Theory Comput.*, 8 (2012) 735–748.
190. S.-Y. Ding, personal communication, manuscript submitted.
191. L. E. Alexander, X-Ray Diffraction Methods in Polymer Science, (1969) Wiley-Interscience, New York, p. 44 and Appendix 1, T-14.
192. http://www.unipress.waw.pl/debyer and http://code.google.com/p/debyer/wiki/debyer.
193. Y. Nishiyama, G. P. Johnson, and A. D. French, Diffraction from nonperiodic models of cellulose crystals, *Cellulose*, 19 (2012) 319–336.
194. A. N. Fernandes, L. H. Thomas, C. M. Altaner, P. Callow, V. T. Forsyth, D. C. Apperley, C. J. Kennedy, and M. C. Jarvis, Nanostructure of cellulose microfibrils in spruce wood, *Proc. Natl. Acad. Sci. USA*, 108 (2011) E1195–E1203.
195. A. Patterson, The Scherrer formula for X-ray particle size determination, *Phys. Rev.*, 56 (1939) 978–982.
196. H. A. Krässig, Cellulose—Structure, Accessibility and Reactivity, (1996) Gordon and Breach, Amsterdam.
197. P. Bansal, M. Hall, M. J. Realff, J. H. Lee, and A. S. Bommarius, Multivariate statistical analysis of X-ray data from cellulose: A new method to determine degree of crystallinity and predict hydrolysis rates, *Bioresour. Technol.*, 101 (2010) 4461–4471.

STRATEGIES IN SYNTHESIS OF HEPARIN/HEPARAN SULFATE OLIGOSACCHARIDES: 2000–PRESENT

Steven B. Dulaney and Xuefei Huang

Department of Chemistry, Michigan State University, East Lansing, Michigan, USA

I. Introduction	96
1. Background	96
2. Challenges in Synthesis of Oligosaccharides of Heparin and HS	98
II. Linear Synthesis	105
1. Solution Phase	105
2. Polymer-Supported Synthesis	107
III. Active–Latent Glycosylation Strategy	113
IV. Selective Activation	118
V. Reactivity-Based Chemoselective Glycosylation	121
VI. Reactivity-Independent, Pre-Activation-Based, Chemoselective Glycosylation	123
VII. Chemoenzymatic Synthesis	125
VIII. Future Outlook	130
Acknowledgments	130
References	130

ABBREVIATIONS

2-NAP, 2-naphthyl; 2-OST, 2-O-sulfotransferase; 3-OST, 3-O-sulfotransferase; 6-OST, 6-O-sulfotransferase; Ac, acetyl; AgOTf, silver triflate; All, allyl; ATIII, antithrombin III; AZMB, 2-(azidomethyl)benzoyl; BAIB, iodobenzene diacetate; Bn, benzyl; BSP, 1-benzenesulfinylpiperidine; Bz, benzoyl; C_5-epi, C_5-epimerase; CAN, ceric ammonium nitrate; Cbz, benzyloxycarbonyl; CSA, camphorsulfonic acid; DBU, 1,8-diazabicyclo[5.4.0]undec-7-ene; DCC, N,N'-dicyclohexylcarbodiimide; DDQ, 2,3-dichloro-5,6-dicyanobenzoquinone; DIAD, diisopropylazodicarboxylate; DIC, N,N'-diisopropylcarbodiimide; DMAP, 4-dimethylaminopyridine; DMTST, dimethyl (methylthio)sulfonium triflate; EDC, N-(3-dimethylaminopropyl)-N'-ethylcarbodiimide; Et₃N, triethylamine; FGF, fibroblast growth factor; Fmoc, fluorenylmethoxycarbonyl; GAG, glycosaminoglycan; D-GlcA, D-glucuronic acid; D-GlcN, D-glucosamine; GlcNAc, N-acetyl-D-glucosamine; HDTC, hydrazine dithiocarbonate; HOBT, hydroxybenzotriazole; HS, heparan sulfate; L-IdoA, L-iduronic acid; KfiA, N-acetylglucosaminyl transferase; Lev, levulinoyl; MP, 4-methoxyphenyl; MPEG, monomethyl polyethylene glycol; NDST, N-deacetylase/N-sulfotransferase; NIS, N-iodosuccinimide; NST, N-sulfotransferase; PAPS, 3'-phospho-5'-adenylyl sulfate; PEG, polyethylene glycol; Piv, pivaloyl; PMB, p-methoxybenzyl ether; PmHS2, heparan synthase-2; p-TolSCl, p-toluenesulfenyl chloride; p-TolSOTf, p-toluenesulfenyl triflate; Pyr, pyridine; RRV, relative reactivity value; SEM, 2-(trimethylsilyl)ethoxymethyl; TBDMS, t-butyldimethylsilyl; TBDMSOTf, t-butyldimethylsilyl triflate; TBDPS, t-butyldiphenylsilyl; TCE, 2,2,2-trichloroethyl; TDS, dimethylthexylsilyl; TEMPO, 2,2,6,6-tetramethylpiperidin-1-oxyl; Tf₂O, triflic anhydride; TFA, trifluoroacetic acid; TMS, trimethylsilyl; TMSOTf, trimethylsilyl triflate; Tol, tolyl; p-TsOH, p-toluenesulfonic acid; TTBP, 2,4,6-tri-t-butylpyrimidine; UDP, uridine 5'-diphosphate

I. INTRODUCTION

1. Background

Heparin, first isolated in 1917, was found to be highly effective as an anticoagulant, and within two decades, it was being used clinically.[1–4] Besides their anticoagulation activities, heparin and the related heparan sulfate (HS) play important roles in a wide range of biological functions such as cell differentiation, viral infection, and cancer metastasis.[5–12]

Heparin is a member of the glycosaminoglycan (GAG) family, which ranges from the unsulfated polymer hyaluronan to chondroitin and dermatan sulfates, and to the most complex examples, heparin and HS.[13] Heparin and HS share the basic disaccharide components, composed of D-glucosamine (GlcN) α-(1 → 4)-linked to a uronic acid (Scheme 1A). The GlcN component has a high degree of variability, as its O-6 and O-3 positions can be free or sulfated, and the amino group can be sulfated, acylated, or unmodified. The uronic acid can be either D-glucuronic acid (D-GlcA) or its C-5 epimer, L-iduronic acid (L-IdoA), both of which can be sulfated at the O-2 position.

Heparin and HS are differentiated by their tissue location and their detailed structures. Heparin has a higher degree of sulfation, with around 2.7 sulfate groups

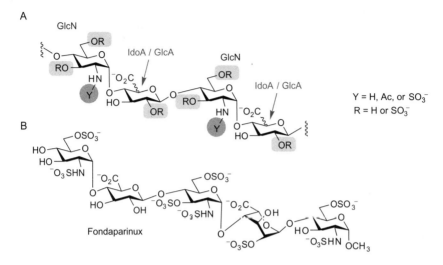

SCHEME 1. (A) Structures of heparin/HS; (B) structure of fondaparinux (Arixtra®). (Note idose and iduronic acid are arbitrarily presented in the $^1C_4$ conformation following common usage in the field. This does not necessarily represent the conformations in solution of the various heparin derivatives depicted throughout the article.) (See Color Insert.)

per disaccharide unit, and contains about 90% of its uronic acid as L-IdoA. Heparin is selectively synthesized in mast cells, whereas HS is omnipresent on cell surfaces and in the extracellular matrix as part of the proteoglycan complex.[14] More prevalent and heterogeneous, HS has on average one sulfate group per disaccharide, but it includes areas of high sulfation and swaths of unsulfated disaccharides.[15] The backbone sequence of HS is also more varied, in that the uronic acid residue is around 40% L-IdoA, with the major entity being D-GlcA.[5,16]

Although the naturally occurring heparin/HS is an exceedingly heterogeneous mixture, its interactions with biological receptors can be highly specific, as is evident from its binding to antithrombin III (ATIII).[17] Thorough structural analysis has demonstrated that the oligosaccharide sequence in heparin responsible for ATIII binding is a rare pentasaccharide fragment that is sulfated at O-3 in the middle GlcN component.[3,17–20] Removal of this O-3 sulfate group diminished its antithrombin affinity 10,000-fold.[21,22] The understanding of this structure–activity relationship led to the development of the drug fondaparinux (trade name: Arixtra, Scheme 1B), a fully synthetic pentasaccharide approved by the US Food and Drug Administration for the treatment of deep-vein thrombosis.[21]

Despite the success in establishing the ATIII-binding site, the heterogeneities of heparin and HS from natural sources present a major challenge in obtaining sufficient quantities of pure materials for the determination of detailed structure–activity relationships. To overcome this limitation, a frequently employed strategy has involved chemical modification of natural heparin and HS. However, this approach can give a complex mixture of many partially modified products from incomplete reactions.[23,24] Thus, the synthesis of pure and homogenous oligosaccharide sequences of the parent heparin and HS polysaccharides becomes crucial in facilitating biological studies. Commercially, Arixtra has been prepared by pure chemical synthesis, which is an impressive accomplishment considering there are over 50 synthetic steps. For Arixtra synthesis and other synthetic work prior to 2000, the reader should refer to several excellent reviews.[21,25–29] This article focuses on the advancement of heparin and HS component synthesis since 2000.

2. Challenges in Synthesis of Oligosaccharides of Heparin and HS

The synthesis of heparin/HS oligosaccharides presents a major challenge. Multiple factors must be considered for a successful synthetic design. These include (a) synthetic access to L-iduronic acid and L-idose; (b) the choice of uronic acid or the corresponding pyranoside as building blocks; (c) formation of the 1,2-*cis* linkage from the GlcN donor; (d) suitable protecting-group strategy to install sulfate groups at desired locations; and (e) methods used for elongation of the backbone sequence.

a. Preparation of L-Iduronic Acid and L-Idose.—L-Iduronic acid (L-IdoA) and the corresponding idopyranosides are not available from natural sources in large quantities and must be synthesized. There has been much research in order to access L-IdoA and its derivatives efficiently. Many approaches start from the commercially available 1,2:5,6-di-*O*-isopropylidene-α-D-glucofuranose (**1**), followed by the inversion of the configuration at C-5 through formation of an L-*ido* epoxide as in **3** (Scheme 2A).[30–32] Other routes employing compound **1** involve oxidation of the 5-hydroxyl group to aldehyde **4** through a three-step process, followed by stereoselective addition of a cyano group (Scheme 2B) or elimination of the primary hydroxyl group with subsequent hydroboration to invert the stereochemistry at C-5 (compound **6** in Scheme 2C).[33–36]

Alternative routes to L-IdoA have been reported.[32–45] As an example, Seeberger and coworkers have spearheaded research in the *de novo* synthesis of L-IdoA. Early work from their laboratory started from L-arabinose, but the low selectivity in the Mukaiyama aldol reaction with aldehyde **7** resulted in a low overall yield (6%) (Scheme 3A).[39] Starting from D-xylose and switching the aldol reaction to a more-selective cyanation furnished the L-IdoA building block **11** in 24% overall yield

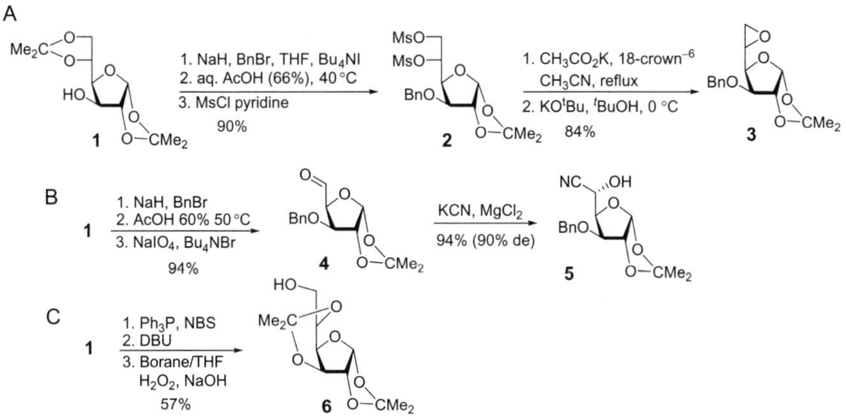

SCHEME 2. Various routes for inverting D-glucose to L-idose derivatives.

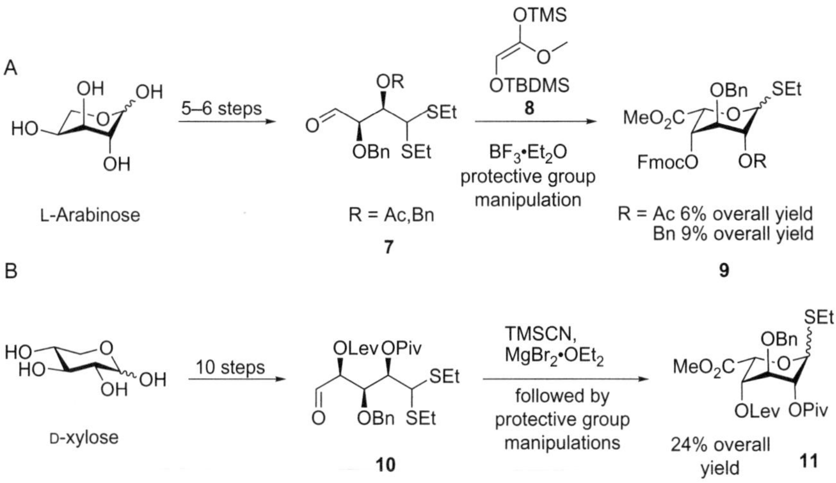

SCHEME 3. Recent routes to monosaccharide precursors of L-IdoA.

(Scheme 3B).[38] However, despite the many routes developed toward the preparation of L-IdoA or L-idose,[32–45] the synthesis of heparin/HS oligosaccharides remains difficult as long as 8–12 synthetic steps are required for the preparation of a single monosaccharide building block.

b. The Choice of Uronic Acid Versus Pyranoside as Building Blocks.—As glycosyl donors based on uronic acids can potentially be epimerized during their preparation, and they are typically less reactive than the corresponding glycopyranosides, the latter are commonly used as surrogate glycosyl donors. However, this approach requires adjustment of the oxidation state on the oligosaccharide after its assembly. As the size of the oligosaccharide increases, high-yielding oxidation can become very difficult.[46,47] Early syntheses relied on the Jones oxidation or the use of similar chromium reagents, which are toxic and frequently give low yields of the desired products (Scheme 4A).[31,48–51] This problem was subsequently overcome by using the mild TEMPO-mediated oxidations, which are typically effected with a co-oxidant such as NaOCl[52–54] or iodobenzene diacetate (BAIB; Scheme 4B),[55,56] and this can be followed by Pinnick oxidation to achieve high yields (Scheme 4C).[46,47,57,58] Alternatively, glycopyranosides could be used to prepare disaccharide intermediates as precursors for longer oligosaccharides by taking advantage of the high anomeric reactivity of the pyranoside donors. Adjustment of the oxidation state can then be performed on the disaccharide through oxidation at C-6 of the nonreducing end to the uronic acid, thus avoiding a late-stage oxidation of the more-valuable larger oligosaccharides (Scheme 4D).[59,60]

The monosaccharide glucuronic and iduronic acids, suitably derivatized, can be used directly as donors. Sinaÿ's synthesis of the ATIII-binding pentasaccharide used uronic acid-based glycosyl bromide donors, which gave glycosylation yields typically around 50%.[30,31] The availability of newer glycosylation methods and an understanding of the effects of protecting groups on anomeric reactivities have potentially circumvented this issue.[61,62] Bonnaffé and coworkers synthesized the disaccharide building block **22** in 75% yield by using the bromide donor **20** (Scheme 5A). The yield was increased to 91% employing the trichloroacetimidate donor **23** (Scheme 5B).[41,63] The resultant disaccharide was then used in a highly convergent manner to afford a dodecasaccharide derivative that was used for the synthesis of an HS proteoglycan analogue (see Scheme 18).[64]

c. Stereochemical Control in Glycosylation.—Stereochemical control is a crucial issue in the synthesis of heparin/HS components. The 1,2-*trans* linkage from the uronic acid to glucosamine is usually achieved through use of a participating group at the 2-position of the uronic acid. However, formation of the 1,2-*cis* linkage from the glucosamine donor can be difficult to control. The azido group, as a nonparticipating functionality, is widely employed as a precursor for the nitrogen atom at C-2 of glucosamine.[65] Such 2-azido glucosamine precursors can lead to the thermodynamically more-stable α glycosides.[29] This route generally provides high stereoselectivities in reactions with L-idosyl acceptors. However, for D-glucuronic acid-based acceptors,

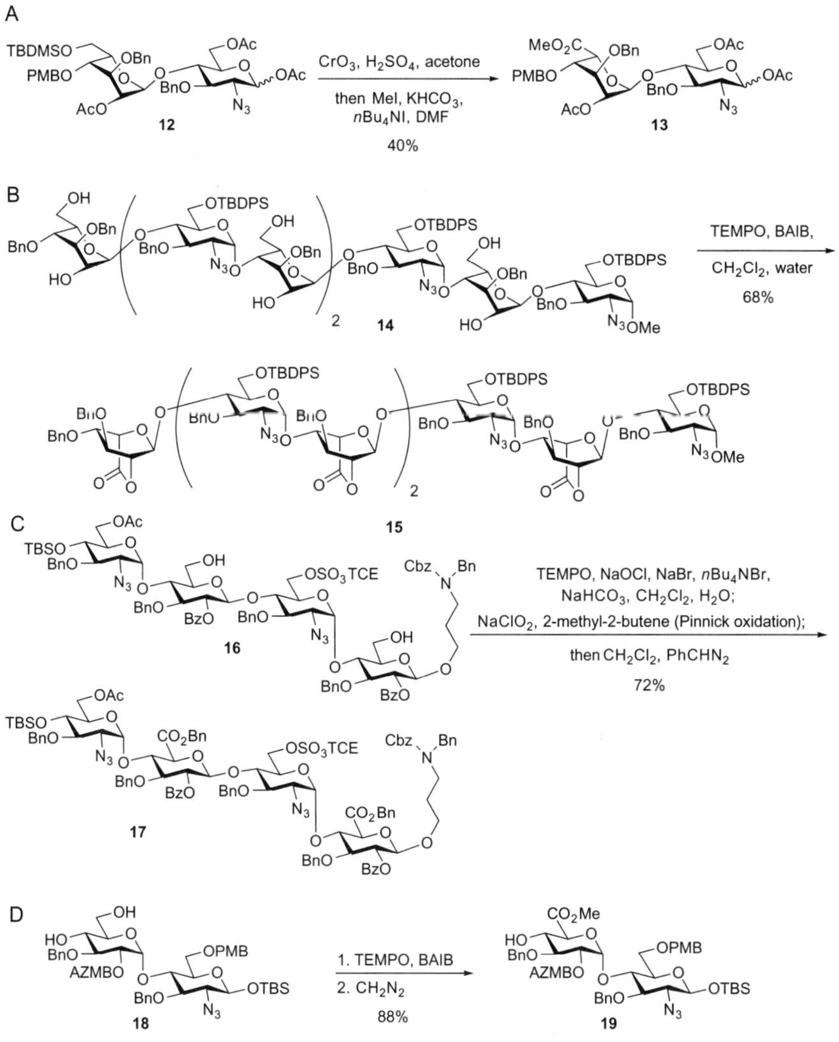

STRATEGIES IN SYNTHESIS OF HEPARIN/HEPARAN SULFATE OLIGOSACCHARIDES 101

SCHEME 4. Conversion of glycopyranosides into uronic acids in synthesis of heparin/HS oligosaccharides.

anomeric mixtures often result from the glycosylation, and this requires fine tuning of protecting groups to achieve high stereoselectivities.[58,66] For example, substituting the 4-benzyl ether in donor **26** by a 4-*t*-butyldimethylsilyl ether (donor **29**) led to formation of the α-linked disaccharide **28b** exclusively (Scheme 6A and B).[58] Bulky protecting

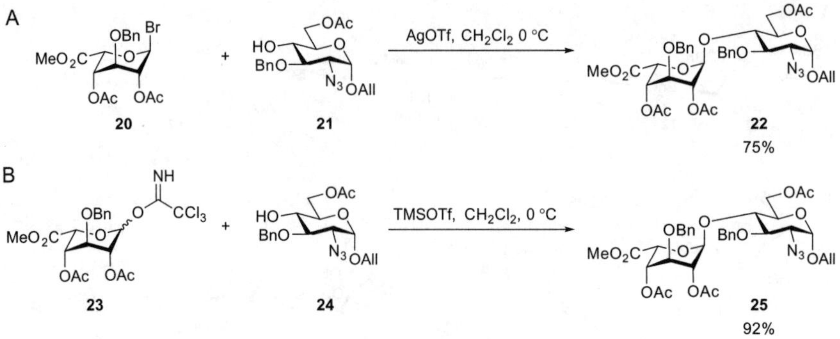

SCHEME 5. Comparison of glycosyl bromide and trichloroacetimidate donors in glycosylation.

groups at O-6 of the glucosamine component have also been explored to decrease the proportion of β anomer formed.[55] In addition to the protecting groups, the conformation of the acceptor can play an important role in determining the stereochemical outcome of the glycosylation. While glycosylation of pentenyl glycoside **31** with trichloroacetimidate **30** gave the disaccharide derivative **32** with an α:β ratio of 3:1 (Scheme 6C), locking the glucuronic acid component into the $^1C_4$ conformation (**33**) led to exclusive α selectivity (Scheme 6D).[67,68] However, caution needs to be taken in extrapolating these results to the assembly of larger oligosaccharides. Thus, glycosylation of the L-idosyl-configured disaccharide derivative **36** by tetrasaccharide donor **35** led to hexasaccharide **37** as an inseparable anomeric mixture (Scheme 6E).[69] The stereochemical outcome of the glycosylation reaction needs to be investigated individually, especially in the formation of large oligosaccharides.

d. Protecting-Group Strategy.—In addition to their roles in dictating stereochemistry, protecting groups are widely used to control the location of sulfate groups. With the high level of functionality in heparin/HS oligosaccharides, and the large number of protecting groups employed, syntheses must be suitably designed to prevent the premature removal of a protecting group.

To establish protecting groups suitable for regioselective sulfation, the Hung group explored the possibility of synthesizing all 48 possible heparin/HS disaccharide structures (disaccharide derivatives **41** and **42**), starting from eight monosaccharide building blocks (**38**–**40**) that are strategically protected.[70] The benzoyl group was used to protect those hydroxyl groups to be sulfated, and benzyl ethers were employed as persistent protecting groups for hydroxyl groups that would remain free in the final oligosaccharide products. The TBDPS substituent temporarily masked the primary hydroxyl group on compound **39** to permit subsequent oxidation to glucuronic acid.

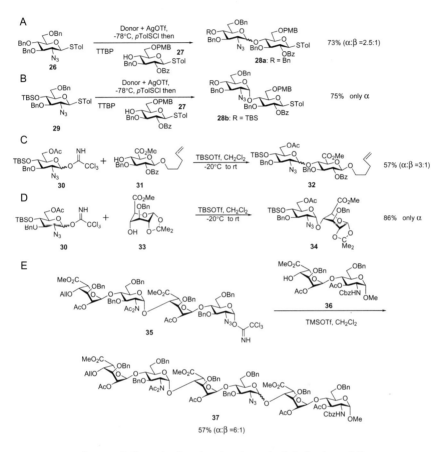

SCHEME 6. Strategies for enhancing stereoselectivity in glycosylation.

The azido group could be selectively reduced by Staudinger reduction and then either acetylated or sulfated, while the benzyloxycarbonylamino (Cbz) group could be deprotected to generate the free amine upon the final hydrogenolysis step. This panel of 48 disaccharide derivatives (compounds **41** and **42**) will be very useful for the assembly of heparin/HS libraries (Scheme 7).

Instead of preparing multiple monosaccharides, Wei and coworkers synthesized the glucuronic acid-containing HS disaccharide **43** having each hydroxyl group orthogonally protected (Scheme 8). Each protecting group could be removed selectively without affecting others. The newly liberated hydroxyl group was sulfated, and other protecting groups were then removed to ensure that the sulfate groups were stable under each set of deprotection conditions.[71] This strategy allowed the divergent

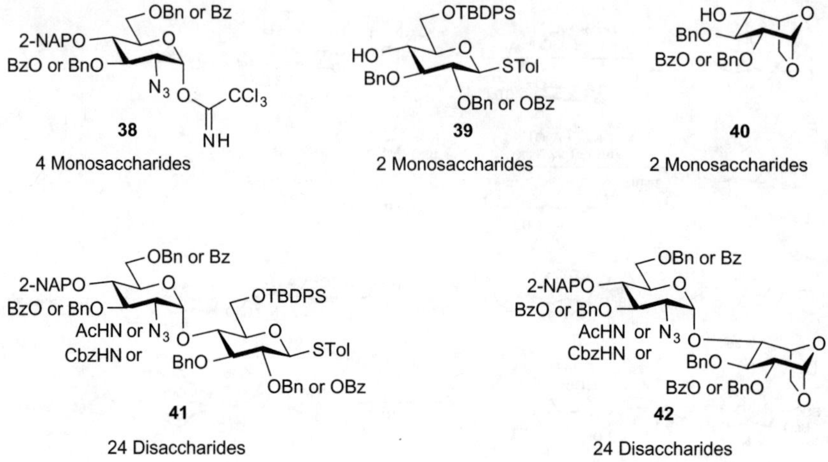

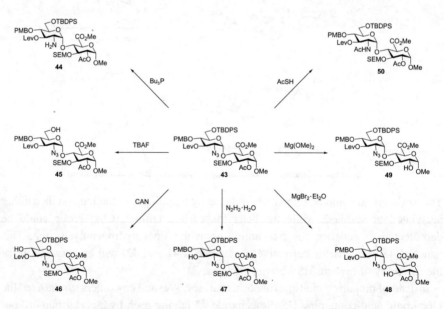

SCHEME 7. Synthesis of all potential heparin/HS disaccharides from eight monosaccharide precursors.

SCHEME 8. Disaccharide derivative **43** can be orthogonally deprotected for sulfation at various locations.

synthesis of multiple sulfation patterns from a single backbone but required more synthetic steps to remove the various protecting groups remaining after sulfation. As the biologically active heparin/HS domains typically are pentasaccharides or longer, these protecting-group strategies need to be extended to the synthesis of longer oligosaccharides.

Sulfate groups have traditionally been installed after assembly of the oligosaccharide backbone. However, late-stage sulfation, especially with larger oligosaccharides, can be quite capricious and challenging. Low yields[72] and incomplete reactions[64,73] are common. As an alternative, the sulfate groups can be installed on building blocks as protected esters prior to glycosylation. Numerous sulfate esters have been developed,[74-76] and Huang and coworkers investigated the utility of 2,2,2-trichloroethyl (TCE) sulfates[57] as developed by the Taylor group. TCE sulfates are stable to common transformations encountered in oligosaccharide synthesis, and the deprotection conditions are very mild.[77] An additional benefit of using TCE was that both sulfated and unsulfated building blocks can be derived from a common intermediate, thus increasing the efficiency of the overall process. For example, deprotection of the primary O-acetyl group in disaccharide derivative **51** followed by treatment with the sulfuryl imidazolium salt **52** provided the sulfate ester **53** (Scheme 9A).[57] The disaccharide derivative **51** was also used for conversion into the nonsulfated acceptor **57** (Scheme 9B). The presence of sulfate ester groups in the building blocks did not significantly affect the glycosylation yield, as reaction of the sulfate ester donor **54** with the acceptor **57** gave tetrasaccharide derivative **58** in 82% yield (Scheme 9C). The sulfate ester-containing tetrasaccharide **59** also functioned as a competent acceptor, as it underwent glycosylation by donor **54** in 70% yield (Scheme 9D). The tetrasaccharide **59** was successfully deprotected, giving rise to the HS tetrasaccharide component **61** (Scheme 9E), demonstrating the compatibility of TCE sulfate esters in the synthesis of heparin/HS oligosaccharides.

With the foregoing general understanding of synthesis of heparin/HS components, the following sections focus on the recent development of strategies to form and extend the heparin/HS oligosaccharide backbone. The discussions are grouped according to the strategy utilized.

II. Linear Synthesis

1. Solution Phase

The linear approach is one of the earliest strategies in oligosaccharide synthesis and is the route employed by Nature to produce heparin/HS.[13] Chemical glycosylation involves the activation of a donor, followed by nucleophilic attack of the activated

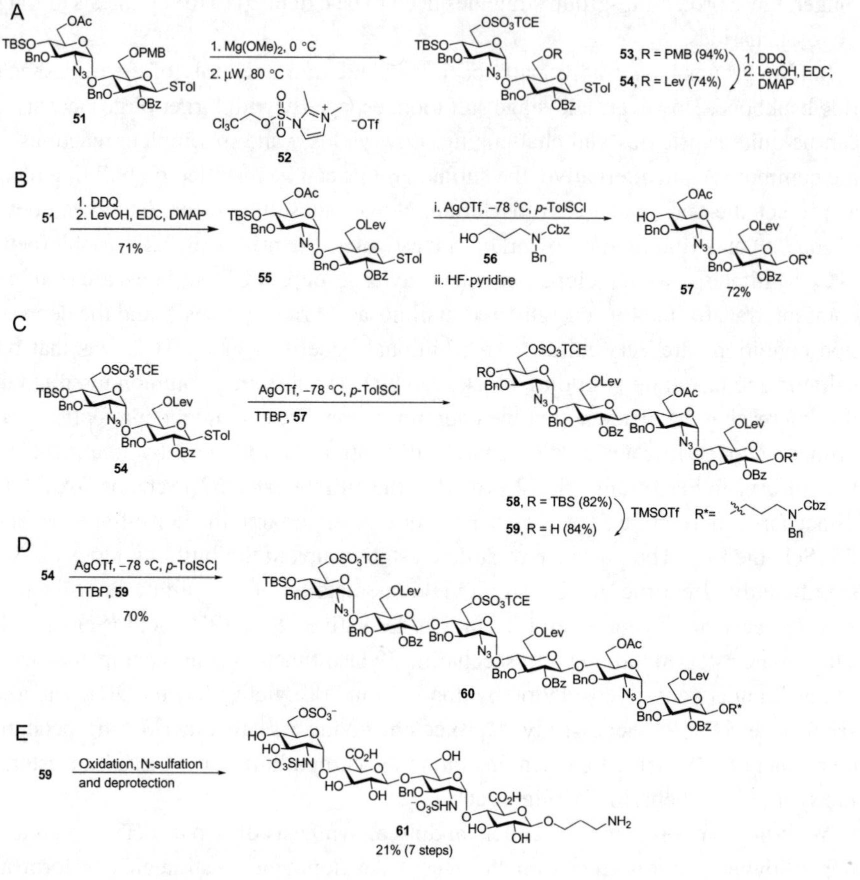

SCHEME 9. Evaluation of TCE sulfate ester-containing donors and acceptors in glycosylation reactions.

donor on the acceptor to form a new glycosidic linkage. In the linear approach toward heparin/HS oligosaccharides, the protecting group on the 4-hydroxyl group at the nonreducing end of the newly formed disaccharide is selectively removed, leading to a new acceptor, which undergoes further glycosylation, extending the chain from the reducing end to the nonreducing end, and producing the heparin/HS oligosaccharide backbone (Scheme 10).

Overall, the number of synthetic steps in linear synthesis is high because of the number of oligosaccharide intermediates generated and the deprotection step required after each glycosylation. Linear synthesis of oligosaccharides is therefore performed mainly for preparing shorter oligosaccharide sequences. Függedi used the linear

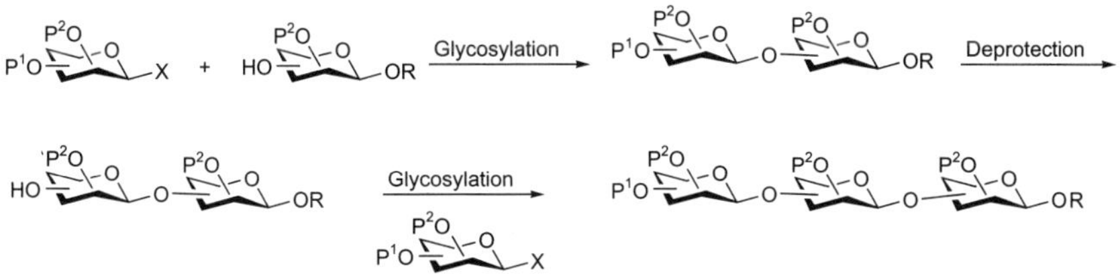

SCHEME 10. Linear synthesis of oligosaccharides from the reducing end.

strategy to synthesize HS trisaccharides considered to be responsible for interactions of HS with the fibroblast growth factors FGF-1 and FGF-2.[32,78] Glycosylation of acceptor **63** with thioglycoside **62**, using the thiophilic promoter dimethyl (methylthio)sulfonium triflate (DMTST), followed by removal of the chloroacetyl protecting group with hydrazine dithiocarbonate (HDTC) furnished the disaccharide acceptor **64** (Scheme 11). A second round of DMTST-mediated glycosylation using donor **65** produced the trisaccharide **66**. After removal of the benzoyl, the 4-methoxyphenyl (MP), and the *t*-butyl group, sulfation of the newly liberated hydroxyl groups was performed with the sulfur trioxide–pyridine complex in DMF. Hydrogenation and selective N-sulfation with the sulfur trioxide–trimethylamine complex under basic conditions furnished the final product **70**.[79] Following the same reaction sequence, except for reversing the steps of MPh-group removal and O-sulfation, generated the trisaccharide **71**, which bore sulfation patterns different from those in compound **70** (Scheme 11).

Boons and coworkers used the linear approach to synthesize a trisaccharide, using the monosaccharide building blocks **72** and **73**.[52] Glycosylation of vinyl donor **72a** with acceptor **73** was followed by removal of the *p*-methoxybenzyl (PMB) group at the nonreducing end by TFA, which generated the disaccharide acceptor **75** (Scheme 12). The trichloroacetimidate donor **72b** was found to be superior to the corresponding vinyl donor **72a**. The monosaccharide derivative **73** was benzoylated to produce vinyl donor **74**, which glycosylated the acceptor **75** to furnish trisaccharide **76**. Deprotection of **76** led to the unsulfated HS trisaccharide derivative **78**. The synthesis, while linear, could be modified into a modular active–latent approach, as discussed in Section III.

2. Polymer-Supported Synthesis

The Holy Grail in oligosaccharide synthesis would be the availability of a general and fully automated system having the synthetic efficiency of the established

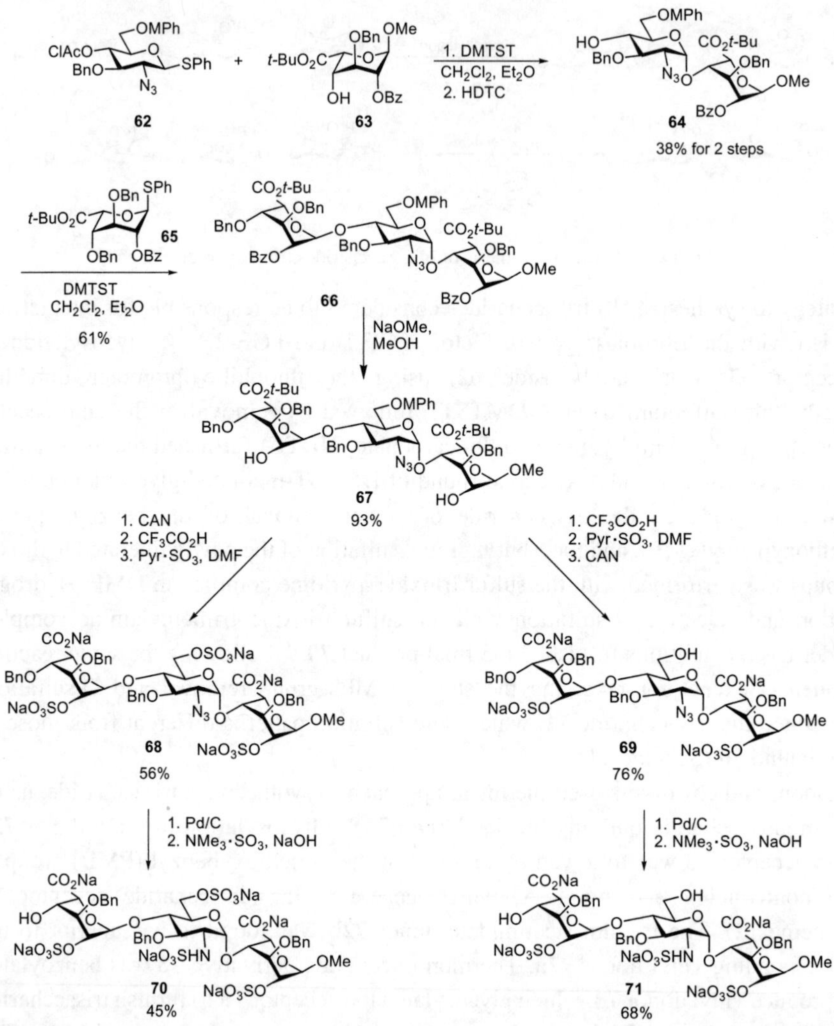

SCHEME 11. A linear synthesis of the heparin/HS trisaccharides responsible for binding with the fibroblast growth factors FGF-1 and FGF-2.

automated systems for peptide synthesis. Toward this goal, the Seeberger group has adapted an automated peptide synthesizer for the synthesis of complex oligosaccharides.[80,81] Thus far, the automated synthesis of heparin/HS oligosaccharides has not been achieved, because of the difficulties in translating solution-phase synthesis to high-yielding polymer-supported synthesis.

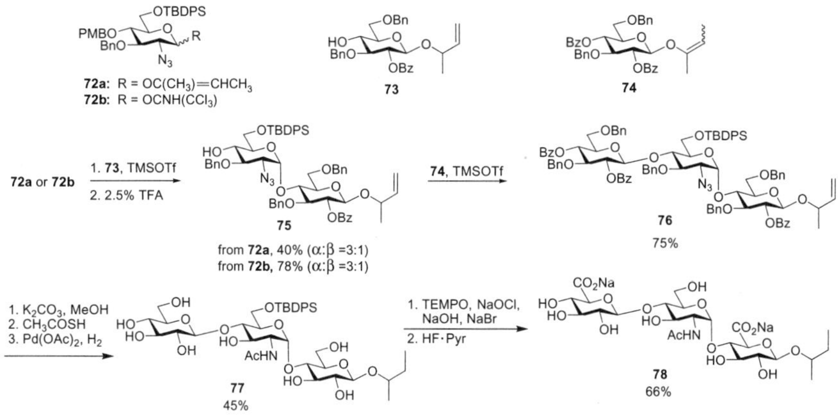

SCHEME 12. Linear synthesis and late-stage oxidation to generate trisaccharide **78**.

To determine the influence of polymer supports on the synthesis of heparin/HS components, Martin-Lomas and coworkers evaluated the use of various polymer supports and linkers on the glycosylation process.[82–84] Through a succinic acid linker, disaccharide derivative **79** was grafted onto the polystyrene resin ArgoGel™, which was then further functionalized, leading to the polymer-bound acceptor **82** (Scheme 13). Glycosylation of disaccharide derivative **82**, mediated by TMSOTf, was performed using 3.7 equiv. of the disaccharide donor **80**, and the excess of activated donor and reagent was removed after the reaction by multiple washes of the resin. As the reactivity of the polymer-bound acceptor was low, the glycosylation reaction was repeated two more times. The resin was then treated with hydrazine to cleave off the product, affording tetrasaccharide derivative **83** in 89% yield. This method was further extended to the synthesis of octasaccharide derivative **90** through successive iterations of deprotection and glycosylation. However, the overall yield of the octasaccharide was very low (∼10%), presumably because of the low reactivity of the larger glycosyl acceptor caused by steric hindrance posed by the insoluble polymer.

To increase the flexibility of the polymer, a water-soluble polymer, monomethyl polyethylene glycol (MPEG), was tested as a support. The succinylated disaccharide derivative **81** was linked to MPEG in a manner similar to that employed with ArgoGel™ (Scheme 14). Following TMSOTf-catalyzed glycosylation with the imidate donor **80**, the polymer was precipitated with diethyl ether and isolated by filtration. The glycosylation reaction was repeated three more times, and the tetrasaccharide derivative **92**, released from the polymer support by hydrazinlolysis, was obtained in 20% overall yield from the polymer-bound disaccharide derivative **91**.

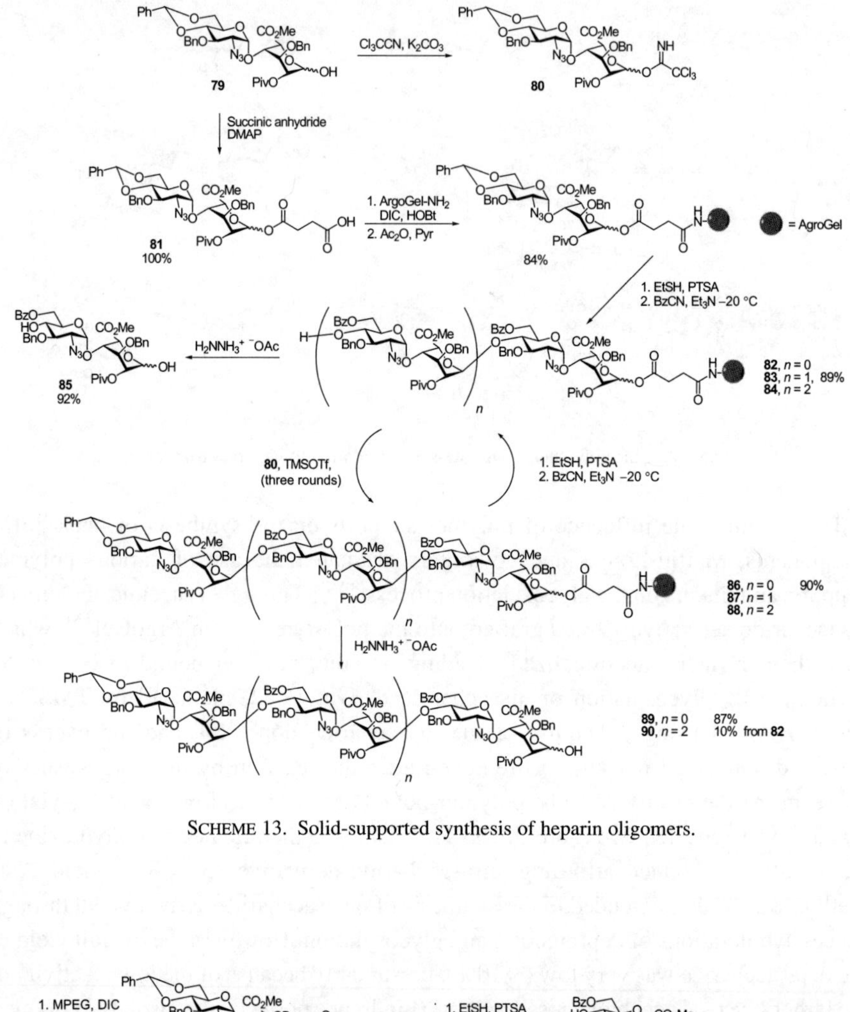

SCHEME 13. Solid-supported synthesis of heparin oligomers.

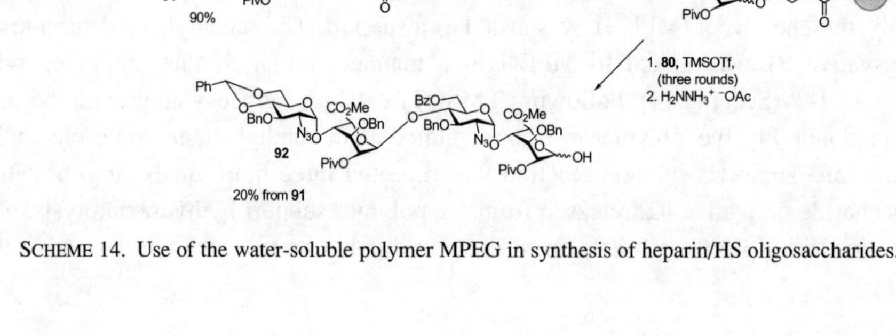

SCHEME 14. Use of the water-soluble polymer MPEG in synthesis of heparin/HS oligosaccharides.

The authors proposed that the yield differences in using MPEG versus ArgoGel could be attributed to the inefficiency of MPEG precipitation, as small losses compounded could become significant over multiple steps of manipulation.

One complication in using the anomeric position to link with the polymer is the production of anomeric mixtures upon release from the polymer. To avoid this, the 6-position of the glucosamine precursor was tested as the site of attachment to MPEG. However, conjugation to the polymer was only 40% effective, and with three rounds of glycosylation and subsequent detachment of the polymer, only 36% of the desired tetrasaccharide was obtained.[83] The carboxylate position of iduronic acid was next evaluated as the site of attachment. This position was ideal for attachment of the polymer as it avoided blocking a potential sulfation site, and cleavage from the polymer support could afford an anomerically pure product. The disaccharide acceptor **93** was bound to the MPEG polymer through the carboxylate site of its iduronic residue, and this was subjected to glycosylation by disaccharide donor **80**. Each backbone–elongation cycle consisted of four rounds of glycosylation (Scheme 15). After each glycosylation, to avoid the loss of desired product through incomplete MPEG precipitation, the nonconsumed acceptor was scavenged by carboxylic acid-functionalized, insoluble Merrifield resins, which were removed by simple filtration. Through this procedure, hexasaccharide product **94** was isolated in 37% overall yield

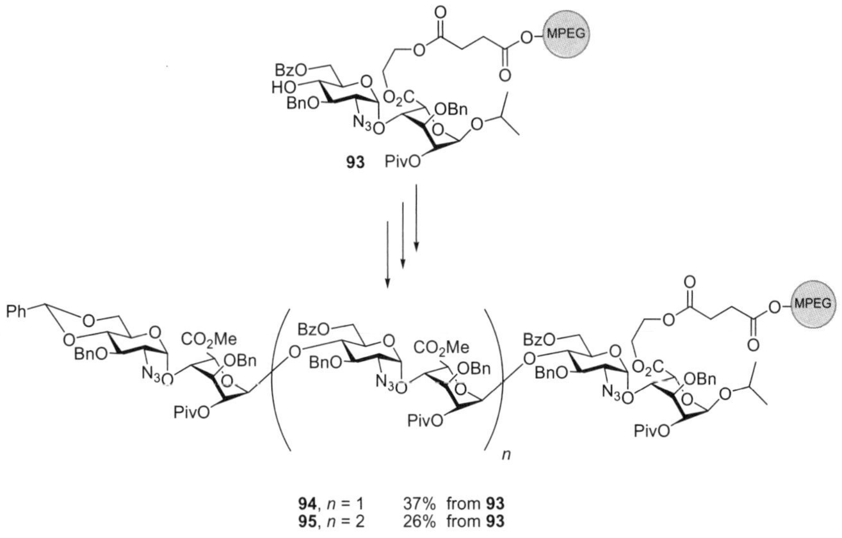

94, $n = 1$ 37% from **93**
95, $n = 2$ 26% from **93**

SCHEME 15. Soluble polymers anchored through the carboxylate group of the iduronic component.

from disaccharide derivative **93**. One more round of elongation gave the octasaccharide derivative **95** in 26% overall yield from compound **93**.[83] Based on yields of product obtained, this route was more efficient than previous MPEG-supported synthesis.

In addition to the succinic acid linker, the Martin-Lomas group designed a novel linker that immobilized an idose-based acceptor (**97**) onto MPEG (Scheme 16). Five iterations of glycosylation of **97** by the imidate **96**, followed by deprotection, gave the disaccharide derivative **99** in 82% yield. Further elongation of the chain by imidate **96** produced the polymer-bound trisaccharide, from which the polymer was cleaved off under basic conditions to yield oligosaccharide **100** bearing a protected amino group

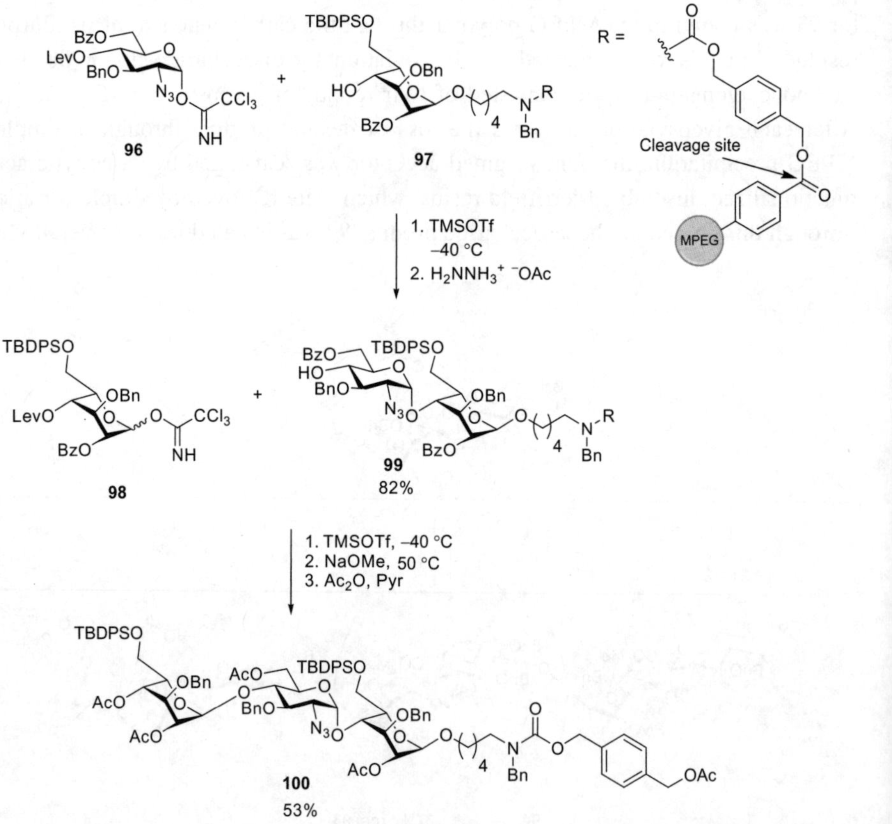

SCHEME 16. Protected amino linker used in conjunction with monosaccharide building blocks used in solid-supported synthesis of heparin/HS oligosaccharides.

in the linker, in 53% yield. The free amino group could be released by hydrogenolysis, which is useful for bioconjugation or immobilization of the oligosaccharides onto glycan microarrays.

In summary, the yields of heparin/HS oligosaccharide by glycosylation through polymer-supported synthesis decrease drastically as the chains grow longer. This is a serious challenge to any efforts at automation. Novel chemistry needs to be developed to significantly enhance the glycosylation yields on polymer support without resorting to the use of large excesses of donors. Until this becomes reality, solution-phase synthesis remains the preferred method for preparing complex heparin/HS oligosaccharides.

III. Active–Latent Glycosylation Strategy

The active–latent strategy is a solution-based method that builds oligosaccharides from the nonreducing end to the reducing end. In this approach, the acceptor carries a latent aglycon, which is inert to the conditions of glycosyl-donor activation. After glycosylation, the resultant oligosaccharide is transformed into an active donor for further chain elongation (Scheme 17).

Allyl glycosides are used widely for the active–latent glycosylation strategy. Inert to many of the conditions used for donor activation, allyl glycosides can be readily transformed into vinyl glycosides, which serve directly as active glycosyl donors in Lewis acid-catalyzed glycosylations. However, the vinyl donors typically give low yields in glycosylation reactions (see Scheme 12).[85] To circumvent this problem, the vinyl aglycon can be cleaved to generate the hemiacetal, which can then be transformed into trichloroacetimidate donors, which are much more reactive.[85–87]

The Bonnaffé group developed an impressive synthesis of a heparin dodecamer by the active–latent strategy, using the allyl glycoside and glycosyl trichloroacetimidate

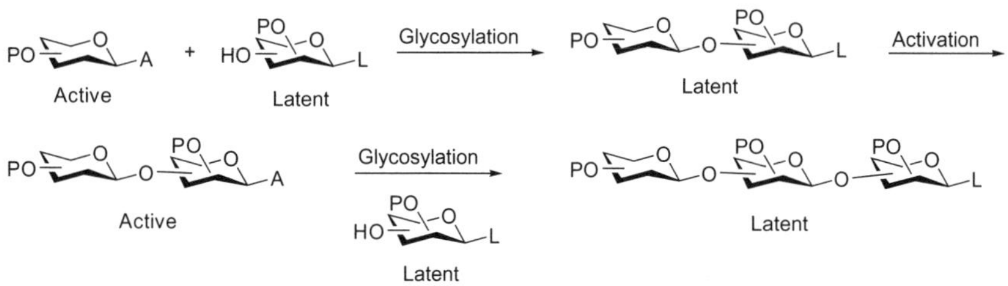

SCHEME 17. The active–latent glycosylation strategy.

combination.[64] To improve the overall synthetic efficiencies, a PMB group was employed to protect the 4-position at the nonreducing end of the oligosaccharide intermediate. This substituent could be removed selectively to expose a free hydroxyl group for further elongation of the chain. In this synthesis, the PMB-derivatized latent allyl disaccharide **103** was first transformed into a trichloroacetimidate donor, **105**, and the allyl disaccharide acceptor **104** (Scheme 18A).[86] Glycosylation of acceptor **104** by imidate **105** generated the latent allyl tetrasaccharide, which was then modified to an active trichloroacetimidate donor **106** (Scheme 18B). The reaction of **106** with tetrasaccharide **107**, followed by removal of the PMB group at the nonreducing end, and another round of glycosylation, furnished the dodecasaccharide **108** in 45% overall yield from the acceptor **107**. After completion of the backbone, deprotection and sulfation were performed. O-Deacetylation by potassium carbonate, and reduction with 1,3-propanedithiol, followed by simultaneous O- and N-sulfation with the sulfur trioxide–pyridine complex gave the sulfated dodecamer. The simultaneous sulfation with pyridine–SO_3 of the hydroxyl and amino groups did not proceed to completion. A second round of sulfation with pyridine–SO_3 in basified water was necessary to complete the sulfation.[64,73] Hydrolysis of the methyl esters, followed by hydrogenolysis, gave the fully deprotected dodecamer **109**, which is the longest heparin oligosaccharide yet prepared by chemical synthesis.[64]

As two synthetic steps are needed to cleave the allyl groups necessitating the use of expensive transition-metal reagents and toxic mercury salts, silyl protecting groups provide attractive alternatives for masking the anomeric position for the active–latent strategy. Many successful syntheses have used a variety of silyl ethers, such as dimethylthexylsilyl (TDS), *t*-butyldimethylsilyl (TBDMS), and trimethylsilyl (TMS).[37,68,73,84,88–91] As an example, glycosylation of the monosaccharide acceptor **111** by the trichloroacetimidate donor **110**, followed by acetylation, generated the latent disaccharide derivative **112** (Scheme 19A). Removal of the anomeric TDS group from **112**, followed by formation of the trichloroacetimidate, converted compound **112** into the active donor **113** (Scheme 19B).[90] The acceptor **114** was prepared by acid-mediated removal of the 4,6-benzylidene acetal from disaccharide derivative **112** and selective benzoylation of the primary hydroxyl group. Glycosylation by donor **113** of acceptor **114** furnished the tetrasaccharide acceptor **115**. The overall yield was only 40% because of the low glycosylation yield, with 44% of the starting disaccharide derivative **114** being recovered. Hexasaccharide derivative **117** was prepared by the reaction of tetrasaccharide derivative **115** with disaccharide donor **116**. With the removal of its anomeric TDS group, the hexasaccharide derivative **117** was transformed into an active trichloroacetimidate donor **118**, which upon reaction with methanol afforded the methyl glycoside **119** (Scheme 19D).

STRATEGIES IN SYNTHESIS OF HEPARIN/HEPARAN SULFATE OLIGOSACCHARIDES 115

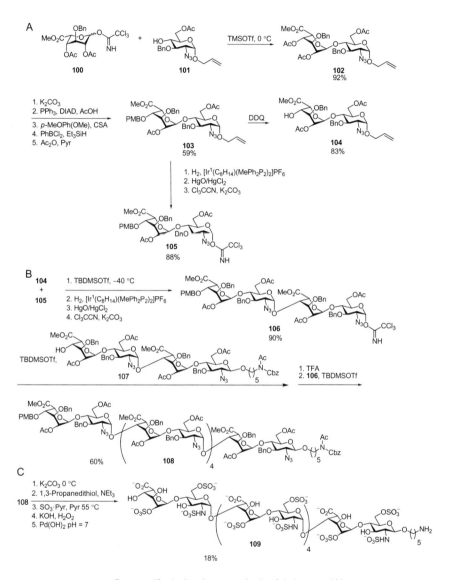

SCHEME 18. Active–latent synthesis of dodecamer **109**.

Besides being compatible with the trichloroacetimidate donors, the silyl protecting group is robust and has also been applied with thioglycosides under various activating systems.[37,59,60] The Boons group used this strategy to prepare disaccharide building blocks for their heparin/HS oligosaccharide synthesis (Scheme 20).[59] Glycosylation

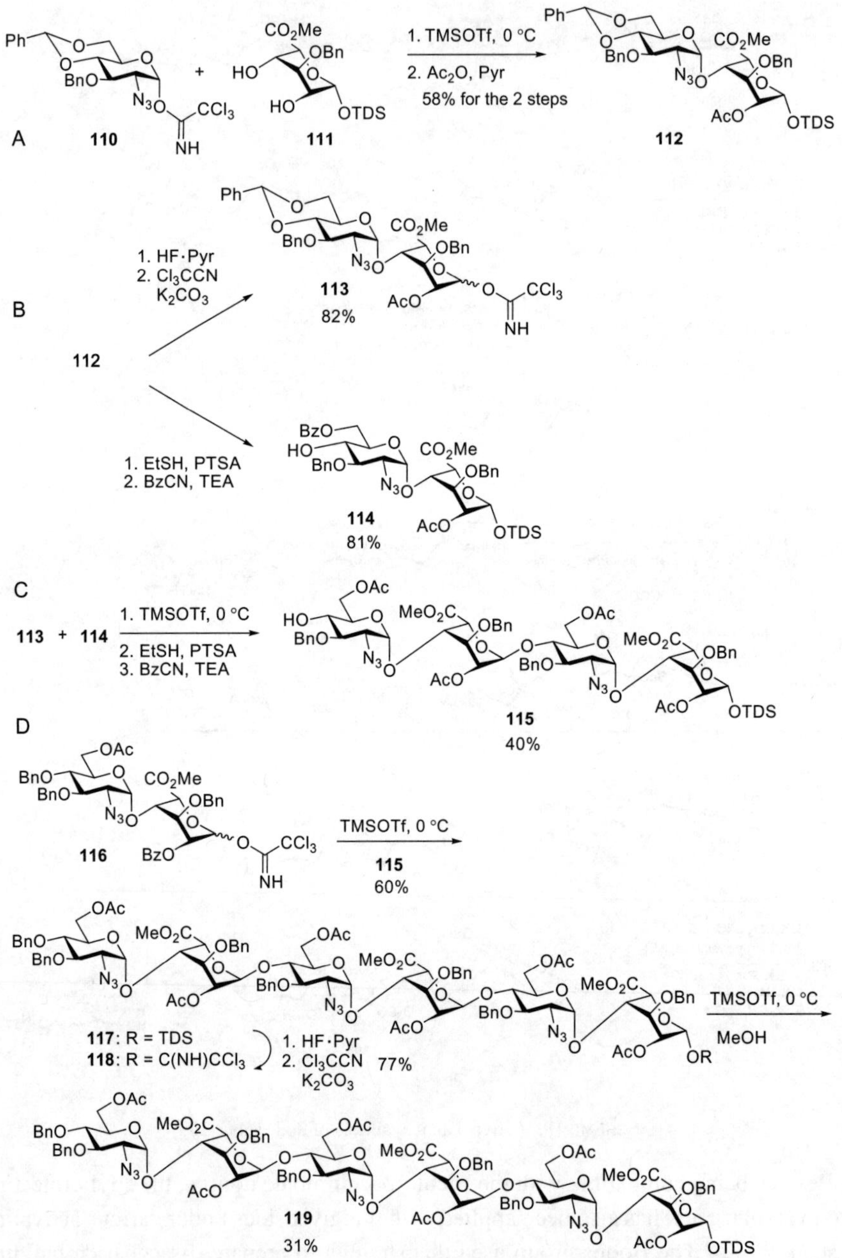

SCHEME 19. Active–latent synthesis with silyl protecting groups.

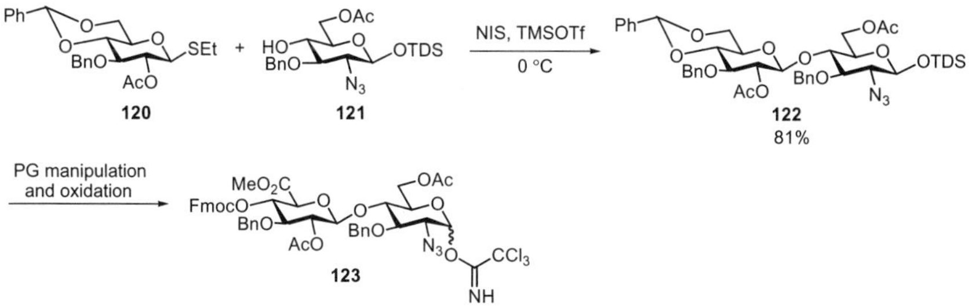

SCHEME 20. Synthesis of one of the eight disaccharide building blocks used by Boons and coworkers to prepare a library of heparin oligosaccharides.

of 1-thioglycoside donor **120** with the TDS-protected acceptor **121** formed the latent disaccharide **122**. After oxidation and protecting group manipulation, the TDS group in **122** was removed and the resulting hemiacetal was converted into the trichloroacetimidate disaccharide donor **123**. Eight disaccharide building blocks were prepared in this manner and were used to construct a panel of 11 heparin/HS tetrasaccharides and 1 hexasaccharide having different backbone structures and sulfation patterns. These tetrasaccharides were used to probe the important structural features of HS for inhibiting β-secretase, a protease considered to be involved in the development of Alzheimer's disease.

In addition to allyl and silyl groups, other functionalities, including isopropylidene acetals[35,69,72] and 1,6-anhydro sugars, have been used to mask the anomeric position of the latent glycosyl donors. The 1,6-anhydro sugars are advantageous to use as they do not require another selectively removable protecting group for the anomeric position. Hung and coworkers developed rapid routes of access to such 1,6-anhydro-L-idose building blocks as compound **125**.[35,36] Glycosylation of anhydro derivative **125** by the glycosyl trichloroacetimidate donor **124** furnished disaccharide **126** (72% yield, α:β = 5.5:1, Scheme 21A). To activate this latent disaccharide, the 1,6-anhydro ring of the α anomer of **126** was cleaved by Cu(OTf)$_2$-catalyzed acetolysis, and the newly installed acetyl group at O-6 was exchanged for the more selectively cleavable levulinoyl ester, followed by formation of the trichloroacetimidate (Scheme 21B). The resulting disaccharide donor was condensed with the glucosaminide precursor **128**, generating trisaccharide derivative **129**. The 2-naphthyl-substituted trisaccharide **129** was selectively deprotected to expose the 4-hydroxyl group at the nonreducing end, where it was glycosylated by the disaccharide donor **127**. Repetition of these deprotection and glycosylation sequences two more times led to formation of the HS nonasaccharide derivative **132**.[72] The active–latent strategy,

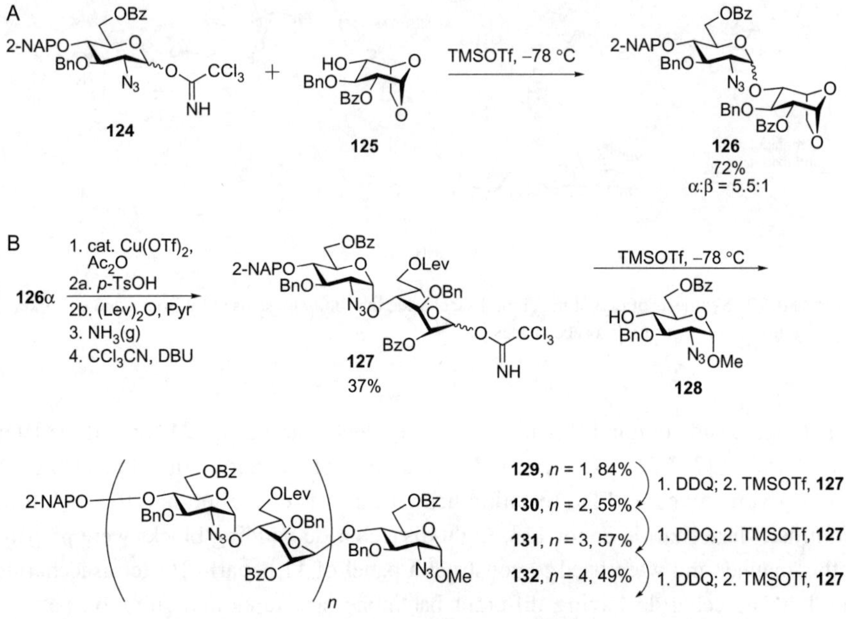

SCHEME 21. The use of the 1,6-anhydro sugars in latent–active strategy.

coupled with the use of a selectively removable protecting group, such as 2-naphthyl, at the 4-hydroxyl group at the nonreducing end, provides additional versatility in comparison to the linear strategy, as oligosaccharides can be built up from both the nonreducing and reducing end. However, multiple synthetic manipulations are still needed on the oligosaccharide intermediates to activate the latent donor.

IV. SELECTIVE ACTIVATION

To decrease the number of steps required for modification of intermediates, as is encountered in the latent–active strategy, the selective-activation method utilizes donors and acceptors having different types of activable aglycons. Upon selective activation of the donor and glycosylation of the acceptor, the resulting disaccharide can be used directly as a donor under a new set of activation conditions, without the need for manipulation of the intermediate (Scheme 22). The most common pairs of glycosyl building blocks in selective-activation methods are glycosyl trichloroacetimidates and thioglycosides, since thioglycosides are stable under the acidic conditions

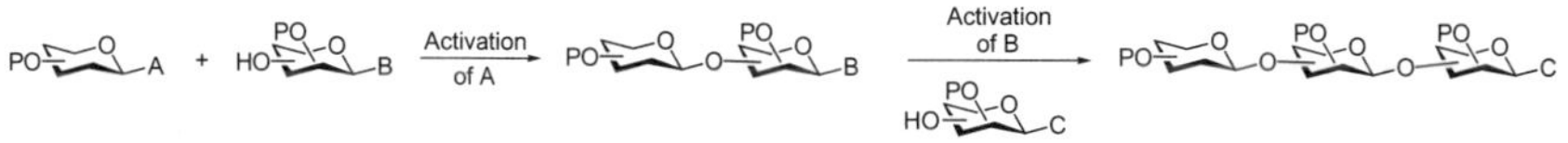

STRATEGIES IN SYNTHESIS OF HEPARIN/HEPARAN SULFATE OLIGOSACCHARIDES   119

SCHEME 22. Glycosylation strategy employing selective activation.

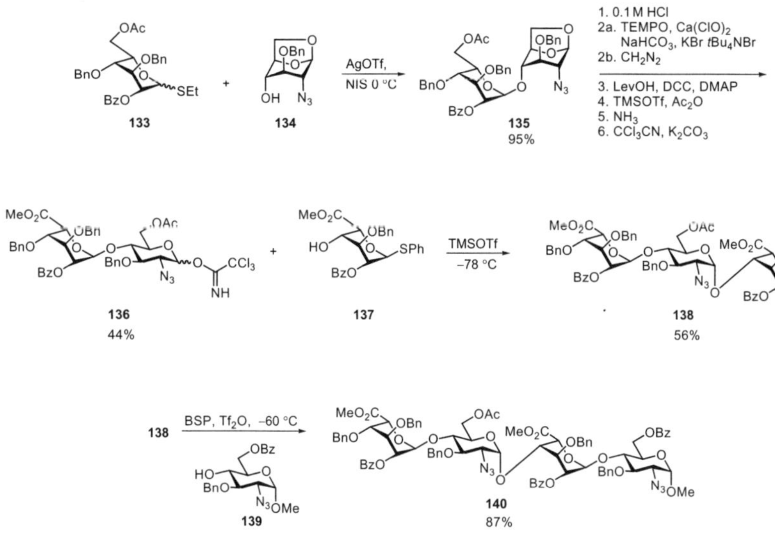

SCHEME 23. Synthesis utilizing the selective activation of trichloroacetimidate donors in the presence of thioglycoside acceptors.

encountered in trichloroacetimidate activation.[55,90,92,93] The selective-activation method can often be combined with the active–latent strategy within a single synthetic operation.

In the preparation of two heparin/HS tetrasaccharides, Yu and coworkers used the active–latent approach to produce disaccharide building blocks and selective activation for extension of the backbone. Glycosylation of the 1,6-anhydro acceptor **134** by the ethyl 1-thio-L-idoside donor **133** was performed with NIS and AgOTf (Scheme 23).[68,90] To convert disaccharide derivative **135** into an active donor, the anhydro ring was opened, followed by protecting-group adjustment, an oxidative manipulation at C-6, and formation of the trichloroacetimidate disaccharide donor **136**. Donor **136** was selectively activated by TMSOTf, with the thioglycoside **137** serving as acceptor, leading to the trisaccharide derivative **138**. Without further synthetic manipulation, trisaccharide **138** was activated by the thiophilic promoter

1-benzenesulfinylpiperidine (BSP) and Tf$_2$O, which glycosylated monosaccharide **139** to produce tetrasaccharide derivative **140**.

Instead of glycosyl trichloroacetimidates and thioglycosides, van der Marel and coworkers explored the utility of free glycoses (glycosyl hemiacetals) and thioglycosides in a selective glycosylation approach toward the pentasaccharide derivative **148**, a fully protected precursor of the heparin backbone. The hemiacetal **141** was selectively activated in the absence of the glycosylthio acceptor **142**, utilizing the preactivation strategy, where the donor was treated with the diphenyl sulfoxide and Tf$_2$O promoter system developed by the Gin laboratory.[94] Upon complete activation, the acceptor **142** was added to the reaction mixture to yield disaccharide derivative **143** (Scheme 24). To extend the chain, the 1,6-anhydro acceptor **134** was glycosylated with disaccharide derivative **143**, producing trisaccharide derivative **144** as a latent

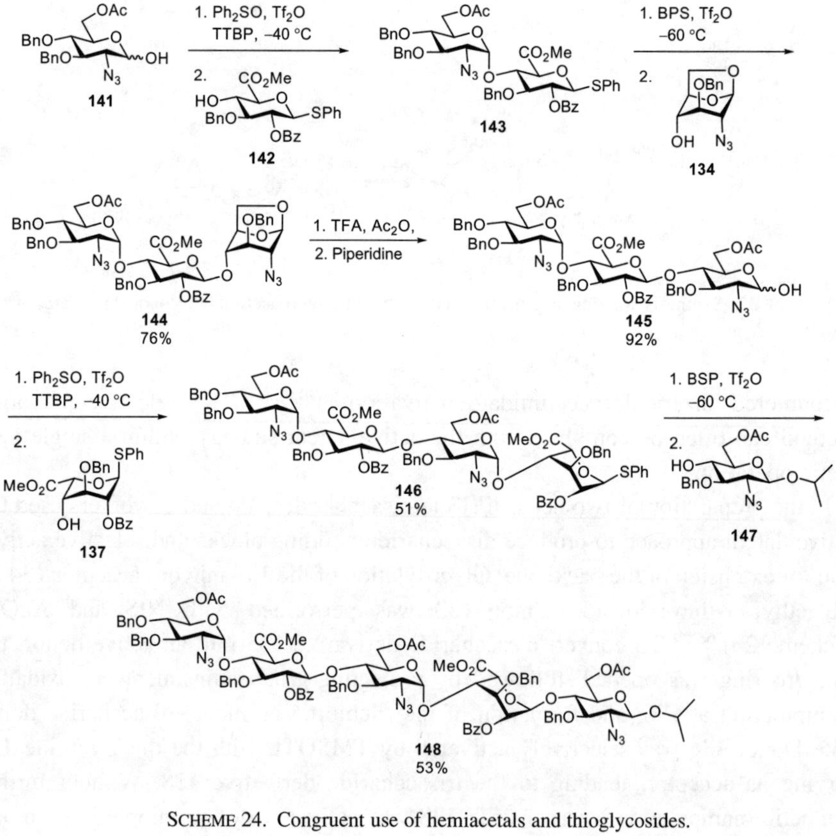

SCHEME 24. Congruent use of hemiacetals and thioglycosides.

donor. The 1,6-anhydro bridge was then opened under acidic conditions to create a trisaccharide hemiacetal donor **145**, which after selective activation was coupled to the glycosylthio acceptor **137**. The resultant tetrasaccharide thioglycoside **146** reacted with the reducing-end acceptor **147** to complete the synthesis.[66] Aided by the 1,6-anhydro ring as a masked hemiacetal, thioglycosides and glycosyl hemiacetals proved to be very effective partners for the selective glycosylation approach. Only one manipulation of an intermediate aglycon was required for preparation of the pentasaccharide derivative **148**.

Although the selective-activation strategy improves synthetic efficiency, it requires two different types of glycosyl donor. To simplify the overall synthetic design, it is desirable for a single type of glycosyl donor to be employed throughout the synthesis, avoiding the need for modification of the aglycon leaving group of the intermediate oligosaccharides. Toward this goal, two chemoselective strategies have been developed. These are the reactivity-based, armed–disarmed method and the reactivity-independent, pre-activation-based method.

V. REACTIVITY-BASED CHEMOSELECTIVE GLYCOSYLATION

In reactivity-based, armed–disarmed glycosylation strategy, glycosyl building blocks, typically thioglycosides, are designed to have different reactivities at the anomeric position. When a mixture of a more-reactive donor (armed) and an acceptor having lower anomeric reactivity (disarmed) is subjected to a limiting amount of promoter, the more-reactive, armed donor is activated preferentially, and this donor glycosylates the acceptor (Scheme 25). The resulting oligosaccharide can function directly as a donor by using the same conditions for further glycosylation with a thioglycoside acceptor that has even lower anomeric reactivity. With suitable design, the anomeric reactivities of various building blocks can be sufficiently different so as to enable multiple glycosylation reactions sequentially in one vessel without the need for purification of the oligosaccharide intermediates.

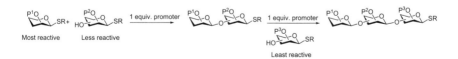

SCHEME 25. The armed–disarmed strategy for chemoselective glycosylation relies on differences in anomeric reactivities of the building blocks.

To achieve the required differentiation of anomeric reactivity, the electronic property or conformational rigidity of the donor can be tuned by strategically placing suitable protecting groups on the glycon ring[95–97] or by modification of the aglycon.[98–100] Reactivities can be quantified as relative reactivity values (RRVs), with the reactivity of *p*-tolyl 2,3,4,6-tetra-*O*-acetyl-α-D-mannopyranoside toward methanol acceptor being set as 1.0.[97]

Wong and coworkers conducted the synthesis of heparin/HS oligosaccharides via the reactivity-based approach. Four monosaccharide building blocks (**149–152**) were prepared and their RRVs measured (Scheme 26A).[101] As the glucoside building block **151** was 30 times more reactive than the acceptor glucosamine precursor **150**, chemoselective activation of **151** was achieved in preference to **150**, leading to disaccharide derivative **153** in excellent (89%) yield (Scheme 26B). Manipulation of protecting groups and oxidation at C-6 of the glucosyl component furnished the new disaccharide building block **154** having a RRV of 18.3. Chemoselective glycosylation of **154** by the azidoglucose donor **149** (RRV = 53.7) was therefore feasible (Scheme 26C) to give the corresponding trisaccharide derivative. The latter was coupled to O-4 of the disaccharide acceptor **155**, and more promoter was added to the reaction mixture. This led to the formation in one vessel of the fully protected HS pentasaccharide precursor **156** in 20% overall yield for the two steps. The modest net

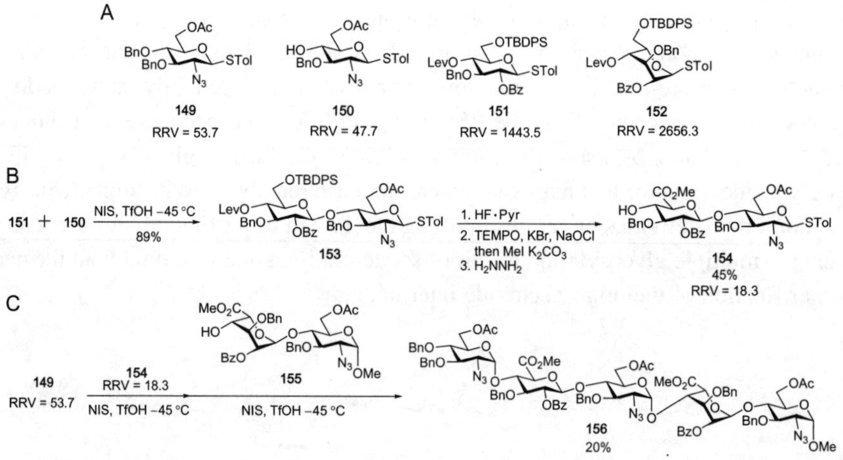

SCHEME 26. (A) Monosaccharide building blocks used in Wong's synthesis of heparin components; (B) preparation of disaccharide building block **154**; (C) one-pot synthesis of heparin pentasaccharide precursor **156** by the armed–disarmed strategy.

yield was most probably attributable to the small reactivity differential between donor **149** and disaccharide **154**.

The RRVs provide general guidance toward the selection of building blocks. However, the RRVs are quantified with reference to methanol as the acceptor, and these values can change according to the structure of the acceptor and the reaction conditions.[102] Accordingly, caution needs to be exercised in relying solely on RRVs to predict the outcome of a reaction. Furthermore, applying the reactivity-based method to the synthesis of longer heparin/HS oligosaccharides could be challenging because the polymeric nature of heparin/HS would require the same glycosyl units to have greatly differing reactivities according to their location in the backbone. The building blocks at the nonreducing end should have higher reactivities than those situated toward the reducing end. This challenge can be overcome by the reactivity-independent, pre-activation based chemoselective strategy for glycosylation.

VI. Reactivity-Independent, Pre-Activation-Based, Chemoselective Glycosylation

The aforementioned glycosylation strategies rely on differences in anomeric reactivity. The acceptor either cannot be activated, as in the case of linear, active–latent, and selective-activation methods, or has a much lower reactivity than the donor in the armed–disarmed reactivity-based approach. The underlying cause for this is the fact that the glycosyl donor and acceptor are both present in the reaction mixture when the promoter is added. Thus, the anomeric reactivities of donors and acceptors must be differentiated to achieve selective activation of the donor. To overcome this limitation, the pre-activation strategy was developed, wherein the donor is activated by a promoter to generate a reactive intermediate in the absence of an acceptor (Scheme 27). The acceptor is then added to react with the reactive intermediate and

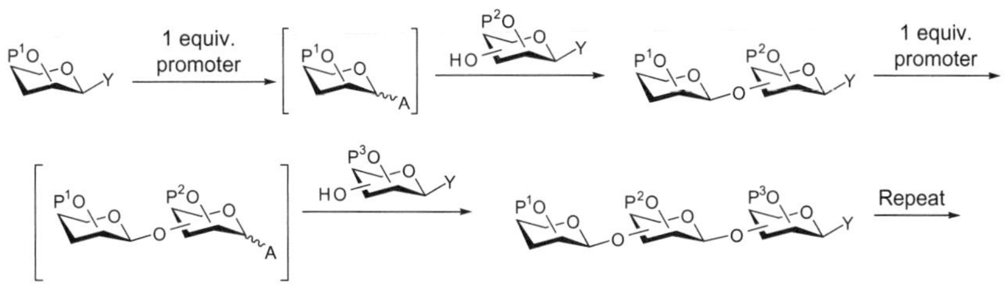

SCHEME 27. Pre-activation-based strategy for glycosylation.

form a new glycosidic bond. Activation of the donor in the absence of the acceptor allows the acceptor to carry the same aglycon group as the donor, negating the need for reactivity tuning. The prerequisite for pre-activation is that the promoter used must be in stoichiometric amount to avoid activation of the acceptor or product, and any side-products from activation of the donor must not be nucleophilic. Several types of glycosyl donor have been used in the pre-activation scheme, and these include hemiacetals,[94] glycals,[103] selenoglycosides,[104] and thioglycosides.[105,106]

Huang and coworkers synthesized a library of 12 heparin/HS hexasaccharides by the reactivity-independent, pre-activation-based strategy. This synthesis employed thioglycoside modules and the powerful promoter p-toluenesulfenyl triflate (p-TolSOTf), which was generated in situ from p-toluenesulfenyl chloride (p-TolSCl) and AgOTf.[106] To simplify the preparation of building blocks, a divergent approach was designed. Starting from three monosaccharide building blocks, two disaccharide derivatives (**162** and **163**) were prepared (Scheme 28A). These compounds were then divergently modified, leading to six disaccharide modules (**164–169**, Scheme **28**B and C).[58] To assemble the hexasaccharide, disaccharide donor **166** was pre-activated with p-TolSCl and AgOTf at −78 °C (Table I). Upon complete activation, the bifunctional 1-thioglycoside acceptor **165** was added to the reaction mixture. The reactive intermediate generated through activation of the donor glycosylated the acceptor **165**, producing a tetrasaccharide. As this tetrasaccharide product already bore an arylthio aglycon, it was activated directly with another equivalent of the promoter and allowed to react with acceptor **167** in the same reaction flask. Hexasaccharide **170** was obtained from this reaction in 54% yield in less than 5 h. Since this synthesis did not require adjustment of the aglycon structure or purification of the intermediate tetrasaccharide, the efficiency of the glycosidic assembly was greatly enhanced.

As the pre-activation method does not require the glycosyl donor to have higher anomeric reactivities than the glycosyl acceptor, the disaccharide building blocks **164–169** could be used in a combinatorial fashion to prepare a library of oligosaccharides (Table I).[58] For example, substituting compound **165** by **168** and then following the same reaction scheme as in the preparation of hexasaccharide **170**, hexasaccharide derivative **171** was formed in 59% yield in a one-pot process. By mixing the disaccharide building blocks **164–169**, six hexasaccharides having systematically varied and precisely controlled backbone structures were produced in 50–62% yields within a few hours (Table I). These hexasaccharides were then deprotected and subsequently sulfated, creating a set of 12 heparin/HS hexasaccharides, which were used to decipher structure–activity relationships in the binding of fibroblast growth factor-2 to heparin.

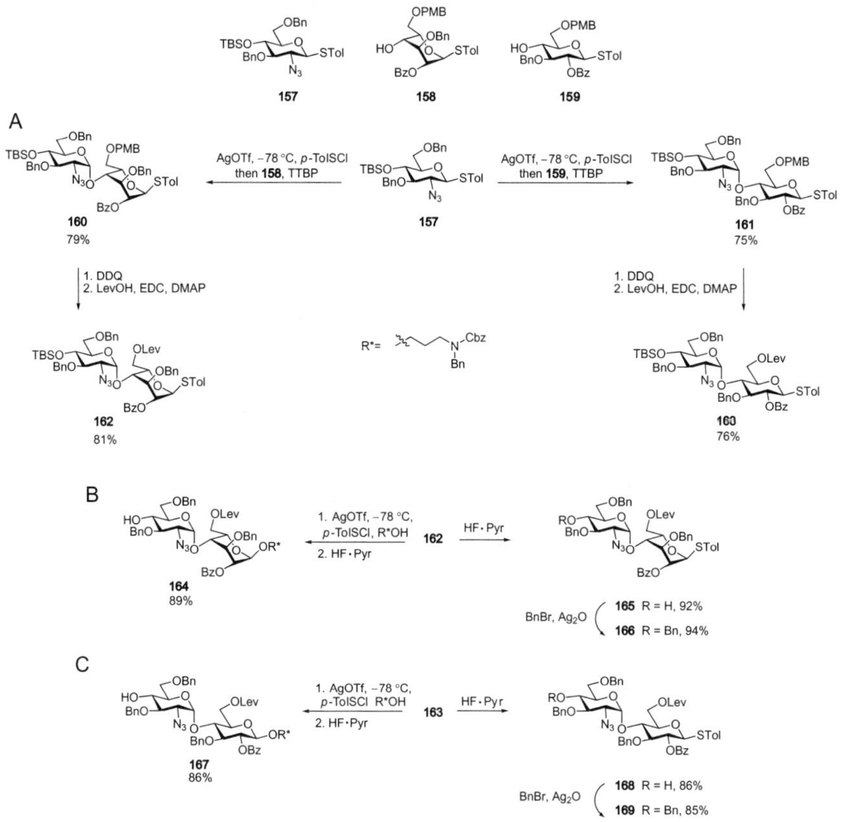

SCHEME 28. Divergent synthesis of the building blocks needed for the assembly of a hexasaccharide library.

In summary, as discussed up to this point, chemical synthesis has been the major path for access to synthetically pure heparin/HS oligomers. Given the length and difficulties in chemical synthesis, several groups have begun to explore the potential of enzymatic synthesis and its integration with chemical methods.

VII. CHEMOENZYMATIC SYNTHESIS

In Nature, the biosynthesis of heparin/HS is performed by multiple enzymes in the Golgi apparatus. Assembly of the HS backbone by glycosyltransferases is followed by such enzymatic modifications as N-deacetylase/N-sulfotransferase (NDST) for removal

TABLE I
One-Pot Preparation of Heparin/HS Hexasaccharides

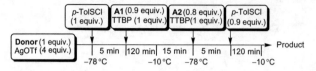

Donor	Acceptor 1	Acceptor 2	Product	Yield (%)
166	165	167	170	54
166	168	167	171	59
166	168	164	172	58
169	165	167	173	62
169	165	164	174	57
169	168	164	175	50

of the *N*-acetyl group and subsequent N-sulfation, C_5-epimerase for isomerization of the uronic acid, and three types of *O*-sulfotransferases, namely 2-OST for sulfating O-2 of IdoA, 3-OST for sulfating O-3 of GlcN, and 6-OST for sulfating O-6 of GlcN. The enzymatic modification of the HS backbone is typically incomplete and thus leads to a wide range of structural variations in naturally occurring heparin and HS.

In order to develop a laboratory synthesis of a pentasaccharide exhibiting strong binding with ATIII, the Rosenberg group explored the enzymatic approach.[107] The backbone of their oligosaccharide was obtained from "heparosan," a polysaccharide from the *Escherichia coli* K5 capsule composed of disaccharide repeating units of [→4)-α-D-GlcNAc-(1→4)-β-D-GlcA-(1→]. The synthesis of pentasaccharide **178** started with N-sulfation of the "heparosan" by incubation with NDST2 and the sulfate-group donor 3′-phospho-5′-adenylyl sulfate (PAPS, Scheme 29). Following N-sulfation, the polymer was depolymerized by heparitinase, and hexasaccharide **176** was isolated by HPLC from the resulting mixture. Sequential epimerization and O-2 sulfation of hexasaccharide **176** by C_5-epimerase and 2-OST1, followed by sulfation at O-6, provided the sulfated hexasaccharide **177**. Removal of the unsaturated uronic acid by $\Delta^{4,5}$-glycosiduronase with subsequent 3-OST-catalyzed sulfation at O-3 produced pentasaccharide **178**, a compound having anticoagulant activity. While this synthesis was groundbreaking, the product **178** was isolated in only microgram quantity and with an overall yield of 1.1%.[107] The low yield was presumably due to the difficulties in purification, particularly in the isolation of hexasaccharide **176** from the complex mixture that arose from cleavage by heparitinase.[108–110] Another obstacle was the low yields of the enzymes expressed from a baculovirus system.

The Liu and Linhardt groups took a different approach for the chemoenzymatic synthesis of heparin/HS oligosaccharides. Instead of relying on the difficult isolation of hexasaccharide **176** from the complex mixture of degradation products resulting from the action of heparitinase on "heparosan," they obtained gram quantities of disaccharide **179** through the complete digestion of "heparosan" by nitrous acid.[109,111] To elongate the chain, two bacterial glycosyltransferases, heparan synthase-2 (PmHS2)[112] and the *N*-acetylglucosaminyl transferase (KfiA) of *E. coli*,[113] were used to transfer GlcA and GlcNAc, respectively (Scheme 30). All of the enzymes for backbone modification, including C_5-epimerase, NDST2, and the *O*-sulfotransferases, were expressed in large quantities in the *E. coli* system. The conversion of the *N*-acetyl group to *N*-sulfate is difficult because of the stability of the acetamido group and the low activity of the *N*-deacetylase. To overcome this, Liu, Linhardt, and coworkers took advantage of the broad substrate specificity of KfiA by incorporating *N*-trifluoro-protected glucosamine (GlcNTFA) into the backbone where N-sulfation is desired.[114,115] Treatment of disaccharide **179** with the glycosyl donor

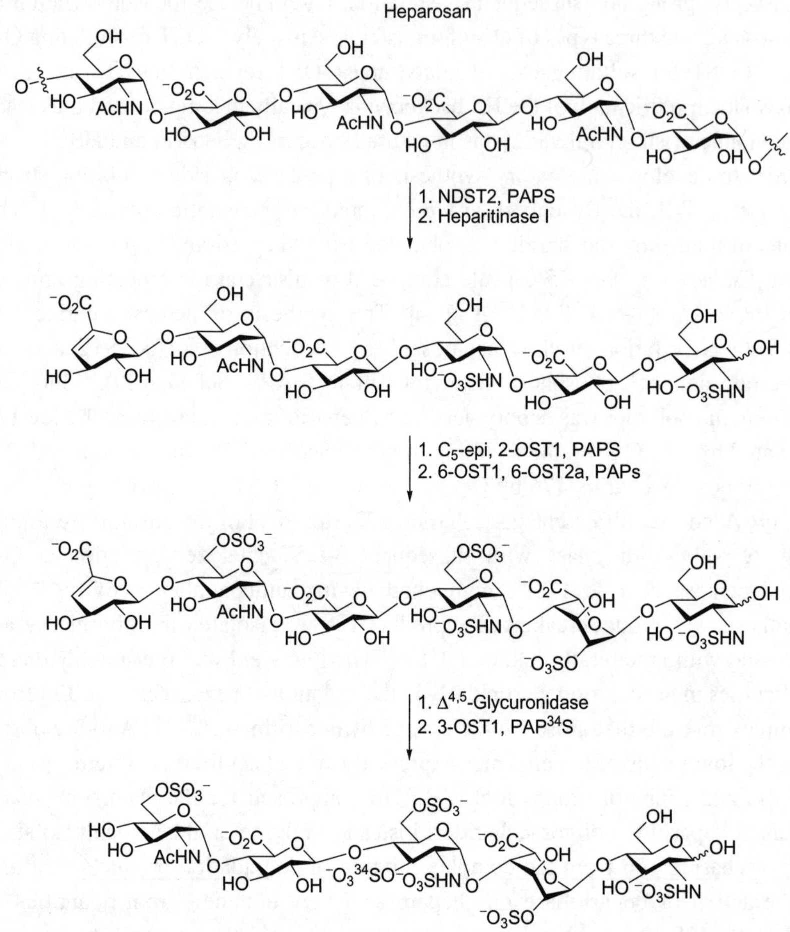

SCHEME 29. Enzymatic synthesis of pentasaccharide **178** from "heparosan."

UDP-GlcNTFA and the transferase KfiA, followed by UDP-GlcA and transferase pmHS2, provided tetrasaccharide **180** in 75% yield. An additional round of elongation with both monosaccharides, followed by removal of the TFA protecting groups with triethylamine, and subsequent N-sulfation by *N*-sulfotransferase (NST) furnished the *N*-sulfated hexasaccharide **181**. Following the addition of another GlcNTFA group, epimerization and sulfation at O-2 were performed in one flask with 2-OST and C$_5$-epimerase to yield the heptasaccharide **182**. The location of enzymatic modification was controlled by the substrate structure. As the C$_5$-epimerase causes GlcA to be

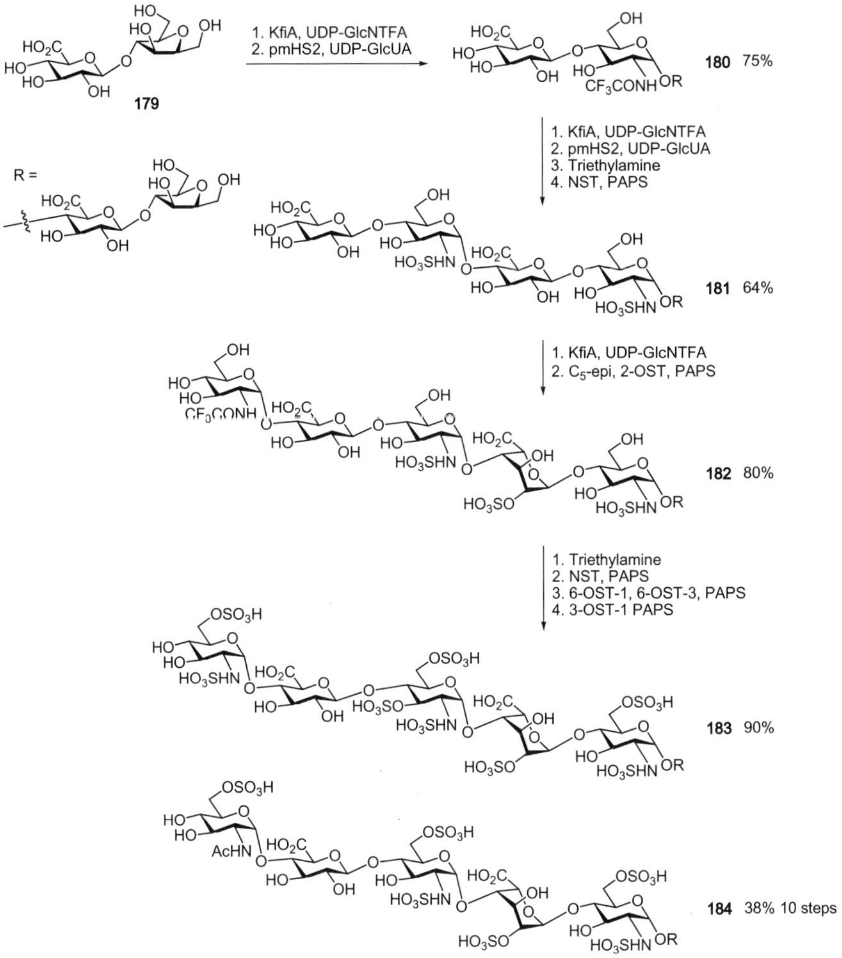

SCHEME 30. The chemoenzymatic synthesis of heparin heptasaccharides **183** and **184**.

modified only when flanked by *N*-sulfated glucosamine groups, the GlcA component closer to the reducing end in **181** alone was epimerized and O-2 sulfated. The last TFA protecting group in **182** was removed with triethylamine, and the product was incubated with NST and PAPS, then PAPS, 6-OST-1, 6-OST-3, and finally PAPS, and 3-OST1 in sequential reactions to provide the final heptasaccharide **183**, which had anticoagulant activity similar to that of the FDA-approved pentasaccharide fondaparinux.[114] In an analogous manner, 49 mg of the heptasaccharide **184** was prepared with an overall yield of 38% from the disaccharide **179**. This work has laid a great foundation for future gram-scale preparation of heparin/HS oligosaccharides.[115] Extending the chemoenzymatic strategy to preparation of fondaparinux will provide an attractive alternative complementing the current complex chemical synthesis of this important molecule.

VIII. Future Outlook

The past decade has seen tremendous advancements in the production of Heparin/HS oligosaccharides. In addition to the more traditional target-oriented synthesis, efforts are being directed toward generating an array of oligosaccharides having diverse patterns of sulfation. In chemical synthesis, multiple strategies have been developed to expedite the glyco-assembly process. Methods are now available for access to tens of oligosaccharides to construct a sample library. However, challenges remain in decreasing the number of synthetic steps required for preparation of building blocks, as well as for establishing a robust method to perform multiple sulfations simultaneously. The enzymatic synthesis of compound **184** at the 49 mg scale is an impressive accomplishment. The substrate specificities of the enzymes may possibly limit the total number of structures that can be generated. Ongoing research has suggested that enzymatic modification can be integrated with chemical synthesis.[116] The combination of the regiospecificity of enzymatic reactions with the flexibility of chemical synthesis can significantly expand our overall synthetic capability, which in turn can greatly aid in the efforts to decipher the exciting biological functions of heparin and HS.

Acknowledgments

We thank the National Institute of General Medical Sciences, NIH (R01GM 72667), and the National Science Foundation (CHE 1111550) for generous financial support of our work.

References

1. R. J. Linhardt, Heparin: An important drug enters its seventh decade, *Chem. Ind.* (1991) 45–47.
2. J. A. Cifonelli and A. Dorfman, Uronic acid of heparin, *Biochem. Biophys. Res. Commun.*, 7 (1962) 41–45.

3. M. Hook, I. Bjork, J. Hopwood, and U. Lindahl, Anticoagulant activity of heparin: Separation of high-activity and low-activity heparin species by affinity chromatography on immobilized antithrombin, *FEBS Lett.*, 66 (1976) 90–93.
4. P. S. Damus, M. Hicks, and R. D. Rosenberg, Anticoagulant action of heparin, *Nature*, 246 (1973) 355–357.
5. N. S. Gandhi and R. L. Mancera, The structure of glycosaminoglycans and their interactions with proteins, *Chem. Biol. Drug Des.*, 72 (2008) 455–482.
6. P. H. Seeberger and D. B. Werz, Synthesis and medical applications of oligosaccharides, *Nature*, 446 (2007) 1046–1051.
7. C. R. Parish, The role of heparan sulphate in inflammation, *Nat. Rev. Immunol.*, 6 (2006) 633–643.
8. A. K. Powell, E. A. Yates, D. G. Fernig, and J. E. Turnbull, Interactions of heparin/heparan sulfate with proteins: Appraisal of structural factors and experimental approaches, *Glycobiology*, 14 (2004) 17R–30R.
9. R. Sasisekharan, Z. Shriver, G. Venkataraman, and U. Narayanasami, Roles of heparan-sulphate glycosaminoglycans in cancer, *Nat. Rev. Cancer*, 2 (2002) 521–528.
10. D. Liu, Z. Shriver, Y. Qi, G. Venkataraman, and R. Sasisekharan, Dynamic regulation of tumor growth and metastasis by heparan sulfate glycosaminoglycans, *Semin. Thromb. Hemost.*, 28 (2002) 67–78.
11. B. Casu and U. Lindahl, Structure and biological interactions of heparin and heparan sulfate, *Adv. Carbohydr. Chem. Biochem.*, 57 (2001) 159–206.
12. R. D. Sanderson, Heparan sulfate proteoglycans in invasion and metastasis, *Semin. Cell Dev. Biol.*, 12 (2001) 89–98.
13. B. Casu, Structure and biological activity of heparin, *Adv. Carbohydr. Chem. Biochem.*, 43 (1985) 51–134.
14. J. D. Esko and U. Lindahl, Molecular diversity of heparan sulfate, *J. Clin. Invest.*, 108 (2001) 169–173.
15. J. T. Gallagher and A. Walker, Molecular distinctions between heparan sulfate and heparin. Analysis of sulfation patterns indicates that heparan sulfate and heparin are separate families of N-sulfated polysaccharides, *Biochem. J.*, 230 (1985) 665–674.
16. U. Lindahl and L. Kjellen, Heparin or heparan sulfate—What is the difference?*Thromb. Haemostasis*, 66 (1991) 44–48.
17. L. H. Lam, J. E. Silbert, and R. D. Rosenberg, The separation of active and inactive forms of heparin, *Biochem. Biophys. Res. Commun.*, 69 (1976) 570–577.
18. U. Lindahl, G. Baeckstroem, M. Hoeoek, L. Thunberg, L.-A. Fransson, and A. Linker, Structure of the antithrombin-binding site in heparin, *Proc. Natl. Acad. Sci. U. S. A.*, 76 (1979) 3198–3202.
19. R. D. Rosenberg and L. Lam, Correlation between structure and function of heparin, *Proc. Natl. Acad. Sci. U. S. A.*, 76 (1979) 1218–1222.
20. M. Petitou, B. Casu, and U. Lindahl, 1976-1983, a critical period in the history of heparin: The discovery of the antithrombin binding site, *Biochimie*, 85 (2003) 83–89.
21. M. Petitou and C. A. A. van Boeckel, A synthetic antithrombin III binding pentasaccharide is now a drug! What comes next?*Angew. Chem. Int. Ed.*, 43 (2004) 3118–3133.
22. M. Petitou, P. Duchaussoy, I. Lederman, J. Choay, and P. Sinaÿ, Binding of heparin to antithrombin III: A chemical proof of the critical role played by a 3-sulfated 2-amino-2-deoxy-D-glucose residue, *Carbohydr. Res.*, 179 (1988) 163–172.
23. M. Ishihara, Y. Kariya, H. Kikuchi, T. Minamisawa, and K. Yoshida, Importance of 2-O-sulfate groups of uronate residues in heparin for activation of FGF-1 and FGF-2, *J. Biochem.*, 121 (1997) 345–349.

24. M. Ishihara, R. Takano, T. Kanda, K. Hayashi, S. Hara, H. Kikuchi, and K. Yoshida, Importance of 6-O-sulfate groups of glucosamine residues in heparin for activation of FGF-1 and FGF-2, *J. Biochem.*, 118 (1995) 1255–1260.
25. C. Noti and P. H. Seeberger, Chemical approaches to define the structure-activity relationship of heparin-like glycosaminoglycans, *Chem. Biol.*, 12 (2005) 731–756.
26. C. Noti and P. H. Seeberger, Synthetic approach to define structure-activity relationship of heparin and heparan sulfate, in H. G. Garg, R. J. Linhardt, and C. A. Hales, (Eds.), *Chemistry and Biology of Heparin and Heparan Sulfate,* Elsevier, Oxford, 2005, pp. 79–142.
27. J. D. C. Codée, H. S. Overkleeft, G. A. van der Marel, and C. A. A. van Boeckel, The synthesis of well-defined heparin and heparan sulfate fragments, *Drug Discov. Today Technol.*, 1 (2004) 317–326.
28. N. A. Karst and R. J. Linhardt, Recent chemical and enzymatic approaches to the synthesis of glycosaminoglycan oligosaccharides, *Curr. Med. Chem.*, 10 (2003) 1993–2031.
29. L. Poletti and L. Lay, Chemical contributions to understanding heparin activity: Synthesis of related sulfated oligosaccharides, *Eur. J. Org. Chem.* (2003) 2999–3024.
30. P. Sinaÿ, J. C. Jacquinet, M. Petitou, P. Duchaussoy, I. Lederman, J. Choay, and G. Torri, Total synthesis of a heparin pentasaccharide fragment having high affinity for antithrombin III, *Carbohydr. Res.*, 132 (1984) C5–C9.
31. C. A. A. Van Boeckel, T. Beetz, J. N. Vos, A. J. M. de Jong, S. F. van Aelst, R. H. van den Bosch, J. M. R. Mertens, and F. A. van der Vlugt, Synthesis of a pentasaccharide corresponding to the antithrombin III binding fragment of heparin, *J. Carbohydr. Chem.*, 4 (1985) 293–321.
32. J. Tatai, G. Osztrovszky, M. Kajtar-Peredy, and P. Fügedi, An efficient synthesis of L-idose and L-iduronic acid thioglycosides and their use for the synthesis of heparin oligosaccharides, *Carbohydr. Res.*, 343 (2008) 596–606.
33. S. U. Hansen, M. Barath, B. A. B. Salameh, R. G. Pritchard, W. T. Stimpson, J. M. Gardiner, and G. C. Jayson, Scalable synthesis of L-iduronic acid derivatives via stereocontrolled cyanohydrin reaction for synthesis of heparin-related disaccharides, *Org. Lett.*, 11 (2009) 4528–4531.
34. A. Dilhas and D. Bonnaffe, Efficient selective preparation of methyl-1,2,4-tri-*O*-acetyl-3-*O*-benzyl-β-L-idopyranuronate from methyl 3-*O*-benzyl-L-iduronate, *Carbohydr. Res.*, 338 (2003) 681–686.
35. S.-C. Hung, S. R. Thopate, F.-C. Chi, S.-W. Chang, J.-C. Lee, C.-C. Wang, and Y.-S. Wen, 1,6-Anhydro-β-L-hexopyranoses as potent synthons in the synthesis of the disaccharide units of bleomycin A2 and heparin, *J. Am. Chem. Soc.*, 123 (2001) 3153–3154.
36. S.-C. Hung, R. Puranik, and F.-C. Chi, Novel synthesis of 1,2:3,5-di-*O*-isopropylidene-β-L-idofurano-side and its derivatives at C6, *Tetrahedron Lett.*, 41 (2000) 77–80.
37. A. Saito, M. Wakao, H. Deguchi, A. Mawatari, M. Sobel, and Y. Suda, Toward the assembly of heparin and heparan sulfate oligosaccharide libraries: Efficient synthesis of uronic acid and disaccharide building blocks, *Tetrahedron*, 66 (2010) 3951–3962.
38. A. Adibekian, P. Bindschaedler, M. S. M. Timmer, C. Noti, N. Schuetzenmeister, and P. H. Seeberger, De Novo synthesis of uronic acid building blocks for assembly of heparin oligosaccharides, *Chem. Eur. J.*, 13 (2007) 4510–4522.
39. M. S. M. Timmer, A. Adibekian, and P. H. Seeberger, Short de novo synthesis of fully functionalized uronic acid monosaccharides, *Angew. Chem. Int. Ed.*, 44 (2005) 7605–7607.
40. H. N. Yu, J. Furukawa, T. Ikeda, and C.-H. Wong, Novel efficient routes to heparin monosaccharides and disaccharides achieved via regio- and stereoselective glycosidation, *Org. Lett.*, 6 (2004) 723–726.
41. O. Gavard, Y. Hersant, J. Alais, V. Duverger, A. Dilhas, A. Bascou, and D. Bonnaffe, Efficient preparation of three building blocks for the synthesis of heparan sulfate fragments: Towards the combinatorial synthesis of oligosaccharides from hypervariable regions, *Eur. J. Org. Chem.* (2003) 3603–3620.
42. W. Ke, D. M. Whitfield, M. Gill, S. Larocque, and S.-H. Yu, A short route to L-iduronic acid building blocks for the syntheses of heparin-like disaccharides, *Tetrahedron Lett.*, 44 (2003) 7767–7770.

43. P. Schell, H. A. Orgueira, S. Roehrig, and P. H. Seeberger, Synthesis and transformations of D-glucuronic and L-iduronic acid glycals, *Tetrahedron Lett.*, 42 (2001) 3811–3814.
44. A. Lubineau, O. Gavard, J. Alais, and D. Bonnaffe, New accesses to L-iduronyl synthons, *Tetrahedron Lett.*, 41 (2000) 307–311.
45. R. Fernandez, E. Martin-Zamora, C. Pareja, and J. M. Lassaletta, Stereoselective nucleophilic formylation and cyanation of α-alkoxy- and α-aminoaldehydes, *J. Org. Chem.*, 66 (2001) 5201–5207.
46. L. Huang and X. Huang, Highly efficient syntheses of hyaluronic acid oligosaccharides, *Chem. Eur. J.*, 13 (2007) 529–540.
47. L. Huang, N. Teumelsan, and X. Huang, A facile method for oxidation of primary alcohols to carboxylic acids and its application in glycosaminoglycan syntheses, *Chem. Eur. J.*, 12 (2006) 5246–5252.
48. E. R. Palmacci and P. H. Seeberger, Toward the modular synthesis of glycosaminoglycans: Synthesis of hyaluronic acid disaccharide building blocks using a periodic acid oxidation, *Tetrahedron*, 60 (2004) 7755–7766.
49. B. La Ferla, L. Lay, M. Guerrini, L. Poletti, L. Panza, and G. Russo, Synthesis of disaccharidic subunits of a new series of heparin related oligosaccharides, *Tetrahedron*, 55 (1999) 9867–9880.
50. C. Tabeur, J. M. Mallet, F. Bono, J. M. Herbert, M. Petitou, and P. Sinaÿ, Oligosaccharides corresponding to the regular sequence of heparin: Chemical synthesis and interaction with FGF-2, *Bioorg. Med. Chem.*, 7 (1999) 2003–2012.
51. J. Kovensky, P. Duchaussoy, F. Bono, M. Salmivirta, P. Sizun, J.-M. Herbert, M. Petitou, and P. Sinaÿ, A synthetic heparan sulfate pentasaccharide, exclusively containing L-iduronic acid, displays higher affinity for FGF-2 than its D-glucuronic acid-containing isomers, *Bioorg. Med. Chem.*, 7 (1999) 1567–1580.
52. M. Haller and G.-J. Boons, Towards a modular approach for heparin synthesis, *J. Chem. Soc. Perkin Trans. 1* (2001) 814–822.
53. A. E. J. de Nooy, A. C. Besemer, and H. van Bekkum, Highly selective nitroxyl radical-mediated oxidation of primary alcohol groups in water-soluble glucans, *Carbohydr. Res.*, 269 (1995) 89–96.
54. N. J. Davis and S. L. Flitsch, Selective oxidation of monosaccharide derivatives to uronic acids, *Tetrahedron Lett.*, 34 (1993) 1181–1184.
55. Y.-P. Hu, S.-Y. Lin, C.-Y. Huang, M. M. L. Zulueta, J.-Y. Liu, W. Chang, and S.-C. Hung, Synthesis of 3-O-sulfonated heparan sulfate octasaccharides that inhibit the herpes simplex virus type 1 host-cell interaction, *Nat. Chem.*, 3 (2011) 557–563.
56. L. J. van den Bos, J. D. C. Codée, J. C. van der Toorn, T. J. Boltje, J. H. van Boom, H. S. Overkleeft, and G. A. van der Marel, Thio-glycuronides: Synthesis and application in the assembly of acidic oligosaccharides, *Org. Lett.*, 6 (2004) 2165–2168.
57. G. Tiruchinapally, Z. Yin, M. El-Dakdouki, Z. Wang, and X. Huang, Divergent heparin oligosaccharide synthesis with pre-installed sulfate esters, *Chem. Eur. J.*, 17 (2011) 10106–10112.
58. Z. Wang, Y. Xu, B. Yang, G. Tiruchinapally, B. Sun, R. Liu, S. Dulaney, J. Liu, and X. Huang, Preactivation-based, one-pot combinatorial synthesis of heparin-like hexasaccharides for the analysis of heparin-protein interactions, *Chem. Eur. J.*, 16 (2010) 8365–8375.
59. S. Arungundram, K. Al-Mafraji, J. Asong, F. E. Leach, III,, I. J. Amster, A. Venot, J. E. Turnbull, and G.-J. Boons, Modular synthesis of heparan sulfate oligosaccharides for structure-activity relationship studies, *J. Am. Chem. Soc.*, 131 (2009) 17394–17405.
60. J. Chen, Y. Zhou, C. Chen, W. Xu, and B. Yu, Synthesis of a tetrasaccharide substrate of heparanase, *Carbohydr. Res.*, 343 (2008) 2853–2862.
61. M. T. C. Walvoort, W. de Witte, J. van Dijk, J. Dinkelaar, G. Lodder, H. S. Overkleeft, J. D. C. Codée, and G. A. van der Marel, Mannopyranosyl uronic acid donor reactivity, *Org. Lett.*, 13 (2011) 4360–4363.
62. Y. Zeng, Z. Wang, D. Whitfield, and X. Huang, Installation of electron-donating protective groups, a strategy for glycosylating unreactive thioglycosyl acceptors using the preactivation-based glycosylation method, *J. Org. Chem.*, 73 (2008) 7952–7962.

63. A. Dilhas and D. Bonnaffe, PhBCl$_2$ promoted reductive opening of 2′,4′-O-p-methoxybenzylidene: New regioselective differentiation of position 2′ and 4′ of α-L-iduronyl moieties in disaccharide building blocks, *Tetrahedron Lett.*, 45 (2004) 3643–3645.
64. F. Baleux, L. Loureiro-Morais, Y. Hersant, P. Clayette, F. Arenzana-Seisdedos, D. Bonnaffe, and H. Lortat-Jacob, A synthetic CD4-heparan sulfate glycoconjugate inhibits CCR5 and CXCR4 HIV-1 attachment and entry, *Nat. Chem. Biol.*, 5 (2009) 743–748.
65. H. Paulsen, Progress in the selective chemical synthesis of complex oligosaccharides, *Angew. Chem. Int. Ed.*, 21 (1982) 155–173.
66. J. D. C. Codée, B. Stubba, M. Schiattarella, H. S. Overkleeft, C. A. A. van Boeckel, J. H. van Boom, and G. A. van der Marel, A modular strategy toward the synthesis of heparin-like oligosaccharides using monomeric building blocks in a sequential glycosylation strategy, *J. Am. Chem. Soc.*, 127 (2005) 3767–3773.
67. H. A. Orgueira, A. Bartolozzi, P. Schell, R. E. J. N. Litjens, E. R. Palmacci, and P. H. Seeberger, Modular synthesis of heparin oligosaccharides, *Chem. Eur. J.*, 9 (2003) 140–169.
68. H. A. Orgueira, A. Bartolozzi, P. Schell, and P. H. Seeberger, Conformational locking of the glycosyl acceptor for stereocontrol in the key step in the synthesis of heparin, *Angew. Chem. Int. Ed.*, 41 (2002) 2128–2131.
69. G. J. S. Lohman and P. H. Seeberger, A stereochemical surprise at the late stage of the synthesis of fully N-differentiated heparin oligosaccharides containing amino, acetamido, and N-sulfonate groups, *J. Org. Chem.*, 69 (2004) 4081–4093.
70. L.-D. Lu, C.-R. Shie, S. S. Kulkarni, G.-R. Pan, X.-A. Lu, and S.-C. Hung, Synthesis of 48 disaccharide building blocks for the assembly of a heparin and heparan sulfate oligosaccharide library, *Org. Lett.*, 8 (2006) 5995–5998.
71. R.-H. Fan, J. Achkar, J. M. Hernandez-Torres, and A. Wei, Orthogonal sulfation strategy for synthetic heparan sulfate ligands, *Org. Lett.*, 7 (2005) 5095–5098.
72. J.-C. Lee, X.-A. Lu, S. S. Kulkarni, Y.-S. Wen, and S.-C. Hung, Synthesis of heparin oligosaccharides, *J. Am. Chem. Soc.*, 126 (2004) 476–477.
73. C. Noti, J. L. de Paz, L. Polito, and P. H. Seeberger, Preparation and use of microarrays containing synthetic heparin oligosaccharides for the rapid analysis of heparin-protein interactions, *Chem. Eur. J.*, 12 (2006) 8664–8686.
74. M. Huibers, A. Manuzi, F. P. J. T. Rutjes, and F. L. van Delft, A sulfitylation-oxidation protocol for the preparation of sulfates, *J. Org. Chem.*, 71 (2006) 7473–7476.
75. L. S. Simpson and T. S. Widlanski, A comprehensive approach to the synthesis of sulfate esters, *J. Am. Chem. Soc.*, 128 (2006) 1605–1610.
76. N. A. Karst, T. F. Islam, and R. J. Linhardt, Sulfo-protected hexosamine monosaccharides: Potentially versatile building blocks for glycosaminoglycan synthesis, *Org. Lett.*, 5 (2003) 4839–4842.
77. L. J. Ingram and S. D. Taylor, Introduction of 2,2,2-trichloroethyl-protected sulfates into monosaccharides with a sulfuryl imidazolium salt and application to the synthesis of sulfated carbohydrates, *Angew. Chem. Int. Ed.*, 45 (2006) 3503–3506.
78. J. Kreuger, M. Salmivirta, L. Sturiale, G. Gimenez-Gallego, and U. Lindahl, Sequence analysis of heparan sulfate epitopes with graded affinities for fibroblast growth factors 1 and 2, *J. Biol. Chem.*, 276 (2001) 30744–30752.
79. J. Tatai and P. Fügedi, Synthesis of the putative minimal FGF binding motif heparan sulfate trisaccharides by an orthogonal protecting group strategy, *Tetrahedron*, 64 (2008) 9865–9873.
80. M. Weishaupt, S. Eller, and P. H. Seeberger, Solid phase synthesis of oligosaccharides, *Methods Enzymol.*, 478 (2010) 463–484.
81. O. J. Plante, E. R. Palmacci, and P. H. Seeberger, Automated solid-phase synthesis of oligosaccharides, *Science*, 291 (2001) 1523–1527.

82. P. Czechura, N. Guedes, S. Kopitzki, N. Vazquez, M. Martin-Lomas, and N.-C. Reichardt, A new linker for solid-phase synthesis of heparan sulfate precursors by sequential assembly of monosaccharide building blocks, *Chem. Commun.*, 47 (2011) 2390–2392.
83. R. Ojeda, O. Terenti, J.-L. de Paz, and M. Martin-Lomas, Synthesis of heparin-like oligosaccharides on polymer supports, *Glycoconj. J.*, 21 (2004) 179–195.
84. J.-L. de Paz, J. Angulo, J.-M. Lassaletta, P. M. Nieto, M. Redondo-Horcajo, R. M. Lozano, G. Gimenez-Gallego, and M. Martin-Lomas, The activation of fibroblast growth factors by heparin: Synthesis, structure, and biological activity of heparin-like oligosaccharides, *Chembiochem*, 2 (2001) 673–685.
85. A. Prabhu, A. Venot, and G.-J. Boons, New set of orthogonal protecting groups for the modular synthesis of heparan sulfate fragments, *Org. Lett.*, 5 (2003) 4975–4978.
86. A. Lubineau, H. Lortat-Jacob, O. Gavard, S. Sarrazin, and D. Bonnaffe, Synthesis of tailor-made glycoconjugate mimetics of heparan sulfate that bind IFN-γ in the nanomolar range, *Chem. Eur. J.*, 10 (2004) 4265–4282.
87. L. Poletti, M. Fleischer, C. Vogel, M. Guerrini, G. Torri, and L. Lay, A rational approach to heparin-related fragments—Synthesis of differently sulfated tetrasaccharides as potential ligands for fibroblast growth factors, *Eur. J. Org. Chem.* (2001) 2727–2734.
88. J. L. de Paz and M. Martin-Lomas, Synthesis and biological evaluation of a heparin-like hexasaccharide with the structural motifs for binding to FGF and FGFR, *Eur. J. Org. Chem.* (2005) 1849–1858.
89. R. Lucas, J. Angulo, P. M. Nieto, and M. Martin-Lomas, Synthesis and structural study of two new heparin-like hexasaccharides, *Org. Biomol. Chem.*, 1 (2003) 2253–2266.
90. J.-L. de Paz, R. Ojeda, N. Reichardt, and M. Martin-Lomas, Some key experimental features of a modular synthesis of heparin-like oligosaccharides, *Eur. J. Org. Chem.* (2003) 3308–3324.
91. R. Ojeda, J. Angulo, P. M. Nieto, and M. Martin-Lomas, The activation of fibroblast growth factors by heparin: Synthesis and structural study of rationally modified heparin-like oligosaccharides, *Can. J. Chem.*, 80 (2002) 917–936.
92. K. Daragics and P. Fügedi, Synthesis of glycosaminoglycan oligosaccharides. Part 5: Synthesis of a putative heparan sulfate tetrasaccharide antigen involved in prion diseases, *Tetrahedron*, 66 (2010) 8036–8046.
93. C. L. Cole, S. U. Hansen, M. Barath, G. Rushton, J. M. Gardiner, E. Avizienyte, and G. C. Jayson, Synthetic heparan sulfate oligosaccharides inhibit endothelial cell functions essential for angiogenesis, *PLoS One*, 5 (2010) e11644.
94. H. M. Nguyen, J. L. Poole, and D. Y. Gin, Chemoselective iterative dehydrative glycosylation, *Angew. Chem. Int. Ed.*, 40 (2001) 414–417.
95. C. McDonnell, O. Lopez, P. Murphy, J. G. F. Bolanos, R. Hazell, and M. Bols, Conformational effects on glycoside reactivity: Study of the high reactive conformer of glucose, *J. Am. Chem. Soc.*, 126 (2004) 12374–12385.
96. H. H. Jensen, L. U. Nordstrom, and M. Bols, The disarming effect of the 4,6-acetal group on glycoside reactivity: Torsional or electronic?*J. Am. Chem. Soc.*, 126 (2004) 9205–9213.
97. Z. Zhang, I. R. Ollmann, X.-S. Ye, R. Wischnat, T. Baasov, and C.-H. Wong, Programmable one-pot oligosaccharide synthesis, *J. Am. Chem. Soc.*, 121 (1999) 734–753.
98. L. Huang, Z. Wang, and X. Huang, One-pot oligosaccharide synthesis: Reactivity tuning by post-synthetic modification of aglycon, *Chem. Commun.* (2004) 1960–1961.
99. P. Pornsuriyasak, U. B. Gangadharmath, N. P. Rath, and A. V. Demchenko, A novel strategy for oligosaccharide synthesis via temporarily deactivated S-thiazolyl glycosides as glycosyl acceptors, *Org. Lett.*, 6 (2004) 4515–4518.
100. M. Lahmann and S. Oscarson, Investigation of the reactivity difference between thioglycoside donors with variant aglycon parts, *Can. J. Chem.*, 80 (2002) 889–893.

101. T. Polat and C.-H. Wong, Anomeric reactivity-based one-pot synthesis of heparin-like oligosaccharides, *J. Am. Chem. Soc.*, 129 (2007) 12795–12800.
102. X. Li, L. Huang, X. Hu, and X. Huang, Thio-arylglycosides with various aglycon para-substituents: A probe for studying chemical glycosylation reactions, *Org. Biomol. Chem.*, 7 (2009) 117–127.
103. T. K. Park, I. J. Kim, S. Hu, M. T. Bilodeau, J. T. Randolph, O. Kwon, and S. J. Danishefsky, Total synthesis and proof of structure of a human breast tumor (Globo-H) antigen, *J. Am. Chem. Soc.*, 118 (1996) 11488–11500.
104. S. Yamago, T. Yamada, O. Hara, H. Ito, Y. Mino, and J.-i. Yoshida, A new, iterative strategy of oligosaccharide synthesis based on highly reactive β-bromoglycosides derived from selenoglycosides, *Org. Lett.*, 3 (2001) 3867–3870.
105. S. Yamago, T. Yamada, T. Maruyama, and J. Yoshida, Iterative glycosylation of 2-deoxy-2-aminothioglycosides and its application to the combinatorial synthesis of linear oligo-glucosamines, *Angew. Chem. Int. Ed.*, 43 (2004) 2145–2148.
106. X. Huang, L. Huang, H. Wang, and X.-S. Ye, Iterative one-pot synthesis of oligosaccharides, *Angew. Chem. Int. Ed.*, 43 (2004) 5221–5224.
107. B. Kuberan, M. Z. Lech, D. L. Beeler, Z. L. Wu, and R. D. Rosenberg, Enzymatic synthesis of antithrombin III-binding heparan sulfate pentasaccharide, *Nat. Biotechnol.*, 21 (2003) 1343–1346.
108. J. Chen, C. L. Jones, and J. Liu, Using an enzymatic combinatorial approach to identify anticoagulant heparan sulfate structures with differing sulfation patterns, *Chem. Biol.*, 14 (2007) 986–993.
109. J. Chen, F. Y. Avci, E. M. Munoz, L. M. McDowell, M. Chen, L. C. Pedersen, L. Zhang, R. J. Linhardt, and J. Liu, Enzymatic redesigning of biologically active heparan sulfate, *J. Biol. Chem.*, 280 (2005) 42817–42825.
110. U. Lindahl, J.-P. Li, M. Kusche-Gullberg, M. Salmivirta, S. Alaranta, T. Veromaa, J. Emeis, I. Roberts, C. Taylor, P. Oreste, G. Zoppetti, A. Naggi, G. Torri, and B. Casu, Generation of "Neoheparin" from *E. coli* K5 capsular polysaccharide, *J. Med. Chem.*, 48 (2005) 349–352.
111. Z. Wang, M. Ly, F. Zhang, W. Zhong, A. Suen, A. M. Hickey, J. S. Dordick, and R. J. Linhardt, *E. coli* K5 fermentation and the preparation of heparosan, a bioengineered heparin precursor, *Biotechnol. Bioeng.*, 107 (2010) 964–973.
112. G. Griffiths, N. J. Cook, E. Gottfridson, T. Lind, K. Lidholt, and I. S. Roberts, Characterization of the glycosyltransferase enzyme from the *Escherichia coli* K5 capsule gene cluster and identification and characterization of the glucuronyl active site, *J. Biol. Chem.*, 273 (1998) 11752–11757.
113. N. Hodson, G. Griffiths, N. Cook, M. Pourhossein, E. Gottfridson, T. Lind, K. Lidholt, and I. S. Roberts, Identification that KfiA, a protein essential for the biosynthesis of the *Escherichia coli* K5 capsular polysaccharide, is an α-UDP-GlcNAc glycosyltransferase. The formation of a membrane-associated K5 biosynthetic complex requires KfiA, KfiB, and KfiC, *J. Biol. Chem.*, 275 (2000) 27311–27315.
114. Y. Xu, S. Masuko, M. Takieddin, H. Xu, R. Liu, J. Jing, S. A. Mousa, R. J. Linhardt, and J. Liu, Chemoenzymatic synthesis of homogeneous ultralow molecular weight heparins, *Science*, 334 (2011) 498–501.
115. R. Liu, Y. Xu, M. Chen, M. Weiwer, X. Zhou, A. S. Bridges, A. P. L. De, Q. Zhang, R. J. Linhardt, and J. Liu, Chemoenzymatic design of heparan sulfate oligosaccharides, *J. Biol. Chem.*, 285 (2010) 34240–34249.
116. Y. Xu, Z. Wang, R. Liu, A. S. Bridges, X. Huang, and J. Liu, Directing the biological activities of heparan sulfate oligosaccharides using a chemoenzymatic approach, *Glycobiology*, 22 (2012) 96–106.

CHEMICAL SYNTHESIS OF GLYCOSYLPHOSPHATIDYLINOSITOL ANCHORS

Benjamin M. Swarts and Zhongwu Guo

Department of Chemistry, Wayne State University, Detroit, Michigan, USA

I. Introduction	138
II. Classic Approaches to GPI Synthesis	142
1. Synthesis of a *Trypanosoma brucei* GPI Anchor by the Ogawa Group (1991)	142
2. Synthesis of a Yeast GPI Anchor by the Schmidt Group (1994)	147
3. Synthesis of a Rat Brain Thy-1 GPI Anchor by the Fraser-Reid Group (1995)	151
4. Synthesis of a *T. brucei* GPI Anchor by the Ley Group (1998)	157
5. Synthesis of a Rat Brain Thy-1 GPI Anchor by the Schmidt Group (1999, 2003)	162
6. Synthesis of a Human Sperm CD52 Antigen GPI Anchor Containing an Acylated Inositol by the Guo Group (2003)	167
7. Synthesis of a *Plasmodium falciparum* GPI Anchor Containing an Acylated Inositol by the Fraser-Reid Group (2004)	172
8. Synthesis of a *P. falciparum* GPI Anchor Containing an Acylated Inositol by the Seeberger Group (2005)	177
9. Synthesis of a *T. cruzi* GPI Anchor by the Vishwakarma Group (2005)	181
10. Synthesis of a Fully Phosphorylated and Lipidated Human Sperm CD52 Antigen GPI Anchor by the Guo group (2007)	185
III. Diversity-Oriented Approaches to GPI Synthesis	187
1. Synthesis of *T. cruzi* GPI Anchors Containing Unsaturated Lipids by the Nikolaev Group (2006)	188
2. Synthesis of a GPI Anchor Containing Unsaturated Lipid Chains by the Guo Group (2010)	193
3. Synthesis of a Human Lymphocyte CD52 Antigen GPI Anchor Containing a Polyunsaturated Arachidonoyl Lipid by the Guo Group (2011)	199
4. Synthesis of "Clickable" GPI Anchors by the Guo Group (2011)	203
5. General Synthetic Strategy for Branched GPI Anchors by the Seeberger Group (2011)	207
IV. Conclusions and Outlook	211
Acknowledgments	212
References	212

Abbreviations

Ac, acetyl; All, allyl; BDA, butanediacetal; Bn, benzyl; Boc, *tert*-butyloxycarbonyl; BRSM, based on recovered starting material; Bz, benzoyl; CA, chloroacetyl; CAM, camphanoyl; CAN, ceric ammonium nitrate; Cbz, benzyloxycarbonyl; "Click" chemistry, [3+2] azide–alkyne cycloaddition; CSA, camphorsulfonic acid; CuAAC, Cu-catalyzed azide–alkyne cycloaddition; DAST, diethylaminosulfur trifluoride; DBU, 1,8-diazabicyclo[5.4.0]undec-7-ene; DCC, N,N'-dicyclohexylcarbodiimide; DCE, dichloroethane; DDQ, 2,3-dichloro-5,6-dicyano-1,4-benzoquinone; DIBAL-H, diisobutylaluminum hydride; DIC, N,N'-diisopropylcarbodiimide; DMAP, 4-dimethylaminopyridine; DMF, N,N-dimethylformamide; DMTrt, 4,4'-dimethoxytrityl; DTBMP, 2,6-di-*tert*-butyl-4-methylpyridine; EE, 1-ethoxyethyl ether; Fmoc, 9-fluorenylmethoxycarbonyl; Gal, galactose; GalNAc, N-acetylgalactosamine; GFP, green fluorescent protein; Glc, glucose; GlcN, glucosamine; GPI, glycosylphosphatidylinositol; KHMDS, potassium hexamethyldisilazide; Lev, levulinoyl; LPG, lipophosphoglycan; m-CPBA, m-chloroperoxybenzoic acid; Man, mannose; Mnt, menthyl; MS, molecular sieves; NAP, 2-naphthylmethyl; NIS, N-iodosuccinimide; NPG, n-pentenyl glycoside; NPOE, n-pentenyl orthoester; p-TsOH, p-toluenesulfonic acid; PA, phenoxyacetyl; PBS, phosphate-buffered saline; PG, protecting group; PI, phosphatidylinositol; PIV, pivaloyl; PMB, p-methoxybenzyl; PMP, p-methoxyphenyl; PMTrt, p-methoxytrityl; PNBSCl, p-nitrobenzenesulfenyl chloride; PPTS, pyridinium p-toluenesulfonate; pyr, pyridine; SEM, 2-trimethylsilylethoxymethyl; TBAF, tetrabutylammonium fluoride; TBDPS, *tert*-butyldiphenylsilyl; TBS, *tert*-butyldimethylsilyl; TCA, trichloroacetyl; TDS, thexyldimethylsilyl; TES, triethylsilyl; Tf, trifluoromethanesulfonate; TFA, trifluoroacetic acid; THF, tetrahydrofuran; THP, tetrahydropyranyl; TIPS, triisopropylsilyl; TMS, trimethylsilyl; tol, toluene; TTBP, 2,4,6-tri-*tert*-butylpyrimidine

I. Introduction

Numerous eukaryotic proteins and glycoproteins are attached to the cell surface by glycosylphosphatidylinositol (GPI) anchors, a structurally complex and diverse class of glycolipids.[1–4] Hundreds of membrane proteins are GPI-anchored,[5] and recent genomic investigations predict that as much as 0.5–1.5% of the eukaryotic proteome may be post-translationally modified and thus anchored to the cell surface by GPIs.[6] The GPIs and GPI-anchored molecules have numerous crucial biological functions, including cell recognition and adhesion,[7] signal transduction,[8,9] pathogenic infections and other disease-causing conditions,[10,11] cell-surface enzymatic activity,[12] and operating as cellular markers.[4]

During the 1970s and 1980s, GPI anchoring was discovered to be a unique type of membrane-protein binding,[1] and this work was highlighted by Ferguson's complete structural elucidation of the *T. brucei* variant surface glycoprotein GPI anchor in 1988.[13] Since then, dozens of GPIs have been identified and characterized, all of which share the following conserved core structure: $H_2N\text{-}(CH_2)_2\text{-}(P)\text{-}6\text{-}\alpha\text{-Man-}(1\rightarrow2)\text{-}\alpha\text{-Man-}(1\rightarrow6)\text{-}\alpha\text{-Man }(1\rightarrow4)\text{-}\alpha\text{-GlcN-}(1\rightarrow6)\text{-}myo\text{-inositol-1-}(P)\text{-glycerolipid}$ (Fig. 1). The C-terminus of the protein is covalently linked to the GPI glycan through a bridging phosphoethanolamine, and the entire protein–GPI structure is anchored to the cell surface by insertion of the fatty acid chains of the phosphatidylinositol (PI) into the outer leaflet of the membrane bilayer. Amongst known GPIs, the tetrasaccharide core exhibits high structural diversity, primarily in the form of additional carbohydrate and phosphoethanolamine units linked to various positions. For

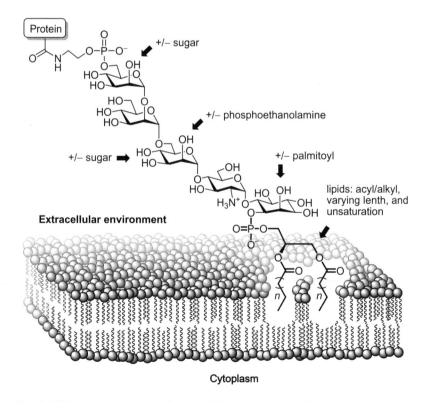

FIG. 1. GPI core structure, points of structural diversity (arrows), and protein anchoring function.

example, mannose and galactose mono- or oligomers may be attached to the O-2 position of Man-III and the O-3 or O-4 positions of Man-I, respectively, while in mammalian GPIs the O-2 position of Man-I frequently carries a phosphoethanolamine group. Additionally, palmitoylation (hexadecanoylation) at the O-2 position of inositol is frequent, and the phosphoglycerolipid can undergo lipid remodeling,[14] which can generate acyl- or alkyl-linked lipids that exhibit varying chain-length and patterns of unsaturation.

Traditional and emerging techniques in biochemistry and in molecular and cell biology have enabled impressive advances in unraveling the biosynthesis, structure, and function of GPI anchors and GPI-anchored proteins.[15] However, many aspects of these molecules remain poorly understood. For example, the significance of structural complexity and diversity of GPI anchors is not well understood and prompts studies on structure–activity relationships. GPIs are expected to have therapeutic value, which is perhaps best exemplified by the recent use of GPI partial structures as antimalarial vaccines,[16,17] but new technologies are needed to fully explore their therapeutic potential. Visualization of GPI anchors and GPI-anchored proteins by fluorescence microscopy can provide insight into their expression, distribution, and endocytosis.[18] "GPIomics" is another emerging field of GPI research which entails the development of tools for identifying post-translational modification of proteins by GPIs.[19]

To accelerate progress in these and other areas of GPI research, structurally defined GPIs, GPI derivatives, and functionalized GPIs must be accessible in sufficient purity and quantity, a requirement that can be addressed by chemical synthesis. The GPIs are among the most complex of natural products, as they combine carbohydrate, lipid, phosphate, and inositol groups, the combination of which makes their synthesis very challenging. Difficult tasks in GPI synthesis include the preparation of enantiomerically pure and properly discriminated inositol derivatives, stereoselective formation of glycosidic bonds, and regiocontrolled introduction of side chains. A general convergent approach for GPI assembly is shown in Scheme 1, where the target GPI anchor **I** is typically accessed from a fully protected intermediate **II** bearing orthogonal protecting groups at sites for late-stage introduction of the phosphoglycerolipid and phosphoethanolamine moieties. In most cases, structure **II** is disconnected at the glucosamine–mannose glycosidic bond, leading back to trimannose and pseudodisaccharide fragments **III** and **IV**, which are in turn generated from their corresponding monomeric subunits.

In this article, the field of GPI anchor synthesis is covered chronologically, starting with the first synthesis of a GPI anchor by the Ogawa group in 1991. In the ensuing two decades, numerous impressive total syntheses of naturally occurring GPI anchors were completed by various research groups, and these works are discussed in the first

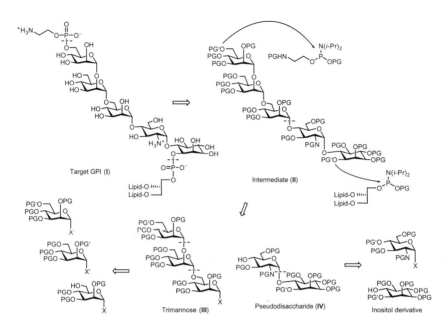

SCHEME 1. General retrosynthesis of a GPI anchor. PG = protecting group, PG' = orthogonal protecting group, X = activatable leaving group, X' = orthogonally activatable leaving group.

section: "Classic Approaches to GPI Anchor Synthesis." Recently, more effort has been focused on the development of novel synthetic strategies that emphasized flexibility, which has enabled access to GPI anchors containing biologically important unsaturated lipids, versatile "click chemistry" tags, and branched structures. These syntheses are surveyed in the second section: "Diversity-Oriented Approaches to GPI Anchor Synthesis." It is expected that this article will have some overlap with previous reviews on the subject,[20–22] but will offer coverage of new material as well as a contemporary and unique perspective.

To keep this article within reasonable constraints, detailed discussion of research involving GPI partial structures has been omitted. Notable works in this category include syntheses of non-lipidated GPIs by Martín-Lomas[23] and Schmidt,[24] lipophosphoglycans (LPGs) by Konradsson and Oscarson,[25,26] GPI-based antimalarial candidates by Seeberger,[17,27,28] and GPI-peptide/glycopeptide constructs by Seeberger[29] and Guo.[30,31] Also excluded from this article is recent work from the Bertozzi group using green fluorescent protein (GFP)-modified GPI partial structures to explore their behavior in live cells[18,32] and work from the Guo laboratory focused on chemoenzymatic ligation of GPI analogues and proteins.[33–35]

II. Classic Approaches to GPI Synthesis

1. Synthesis of a *Trypanosoma brucei* GPI Anchor by the Ogawa Group (1991)

The Ogawa group reported the chemical synthesis of a *T. brucei* GPI anchor in 1991,[36,37] 3 years after Ferguson and coworkers published its complete structural determination. Ogawa's reports, which were preceded by synthetic studies on GPI-related pseudooligosaccharides in the Fraser-Reid,[38] Ogawa,[39] and van Boom[40] laboratories, marked the first total synthesis of a GPI anchor. In accordance with the natural structure of the *T. brucei* GPI anchor, an additional α-Gal-(1→6)-α-Gal-(1→3) component located at Man-I was incorporated into the GPI core structure.

Scheme 2 depicts the retrosynthesis of *T. brucei* GPI anchor **1**. While most modern GPI syntheses involve a convergent assembly featuring formation of the mannose–glucosamine linkage as the key and final glycosidic bond formation (see Scheme 1), Ogawa's pioneering effort relied on a different approach. The assembly strategy had both linear and convergent components, as benzyl-protected GPI precursor **2**, which contained orthogonal protection for late-stage phosphorylation with *H*-phosphonates **3** and **4**, was disconnected to give digalactosyl fluoride **5**, mannosyl halide donors **6** and **7**, and pseudotrisaccharide **8**. The latter compound was further

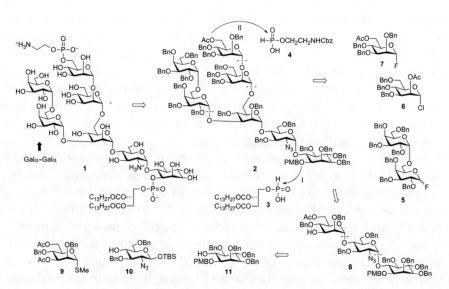

SCHEME 2. Retrosynthesis of *T. brucei* GPI anchor **1** by the Ogawa group.

disconnected into the monosaccharide and inositol building blocks **9–11**. For stereoselective glycosylation reactions, the Ogawa group primarily employed the Suzuki method, using glycosyl fluoride donors,[41] while also making use of Lemieux's halide ion-catalyzed glycosylation[42] and the classical Koenigs–Knorr reaction.[43] The H-phosphonate method[44] was used to introduce the phosphoglycerolipid and phosphoethanolamine groups.

The preparation of digalactosyl fluoride **5** is shown in Scheme 3. Penta-O-acetyl-α-D-galactopyranose (**12**) underwent treatment with p-methoxyphenol in the presence of trimethylsilyl trifluoromethanesulfonate (TMSOTf) in dichloroethane (DCE) to give a p-methoxyphenyl (PMP) glycoside, which was subsequently deacetylated and selectively protected at O-6 with 4,4′-dimethoxytrityl chloride (DMTrtCl) to provide **13**. After the remaining hydroxyl groups had been protected by benzyl ether groups, acid catalyzed methanolysis of the DMTrt group was performed to give the glycosyl acceptor **14**. This compound underwent coupling with the previously described 1-thiogalactoside **15**[45] under Lemieux conditions,[42] namely α-stereoselective halide ion-catalyzed glycosidation of the *in situ*-generated glycosyl bromide. This procedure generated α-galactoside **16** in 67% yield (the β anomer was isolated in 10% yield), and subsequent conversion into glycosyl fluoride **5** was accomplished by oxidative hydrolysis of the anomeric PMP group with ceric ammonium nitrate (CAN), followed by treatment with diethylaminosulfur trifluoride (DAST).

The main fragment in Ogawa's synthesis, pseudotrisaccharide **8**, was synthesized from building blocks **9–11** (Scheme 4). The mannosyl thioglycoside **9** was obtained from compound **18**[46] through acetylation, anomeric acetolysis, and finally treatment with methyl tributyltin sulfide in the presence of tin(IV) chloride. The azidoglucose acceptor **10** was synthesized from triol **19**,[47] which required acid-catalyzed

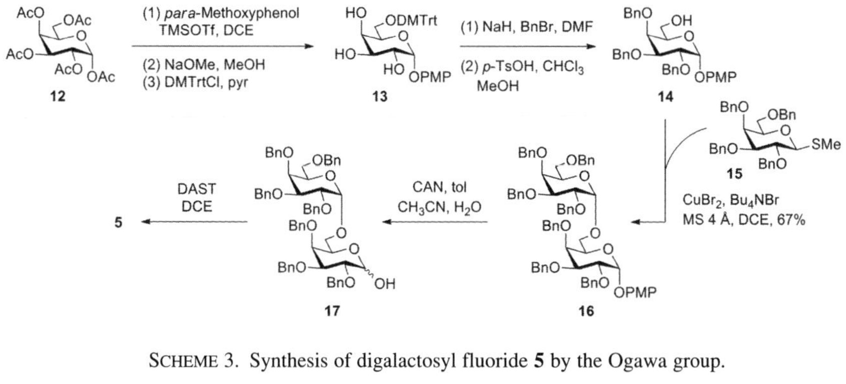

SCHEME 3. Synthesis of digalactosyl fluoride **5** by the Ogawa group.

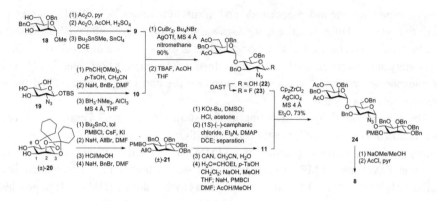

SCHEME 4. Synthesis of building blocks **9–11** and their elaboration to **8** by the Ogawa group.

4,6-benzylidenation with PhCH(OMe)$_2$, benzylation at O-3, and regioselective benzylidene ring opening with BH$_3$·NMe$_3$ in the presence of AlCl$_3$.[48]

Preparation of the enantiomerically pure inositol derivative **11** was lengthy. The synthesis began from racemic 2,3;4,5-di-*O*-cyclohexylidene-*myo*-inositol (±)-**20**,[49] Garegg's inositol derivative, a common starting point for inositol synthesis. Stannylene acetal-directed *p*-methoxybenzylation of (±)-**20** favored the O-6 position over the O-1 position (ratio = 3:1), leaving the latter site free for protection with an allyl group. After exchanging the cyclohexylidene acetals for benzyl groups to give (±)-**21**, the allyl group at O-1 was removed, exposing this position for reaction with the chiral resolving reagent (1*S*)-(−)-camphanic chloride, which generated separable diastereomeric inositol derivatives. At this stage, the absolute structure of the separated inositols was assigned by conversion into the previously characterized compound 2,3,4,5-penta-*O*-benzyl-*myo*-inositol. Unfortunately, the specific rotations for the D and L enantiomers of this compound were incorrectly reported in 1987,[50] an error whose correction in 1991[51] was undetected by the Ogawa group prior to publication of their work.[52,53] Consequently, the wrong inositol enantiomer was chosen, which led to the total synthesis of an unnatural stereoisomer of the *T. brucei* GPI anchor, a feat made no less spectacular by this minor mishap.

Moving forward, the inositol intermediate was converted into **11** through a series of manipulations at the 1 and 6 positions. After CAN-mediated oxidative cleavage of the 6-*O*-PMB group, four steps were performed consecutively without chromatographic purification, including acid-catalyzed reprotection with a 1-ethoxyethyl ether (EE) group, exchange of the (−)-CAM group for a PMB group at O-1, and finally methanolysis of the 6-EE group by use of AcOH/MeOH.

The assembly of pseudotrisaccharide **8** began by the coupling of compounds **9** and **10** (Scheme 4). The glycosylation occurred efficiently under the modified Lemieux conditions previously developed by Ogawa, specifically the use of AgOTf as an additional promoter in the presence of $CuBr_2/Bu_4NBr$ to improve reactivity with poor glycosyl acceptors such as **10**.[54] The α-disaccharide was generated in 90% yield and subsequently converted into glycosyl fluoride **23** by tetrabutylammonium fluoride (TBAF)-mediated anomeric desilylation followed by treatment of intermediate hemiacetal **22** with DAST. Compound **23**, bearing a nonparticipating 2-azido group adjacent to the leaving group, then underwent glycosylation with the inositol acceptor **11** by Suzuki's method employing Cp_2ZrCl_2 and $AgClO_4$[41] to form the crucial 1,2-*cis* glycosidic bond between the glucosamine and inositol components. The α anomer **24** was obtained in 73% yield, whereas the amount of the β anomer was minor (α:β ratio = 3.7:1), probably because of solvent participation from diethyl ether.[55] To expose the O-3 position of Man-I for addition of the digalactose moiety, compound **24** was deacetylated at the O-2 and O-6 positions of Man-I with NaOMe/MeOH and selectively reacetylated at the O-6 position of Man-I to give **8**.

Pseudotrisaccharide **8** was carried forward to the fully protected GPI precursor **2** as shown in Scheme 5. First, the digalactosyl group was installed at O-3 of Man-I by

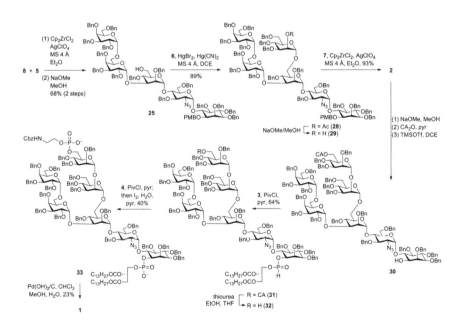

SCHEME 5. Completion of *T. brucei* GPI anchor **1** by the Ogawa group.

reaction of **8** with the digalactosyl fluoride **5** in the presence of $Cp_2ZrCl_2/AgClO_4$ in diethyl ether. An α,β mixture was obtained in 76% yield (α:β ratio = 9:1), and separation of the two anomers was performed following deacetylation at O-6 of Man-I with NaOMe in MeOH to give **25** (68% over two steps). During the glycosylation, pseudotetrasaccharide **26** and 1,6-anhydro sugar **27** were also formed (Scheme 6), suggesting that a portion of glycosyl fluoride **5** underwent decomposition via ring flip, cleavage of the galactose at O-6, and intramolecular attack of the liberated primary hydroxyl group on the oxocarbenium cation. Compound **25**, bearing a primary hydroxyl group at O-6 of Man-I, was coupled with mannosyl chloride **6**[56] in the presence of mercury(II) bromide and mercury(II) cyanide to generate the branched pseudohexasaccharide **28** with excellent α-stereoselectivity (89% yield). Deacetylation of **28** gave compound **29**, which reacted with mannosyl fluoride **7** (generated from its corresponding anomeric acetate[57] by deacetylation and treatment with DAST) in the presence of $Cp_2ZrCl_2/AgClO_4$ in diethyl ether to provide pseudoheptasaccharide **2** in 93% yield. Because of protecting-group complications, the acetyl group at O-6 on Man-III was exchanged for a chloroacetyl (CA) group, followed by removal of the PMB group at O-1 of the inositol with TMSOTf to give compound **30**, which was set up for both late-stage phosphorylation events.

First, the *H*-phosphonate method[44] was used to install a diacylglycerophosphate moiety to the O-1 position of inositol by reaction of **30** with *H*-phosphonate **3** and pivaolyl chloride (PivCl) in pyridine, affording intermediate **31** in 64% yield. Following removal of the 6-chloroacetyl group on Man-III with thiourea, this site was then allowed to react with *H*-phosphonate **4** in the presence of PivCl in 40% yield. After the intermediate had been oxidized with aqueous iodine, product **33** was subjected to global deprotection by hydrogenolysis, using $Pd(OH)_2/C$ as catalyst, which afforded the target GPI anchor **1** in 23% yield.

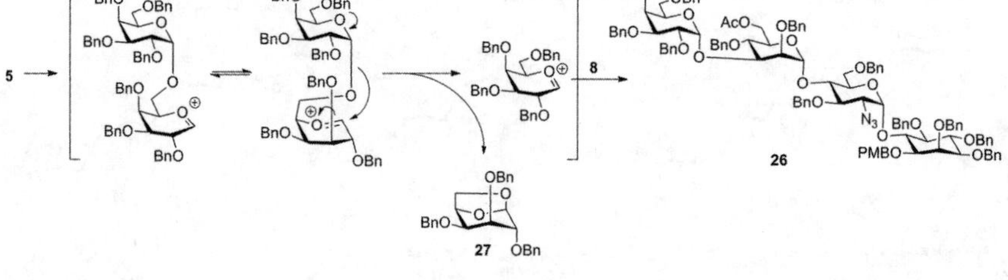

SCHEME 6. Generation of pseudotetrasaccharide **26** and anhydro sugar **27** in a side reaction during the coupling of **5** and **8** by the Ogawa group.

In summary, the Ogawa group accomplished the first synthesis of a GPI anchor by using a combined linear–convergent strategy in concert with classical methods for formation of glycosidic bonds together with *H*-phosphonate chemistry. Although it was an impressive starting point for GPI synthesis, future improvements were called for in the areas of assembly sequence, inositol preparation, orthogonal protecting-group selection, global-deprotection efficiency, and usage of more environmentally acceptable reagents.

2. Synthesis of a Yeast GPI Anchor by the Schmidt Group (1994)

In 1994, Schmidt and coworkers reported the total synthesis of a GPI anchor from yeast (*Saccharomyces cerevisiae*).[58,59] This anchor contains a ceramide lipid region rather than the glycerolipid commonly found in other organisms. Synthesis of the target GPI **34** made use of a fully convergent strategy (Scheme 7). As with Ogawa, Schmidt's laboratory performed two late-stage phosphorylations on a benzyl-protected GPI intermediate (**35**) to install the phosphoethanolamine and ceramide components.

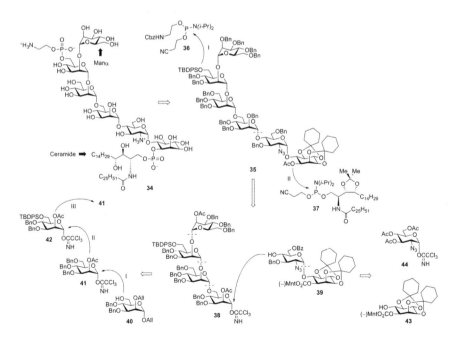

SCHEME 7. Retrosynthesis of yeast GPI anchor **34** by the Schmidt group.

For this purpose, the phosphoramidite method,[60] employing building blocks **36** and **37**, was used as an alternative to the *H*-phosphonate method, allowing incorporation of fully protected phosphate groups. Compound **35** was formed in a key glycosylation between the tetramannosyl trichloroacetimidate **38** and pseudodisaccharide **39**. The former compound **38** was built up from mannose derivatives **40–42**, while the latter was synthesized from inositol acceptor **43** and azidoglucosyl trichloroacetimidate **44**. Schmidt's own powerful method,[61] employing glycosyl trichloroacetimidate donors, was exclusively used for glycosidic bond formation, and acyl protecting groups at O-2 on the glycosyl donors aided the α-stereoselective formation of glycosidic bonds.

For synthesis of the tetramannosyl donor **38**, all required building blocks (**40–42**) were accessed from the known orthoester **45** (Scheme 8).[56] En route to the Man-I monomer **40**, orthoester **45** was treated with *tert*-butylchlorodiphenylsilane in the presence of imidazole to selectively protect O-6, leaving the remaining sites for protection with benzyl groups to give **46**. Acid-catalyzed hydrolysis of the 1,2-orthoester was followed by acetylation and regioselective hydrolysis of the anomeric acetate with $(NH_4)_2CO_3$. Subsequent treatment of the hemiacetal intermediate with trichloroacetonitrile and the nonnucleophilic base 1,8-diazabicyclo[5.4.0]-undec-7-ene (DBU) gave the Man-III glycosyl donor **42**. This compound was also elaborated to Man-I derivative **40** in four steps, including installation of allyl groups at O-1 and O-2 followed by 6-desilylation with TBAF/AcOH. The Man-II donor **41** was

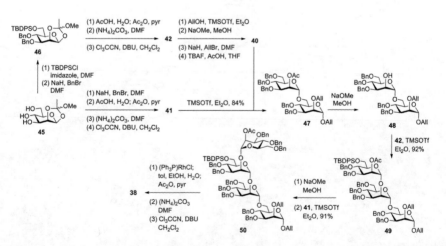

SCHEME 8. Synthesis of building blocks **40–42** and their elaboration to tetramannosyl donor **38** by the Schmidt group.

accessed from orthoester **45** by successive per-*O*-benzylation, orthoester hydrolysis, acetylation, selective anomeric deacetylation, and trichloroacetimidation, as just described for compound **42**.

Coupling of the Man-I acceptor **40** and Man-II trichloroacetimidate donor **41** in the presence of catalytic TMSOTf in Et_2O provided dimannoside **47** in 84% yield with complete α-stereoselectivity. Following removal of the temporary stereodirecting Man-II 2-*O*-acetyl group with NaOMe/MeOH, the acceptor **48** was allowed to react with Man-III trichloroacetimidate donor **42** under similar conditions, affording trimannoside **49** in 92% yield, again with complete α-stereoselectivity through neighboring-group participation. A final iteration of this process, using building block **41**, installed the Man-IV unit in 91% yield. Conversion of the allyl tetramannoside **50** into trichloroacetimidate **38** required four steps, including double deallylation with Wilkinson's catalyst, acetylation of the intermediate diol, regioselective anomeric deacetylation with $(NH_4)_2CO_3$, and treatment with trichloroacetonitrile and DBU. This sequence served to activate the anomeric center as a trichloroacetimidate and install a participating acetyl group at the O-2 position of Man-I to facilitate α-stereoselectivity in the key glycosylation.

The synthesis of pseudodisaccharide **39** featured efficient preparation of the inositol derivative **43**, followed by a highly stereoselective glycosylation reaction with the azidoglucosyl donor **44** (Scheme 9). Synthesis of **43** was accomplished by enantiomeric resolution of inositol diketal (±)-**20**,[49] which underwent treatment with bis(tributyltin)oxide followed by reaction with the chiral reagent (−)-menthyl chloroformate [(−)-MntOCOCl] to give the diastereomeric carbonates **43** and **51**,[62] which were separated by recrystallization. Compound **43** was directly glycosylated by the azidoglucosyl trichloroacetimidate **44**[63] in the presence of TMSOTf to give exclusively the α-stereoisomer **52** in 85% yield. The authors attributed the exceptional stereoselectivity observed in this coupling to an appropriate choice of protecting

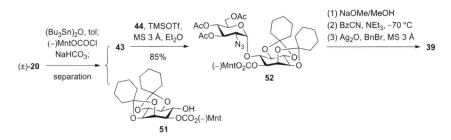

SCHEME 9. Synthesis of pseudodisaccharide **37** by Schmidt.

groups and reaction optimization, which still remains a challenging task for forming the 1,2-*cis* glycosidic bond between azidoglucosyl donors and inositol acceptors.

To expose the O-4 of the azidoglucose unit for the key glycosylation, compound **52** was converted into pseudodisaccharide **39** in three steps, each featuring crucial regiocontrol. First of all, selective deacetylation with NaOMe/MeOH formed an intermediate triol while leaving the (−)-menthyl carbonate intact. The O-6 position of the azidoglucose component was then selectively benzoylated with BzCN in the presence of triethylamine at −70 °C. Finally, selective 3-*O*-benzylation with BnBr promoted by Ag$_2$O gave the pseudodisaccharide **39** in 53% yield over three steps. Overall, the highly economical synthesis of this key intermediate constituted an impressive demonstration of regio- and stereocontrol.

Completion of the synthesis of yeast GPI anchor **34** is shown in Scheme 10. The key glycosylation reaction between tetramannosyl donor **38** and pseudodisaccharide acceptor **39** was accomplished using catalytic TMSOTf in diethyl ether, which afforded the desired α-pseudohexasaccharide intermediate in an excellent 91% yield. Unfortunately, several protecting-group changes were required to access GPI intermediate **35** prior to the two phosphorylation events, including removal of all acyl groups with KCN/MeOH (45%) and reprotection of the exposed positions with benzyl groups, as well as replacement of the 1-*O*-(−)-menthyl carbonate group of the inositol by an acetyl group. These late-stage protecting-group manipulations, which are best avoided on complex intermediates, were the price to be paid for a short pseudodisaccharide synthesis (Scheme 9), where appropriate protection steps were postponed.

Compound **35** was made ready for installation of the phosphoethanolamine group after TBAF-mediated removal of the 6-*O*-TBDPS group on Man-III. Formation of the

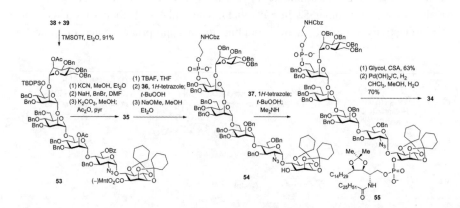

SCHEME 10. Completion of yeast GPI anchor **34** by the Schmidt group.

phosphotriester was achieved by 1*H*-tetrazole-promoted phosphitylation with phosphoramidite **36**, followed by *in situ* oxidation with *t*-BuOOH. Subsequent treatment with NaOMe/MeOH cleaved both the cyanoethoxyl and the inositol 1-*O*-acetyl groups to give phosphodiester **54**, which was poised for attachment of the ceramide moiety. Ceramide phosphoramidite **37**, which was prepared readily from a previously reported azido lipid,[64] reacted with intermediate **54** in the presence of 1*H*-tetrazole, and sequential *in situ* oxidation with *t*-BuOOH, followed by cyanoethoxyl removal with dimethylamine afforded GPI intermediate **55**. Two steps were required for global deprotection, including acid-catalyzed alcoholysis of the acetal protecting groups (63%) and Pd-catalyzed reductive removal of the benzyl/benzyloxycarbonyl groups and reduction of the azido group (70%) to give the target GPI anchor **34**.

Schmidt's total synthesis of yeast GPI anchor **34** was highlighted by incorporation of a ceramide phospholipid, showcasing the trichloroacetimidate "Schmidt" glycosylation method, and perhaps most impressively synthesizing pseudodisaccharide **39** with high efficiency and superb control of regio- and stereoselectivity. Furthermore, global removal of the benzyl groups using Pd-catalyzed hydrogenolysis featured an improved yield (70%) as compared with Ogawa's result in the first total synthesis of a GPI anchor (23%). The versatility of the strategy was further demonstrated by Schmidt's use of building blocks from this synthesis in subsequent syntheses of GPIs from *Toxoplasma gondii*[24] and rat brain Thy-1 (for the latter, see Section II.5).

3. Synthesis of a Rat Brain Thy-1 GPI Anchor by the Fraser-Reid Group (1995)

The Fraser-Reid group published several synthetic studies on GPI partial structures[38,65–69] prior to the 1995 communication by Campbell and Fraser-Reid describing the synthesis of a rat brain Thy-1 GPI anchor, which constituted the first total synthesis of a mammalian GPI and the third total synthesis in general of a GPI.[70] The Thy-1 GPI glycan contains an additional phosphoethanolamine group at the O-2 position of Man-I (a common modification in mammalian GPIs), as well as GalNAc and Man components attached to the O-4 position of Man-I and the O-2 position of Man-III, respectively. These modifications, while numerous, were all effectively incorporated into the "fully phosphorylated" target molecule by Fraser-Reid.

A convergent approach was used to synthesize the Thy-1 GPI (compound **56**), in which a [3+2+2] block strategy was applied (Scheme 11). Because of the additional phosphoethanolamine group present in the target, the benzyl-protected intermediate **57** would undergo three sequential phosphorylation reactions at sites bearing appropriate orthogonal protecting groups. As with Schmidt, Fraser-Reid used

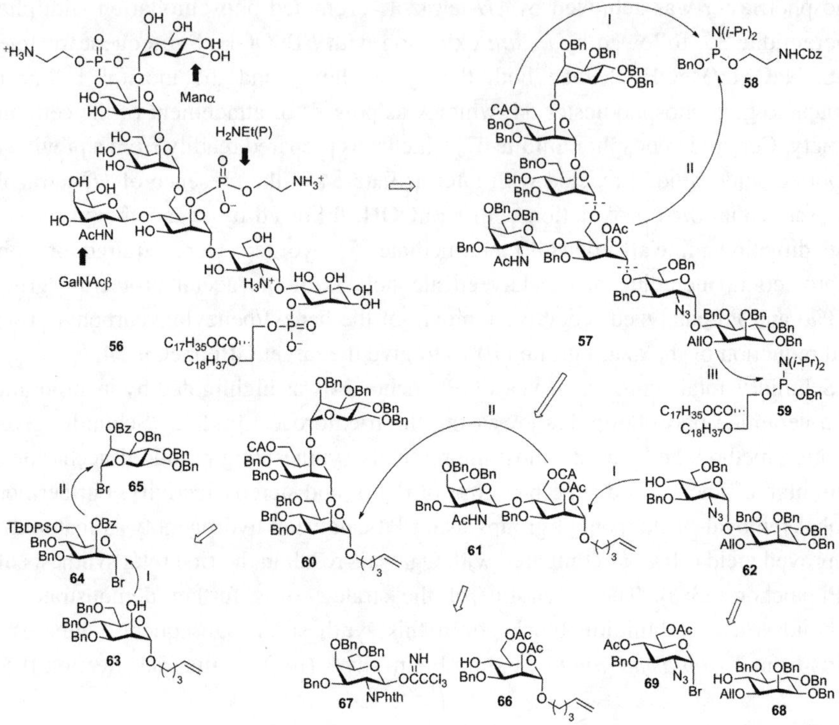

SCHEME 11. Retrosynthesis of Thy-1 GPI anchor **56** by the Fraser-Reid group.

the phosphoramidite method[60,66] with precursors **58** and **59** to install the phosphoethanolamine and phosphoglycerolipid groups. Compound **57** was synthesized from trimannoside **60**, disaccharide **61**, and pseudodisaccharide **62**, which were each accessed from their corresponding sugar and inositol monomers **63–69**. A combination of glycosylation methods were used for stereoselective formation of glycosidic bonds, including the Koenigs–Knorr method,[43] Schmidt's trichloroacetimidate method,[71] and Fraser-Reid's own method employing the versatile *n*-pentenyl glycosides (NPGs).[72]

The synthesis of *n*-pentenyl trimannoside **60** relied on sequential Koenigs–Knorr glycosylations using mannose monomers **63–65**, which were all prepared from the *n*-pentenyl orthoester[73] **70**[68] (Scheme 12). This intermediate allowed for ready conversion into 2-*O*-benzoyl-protected donors, thus facilitating α-stereoselective glycosylations and flexibility with regard to choice of leaving group at the anomeric position. The Man-II component **63** was synthesized from **70** by benzylation, acid-

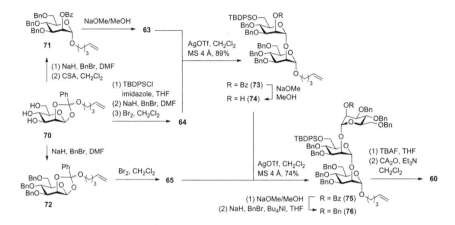

SCHEME 12. Synthesis of *n*-pentenyl trimannoside **60** by the Fraser-Reid group.

catalyzed rearrangement of the *n*-pentenyl orthoester to the corresponding *n*-pentenyl glycoside, and then removal of the newly formed 2-*O*-benzoyl group with NaOMe. Compound **70** was converted into the Man-III component **64** by 6-*O*-silylation with TBDPSCl/imidazole, benzylation, and treatment with bromine. The Man-IV donor **65**, also a glycosyl bromide, was readily generated from **70** by benzylation and treatment with bromine.

The Man-II acceptor **63** and Man-III donor **64** were coupled in the presence of promoter AgOTf, resulting in the formation of α-dimannoside **73** in 89% yield. Removal of the temporary stereodirecting 2-*O*-benzoyl group was accomplished with NaOMe to afford alcohol **74**, which underwent α-stereoselective glycosylation with the Man-IV glycosyl bromide **65**, again under the influence of AgOTf. The product **75** was isolated in 74% yield and subsequently underwent two protecting-group exchanges to provide the *n*-pentenyl trimannoside **60**. First, the stereodirecting 2-*O*-benzoyl group of Man-IV was replaced by a benzyl group, and then the 6-*O*-TBDPS group of Man-III was exchanged for a chloroacetyl group by TBAF-mediated desilylation followed by treatment with chloroacetic anhydride in the presence of triethylamine.

The β-GalNAc-Man disaccharide **61** was prepared from the *n*-pentenyl mannoside **77**[74] and tetraacetate **79**,[75] as shown in Scheme 13. Compound **77**, constituting the Man-I building block, was treated with PhCH(OMe)$_2$ in the presence of pyridinium *p*-toluenesulfonate (PPTS) to generate the 4,6-benzylidene acetal. Stannylene acetal-directed, regioselective 3-*O*-benzylation was followed by acetylation to give compound **78**, which was converted into the Man-I acceptor **66** by acid-promoted

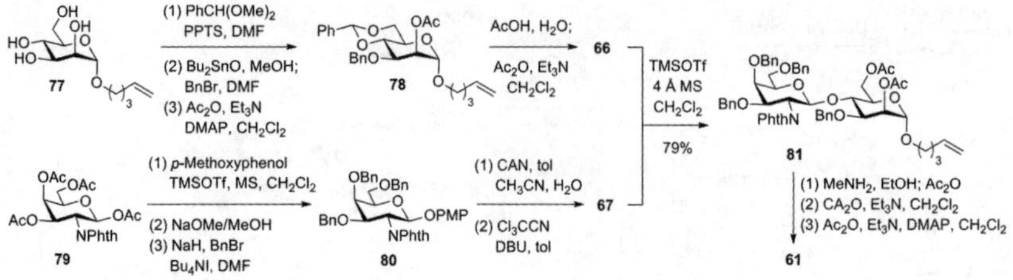

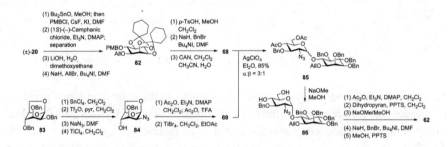

SCHEME 13. Synthesis of β-GalNAc-Man disaccharide **61** by the Fraser-Reid group.

SCHEME 14. Synthesis of pseudodisaccharide **62** by the Fraser-Reid group.

hydrolysis of the benzylidene group and selective acetylation at O-6. The 2-phthalimido galactosyl trichloroacetimidate **67** was accessed from 2-deoxy-2-phthalimido-β-D-galactose tetraacetate (**79**) by successive TMSOTf-promoted glycosidation with p-methoxyphenol, deacetylation, benzylation, oxidative hydrolysis of the anomeric PMP group, and finally reaction with trichloroacetonitrile and DBU.

The TMSOTf-catalyzed Schmidt glycosylation of **66** by **67** was completely β-stereoselective because of the participating effect of the 2-phthalimido group. The β-disaccharide product **81** was isolated in 79% yield, and later, the temporary 2-phthalimido group was replaced by a 2-acetamido group via sequential aminolysis and selective N-acetylation. Differentiation of the two free hydroxyl groups on Man-I was accomplished by selective O-6 chloroacetylation followed by acetylation at O-2, affording building block **61**.

Fraser-Reid's synthesis of the pseudodisaccharide fragment began with Garegg's indispensable inositol derivative (±)-**20**[49] (Scheme 14), as did the previous GPI syntheses by Ogawa and Schmidt. The route was initiated by stannylene-mediated p-methoxybenzylation at O-6, which was followed by esterification of the free O-1 position by the chiral resolution reagent (1S)-(−)-camphanic chloride.[76] After separation, the correct diastereomer was subjected to saponification with LiOH followed by allylation to provide optically active compound **82**. After the cyclohexylidene acetals had been replaced by benzyl groups, CAN-mediated cleavage of the PMB group at O-6 furnished the inositol acceptor **68**. To decrease loss of material, Fraser-Reid also developed a route to convert the 1-*p*-methoxybenzylated inositol regioisomer, generated from (±)-**20** in the first step, into compound **68** (not shown here).[69] The 2-azido glucosyl bromide **69** was synthesized from 1,6-anhydro-2,3,4-tri-*O*-benzyl-β-D-mannopyranose (**83**) via intermediate **84** (a route developed by Hori and coworkers[77]), which after trifluoroacetic acid (TFA)-promoted ring opening/acetylation was treated with $TiBr_4$ to afford **69**.

The Koenigs–Knorr-type glycosylation between inositol acceptor **68** and azido glucosyl bromide **69** proceeded with acceptable α-stereoselectivity when using $AgClO_4$ in diethyl ether. The inseparable α,β mixture **85**, obtained in 85% yield (α:β = 3:1), was subjected to deacetylation at O-4 and O-6, after which the preponderant α-pseudodisaccharide diol **86** was separated out. The remaining task of selective benzylation at O-6 turned out to be quite problematic, as the authors were forced to perform five successive steps, namely selective 6-*O*-acetylation, 4-*O*-tetrahydropyranylation (THP), deacetylation, 6-*O*-benzylation, and acid-promoted methanolysis of the THP group. This sequence blemished an otherwise efficient synthesis of pseudodisaccharide fragment **62**.

Assembly of the pseudoheptasaccharide from subunits **60**–**62** using the [3+2+2] block approach was accomplished with *n*-pentenyl glycoside chemistry (Scheme 15). The *n*-pentenyl disaccharide **61** was chemoselectively activated by *N*-iodosuccinimide (NIS) and triethylsilyl trifluoromethansulfonate (TESOTf), and subsequent reaction with pseudodisaccharide acceptor **62** gave the α-pseudotetrasaccharide **87** as the sole stereoisomer in 55% yield. The remainder of the mass balance was attributed to triethylsilylation of the acceptor hydroxyl group. To prepare for chain extension of the glycan, the 6-chloroacetyl group on Man-I of **87** was removed with thiourea to give **88**, which reacted with *n*-pentenyl trimannoside **60** in the presence of NIS/TESOTf to afford α-pseudoheptasaccharide **57** in moderate (32%) yield [39% based on recovered starting material (BRSM)]. Again, triethylsilylation of the acceptor hydroxyl group accounted for the low yield, although the silylated product could be recovered and transformed back into **88** by treatment with TBAF. This recycling effort would not have been possible without the earlier protecting-group exchange at O-6 of Man-III from TBDPS to chloroacetyl (Scheme 12), which on first glance may have seemed unnecessary. Also of note, Fraser-Reid took advantage of the ability of *n*-pentenyl glycosides to be selectively activated with NIS/TESOTf in the presence of allyl protecting groups in this sequence.

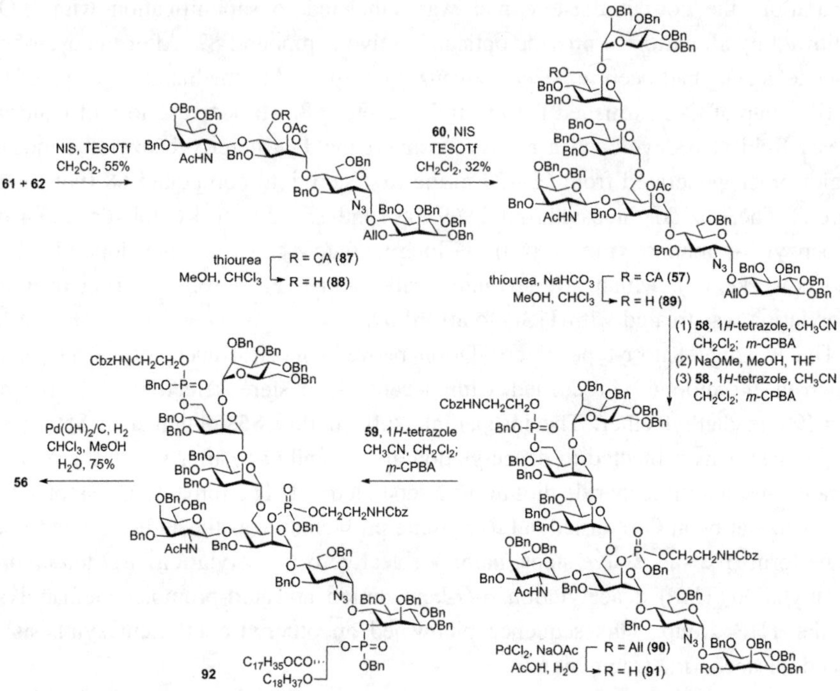

SCHEME 15. Completion of mammalian Thy-1 GPI anchor **56** by the Fraser-Reid group.

Pseudoheptasaccharide **57** was elaborated to the target GPI (**56**) via three phosphorylation events and global deprotection. Phosphoramidite chemistry optimized for GPI synthesis by Fraser-Reid[66] was used to form the appropriate phosphotriester bonds efficiently. For the phosphoethanolamine group on Man-III, dechloroacetylation of **57** was followed by 1H-tetrazole-promoted phosphitylation with **58** and in situ oxidation with m-chloroperoxybenzoic acid (m-CPBA). After deacetylation of the O-2 position of Man-I, the same conditions were used to install a phosphoethanolamine group at this site. Next, compound **90** was subjected to Pd-mediated removal of the inositol O-1 allyl group to give **91**, which was coupled with phospholipid precursor **59**, again using the phosphoramidite method. All three phosphorylation events proceeded smoothly and in high yield (75–90%). Finally, hydrogenolytic global deprotection with Pd(OH)$_2$/C and H$_2$ provided the fully phosphorylated Thy-1 GPI **56** in 75% yield.

The in-depth synthetic studies on GPI synthesis between 1989 and 1995 by the Fraser-Reid group culminated in the first total synthesis of a mammalian GPI anchor, namely Thy-1 GPI (**56**). Through careful choice of assembly strategy and orthogonal protecting groups, the structural modifications present in the natural compound were successfully incorporated into the target molecule, including two sugar groups and an additional phosphoethanolamine group. Fraser-Reid's *n*-pentenyl glycoside chemistry was used prominently, both in the preparation of mannose building blocks and in the final two stereoselective glycosylations to join fragments **60–62**. Where necessary, the glycosylation methods of Koenigs–Knorr and Schmidt were also used for stereoselective formation of glycosidic bonds.

4. Synthesis of a *T. brucei* GPI Anchor by the Ley Group (1998)

In contrast to the combined linear–convergent synthesis of the *T. brucei* GPI anchor **1** reported by Ogawa,[36] the Ley group used a highly convergent route (Scheme 16).[78,79] The target molecule **1** was accessed from a fully protected pseudoheptasaccharide **93** via sequential phosphorylation reactions with **58** and **94** using the phosphoramidite method.[60] In turn, compound **93** was formed in a key glycosylation by two highly elaborated building blocks, namely pentasaccharide donor **95** and pseudodisaccharide **96**. To synthesize these intermediates, the Ley group made elegant use of their bis(dihydropyran) chemistry for selective protection and desymmetrization procedures.[80–82] Rapid preparation of pentasaccharide **95** from building blocks **99–103** (via **97** and **98**) relied on tuning the reactivity of glycosyl donors using judicious protecting-group choices,[83] as well as by taking advantage of selenoglycoside and thioglycoside orthogonality.[84] Pseudodisaccharide **96** was generated from azido glucosyl bromide **104** and inositol acceptor **68**, the latter of which was made chiral by using bis(dihydropyran)-based desymmetrization.

Building blocks **99–103** were generated from seleno- and thioglycosides **105–107** (Scheme 17). The Man-I building block **99** was obtained from thioglycoside **105**[85] in seven steps. Selective installation of a TBS group at O-6 was followed by *cis*-acetonation to protect the O-2 and O-3 positions, which readied O-4 for benzylation. After deacetonation, intermediate **108** underwent trimethylsilylation at O-3, chloroacetylation at O-2, and aqueous HF-mediated desilylation to provide compound **99**. The selenogalactoside **106**[86] was used as a common starting point for the preparation of compounds **100** and **101**. Thus, compound **106** was subjected to 2,3-butanediacetal (BDA) protection with butane-2,3-dione/CSA, selective silylation at O-6 with TBSCl, chloroacetylation of the free axial OH group at O-4, and removal of the TBS group with aqueous HF to give the glycosyl acceptor **100**. Alternatively, compound **106** was readily perbenzylated to give glycosyl donor **101**. Finally, selenomannoside **107**[87]

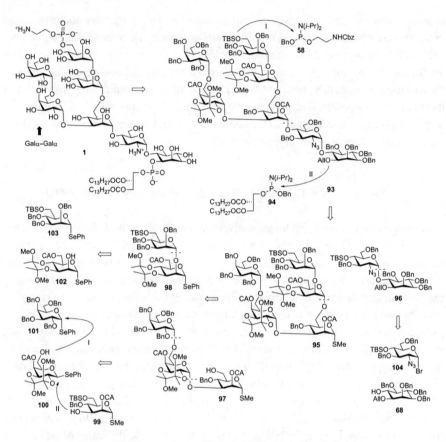

SCHEME 16. Retrosynthesis of *T. brucei* GPI anchor **1** by the Ley group.

was readily converted into the Man-II and Man-III building blocks **102** and **103**, the former being accessed by 3,4-BDA protection followed by tin-mediated chloroacetylation at O-6, and the latter by silylation at O-6 with TBSCl followed by benzylation.

With careful use of "armed" and "disarmed" glycosyl donors, as introduced by Fraser-Reid,[83] as well as by using selectively activatable selenoglycosides and thioglycosides,[84] the preparation of pentasaccharide **95** remarkably took only five steps from building blocks **99–103** (Scheme 18). First, the glycosyl acceptor **100**, whose anomeric selenoacetal group was disarmed by virtue of the deactivating BDA[81] and chloroacetyl groups, was coupled with the benzyl-protected, armed glycosyl donor **101** in the presence of NIS/TMSOTf. The α-digalactoside **111** was obtained in 71%

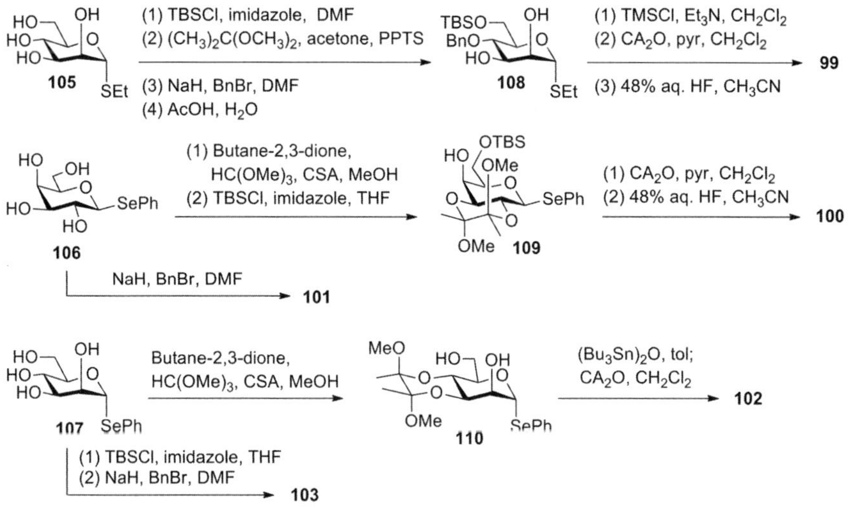

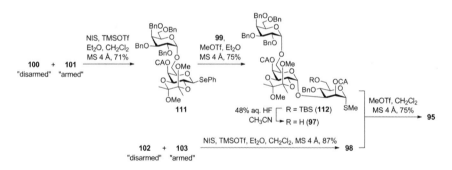

SCHEME 17. Synthesis of building blocks **99–103** by the Ley group.

SCHEME 18. Synthesis of pentasaccharide **95** by the Ley group.

yield, while the β anomer was isolated in 15% yield, and no homocoupling product was observed. Subsequent reaction of **111** with the Man-I thioglycoside acceptor **99** was facilitated by chemoselective activation of the selenoacetal with MeOTf,[88] affording the α-trisaccharide **112** in 75% yield. The α-dimannose fragment **98** was generated in 87% yield by selective coupling of selenoglycosides **102** (disarmed) and **103** (armed) in the presence of NIS/TMSOTf. Finally, compound **98** was combined with **97** (generated from **112** by desilylation), again relying on the higher reactivity of selenoglycosides with MeOTf as compared to thioglycosides (product generated in

75% yield). A large excess of the donor and high concentrations of reactants were used to minimize formation of an anhydro sugar resulting from intramolecular glycosidation of **97** through attack from O-6. Overall, the α-pentasaccharide **95** was efficiently prepared from compounds **99–103** by employing strategies to control anomeric reactivity.

The synthesis of pseudodisaccharide **62** shown in Scheme 19 was highlighted by Ley's efficient preparation of inositol derivative **68** using bis(dihydropyran)-mediated desymmetrization. The known symmetrical tetraol **113**[89] reacted with the chiral bis (dihydropyran) **114** in the presence of CSA to form dispiroketal **115** (ee ≥ 98%) in 88% yield. Positions 2–5 were masked with benzyl groups by deacylation and perbenzylation. The dispiroketal group was removed in a two-step process involving oxidation with *m*-CPBA followed by treatment with $LiN(TMS)_2$. The inositol derivative **68** was then formed by stannylene acetal-directed regioselective allylation at the O-1 position. The azido glucosyl bromide **104** was prepared by conventional methods beginning with thioglycoside **116**,[90] which underwent benzylation at O-3, regioselective cleavage of the benzylidene ring with Et_3SiH/TFA, and exchange of the phthalimido group by an azido group using the diazo-transfer chemistry developed by Vasella and coworkers.[91] This four-step procedure generated compound **117**, which was silylated at O-4 with TBSCl/KHMDS and then treated with bromine to give the glycosyl bromide **104**. Lemieux conditions[42] were employed for α-stereoselective joining of inositol acceptor **68** and glycosyl bromide **104**, which provided compound **96** in 65% yield. Removal of the TBS group at O-4 was accomplished by treatment with TBAF, giving the pseudodisaccharide acceptor **62**.

Completion of the synthesis of *T. brucei* GPI anchor **1** by the Ley group is shown in Scheme 20. The key glycosylation between the pentasaccharide ethylthio donor **95** and pseudodisaccharide **62** was promoted by NIS/TMSOTf and generated the

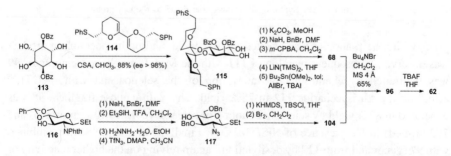

SCHEME 19. Synthesis of pseudodisaccharide **62** by the Ley group.

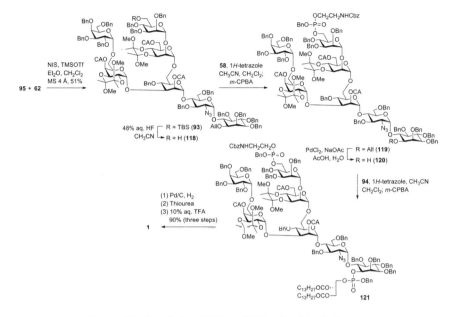

SCHEME 20. Completion of *T. brucei* GPI anchor **1** by the Ley group.

α-pseudoheptasaccharide **93** in 51% yield. Installation of the phosphoethanolamine group was performed by treatment with aqueous HF to expose the O-6 position of Man-III, followed by reaction at this site with the phosphoramidite **58** and *in situ* oxidation with *m*-CPBA. Intermediate **119** underwent Pd-mediated deallylation to deprotect the O-1 position of the inositol, which was followed by phospholipidation with **94** (synthesized using a bis(dihydropyran)-controlled desymmetrization of glycerol), again using phosphoramidite chemistry, to provide compound **121**. The global deprotection required three sets of conditions, including Pd-catalyzed hydrogenolysis, thiourea-mediated dechloroacetylation, and finally TFA-promoted hydrolysis of the BDA groups. This sequence generated the target GPI **1** in an impressive 90% yield over the final three steps.

In summary, Ley and coworkers successfully completed the fourth total synthesis of a GPI anchor, and this was the second effort aimed at the GPI (**1**) of *T. brucei*, following Ogawa's seminal work. This synthesis served as a proving ground for methods developed in the Ley laboratory, namely the usage of bis(dihydropyrans) as reactivity-tuning BDA protecting groups and desymmetrizing chiral reagents. These chemistries were elegantly integrated with strategies for selective activation of the glycosyl donor, specifically Fraser-Reid's armed/disarmed concept and orthogonally activatable anomeric groups (selenoglycosides and thioglycosides), culminating in a highly efficient convergent synthesis of the target molecule.

5. Synthesis of a Rat Brain Thy-1 GPI Anchor by the Schmidt Group (1999, 2003)

Following their 1994 synthesis of a yeast GPI anchor,[58] the Schmidt group reported a synthesis of the Thy-1 GPI anchor (**56**) of rat brain in 1999,[92] a target that had previously yielded to total synthesis by Fraser-Reid in 1995 (see Section II.3).[70] Unsurprisingly, Schmidt's assembly strategy closely resembled that of Fraser-Reid, particularly in the synthetic end-game. We forgo here a detailed discussion of this 1999 report, instead opting to cover a 2003 paper by Schmidt in which the Thy-1 GPI anchor was synthesized by a similar but modified strategy; it emphasized future couplings with peptides/glycopeptides at the phosphoethanolamine group of Man-III,[93] a strategic decision that foreshadowed looming advances in GPI–protein synthetic methods. In terms of synthetic strategy, this work was also closely tied to a 2001 report by Schmidt disclosing the synthesis of a partial structure of a nonlipidated GPI from the parasitic protozoan *T. gondii*.[24]

The retrosynthetic analysis from Schmidt's 2003 synthesis of the Thy-1 GPI anchor is shown in Scheme 21. The precursor to the target molecule, benzyl-protected pseudoheptasaccharide **122**, differed only slightly from the corresponding intermediate in Fraser-Reid's synthesis of the Thy-1 GPI (compound **57**, Scheme 11). However, late-stage phosphorylation events were designed to install orthogonally protected phosphoethanolamine groups (using the phosphoramidites **36** and **123**), which would permit selective amino-group deprotection and therefore the opportunity for future peptide/glycopeptide conjugation at Man-III. Intermediates **125** and **126** were employed in a highly convergent assembly strategy, for which the Schmidt group capitalized on the versatility of their previous GPI syntheses by employing some of the same building blocks. Naturally, the Schmidt trichloroacetimidate method[63] was used for stereoselective formation of glycosidic bonds.

The synthesis of pentasaccharide donor **125** was accomplished by stepwise elongation of the β-Gal-Man disaccharide **127** with mannosyl trichloroacetimidates **41** and **42** (Scheme 22). In turn, disaccharide **127** was prepared from monosaccharides **129** and **131**. First, the allyl mannoside **128**[94] was converted into Man-I derivative **129** via stannylene-mediated benzylation at O-3, *p*-methoxybenzylidenation at O-4 and O-6, allylation at O-2, and regioselective reductive cleavage of the acetal ring with NaBH₃CN/TFA. Preparation of the galactosyl trichloroacetimidate **131**, which

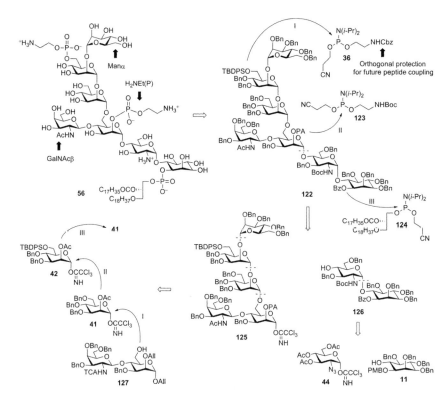

SCHEME 21. Retrosynthesis of Thy-1 GPI anchor **56** by the Schmidt group (2003).

contained a trichloroacetamido group to ensure β-selective glycosylation while also acting as a latent acetamido group, was initiated from the known azido galactose derivative **130**.[71] The anomeric position of **130** was temporarily masked with a thexyldimethylsilyl (TDS) group, followed by exchange of the azide group for a trichloroacetamido group. Conversion into the glycosyl donor **131** was accomplished by anomeric desilylation with TBAF, followed by treatment with trichloroacetonitrile and DBU. Coupling of Man-I acceptor **129** and trichloroacetamido galactosyl donor **131** in the presence of boron trifluoride etherate afforded exclusively the β-linked disaccharide **132**, which was treated with CAN to oxidatively cleave the PMB group to give **127**.

Starting from protected disaccharide **127**, consecutive mannosylations using donors **41** and **42** were effected under standard Schmidt glycosylation conditions to extend the glycan chain. Each of the three glycosylations was high yielding (92–96%)

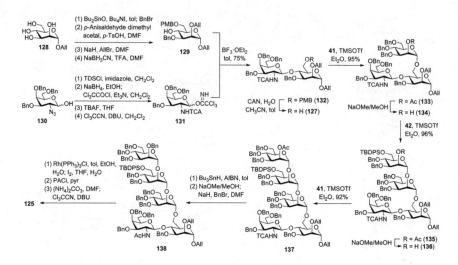

SCHEME 22. Synthesis of pentasaccharide **125** by the Schmidt group.

and highly α-stereoselective because of neighboring-group participation. The pentasaccharide **137** generated through this process required several manipulations to obtain donor **125**, starting with radical reduction of the trichloroacetamido group to form an acetamido group, followed by deacetylation–benzylation of the O-2 position of Man-III. The Man-I component was then adjusted via double deallylation with Wilkinson's catalyst, followed by diacylation with phenoxyacetyl (PA) chloride. Selective anomeric deacylation with $(NH_4)_2CO_3$ generated an intermediate hemiacetal that was treated with trichloroacetonitrile and DBU to provide the donor compound **125**.

Compound **126** differed from the pseudodisaccharide building blocks in previous GPI syntheses because *N-tert*-butyloxycarbonyl (Boc) protection was used rather than the traditional azido group. This was necessary to permit late-stage selective deprotection of the Man-III phosphoethanolamine (Cbz protected) for peptide conjugation. The preparation of **126** began from inositol derivative **11**, which Schmidt generated from the optically active inositol **43** in a few straightforward steps (Scheme 23). Allylation at O-6 was assisted by Ag_2O rather than basic conditions, so as not to affect the carbonate bond. After replacement of the 1-*O*-menthyl carbonate with a PMB group and the cyclohexylidene acetals with benzyl groups, the allyl group of O-6 was removed with Wilkinson's catalyst to give **11**.

Inositol acceptor **11** was then coupled with azido glucosyl trichloroacetimidate **44** in the presence of TMSOTf, which afforded pseudodisaccharide **139** in 70% yield.

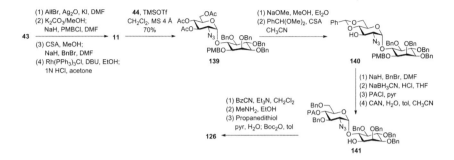

SCHEME 23. Synthesis of pseudodisaccharide **126** by the Schmidt group.

Numerous protecting-group manipulations were required at this point to prepare the azido glucose unit for the remainder of the synthesis. First, deacetylation with NaOMe gave an intermediate triol that was masked with a 4,6-*O*-benzylidene ring to give the pseudodisaccharide derivative **140**, which was converted into compound **141** by benzylation at O-3, regioselective benzylidene cleavage, and phenoxyacetylation at O-4 with PACl in pyridine. Exchange of the inositol 1-*O*-PMB group for a benzoyl group and dephenoxyacetylation with MeNH$_2$ preceded the crucial amino-protection swap, which was accomplished by propanedithiol-assisted azide reduction and treatment with Boc anhydride. The lengthy route to **126** dampened the overall efficiency of this GPI synthesis.

Completion of the Thy-1 GPI anchor synthesis by Schmidt is shown in Scheme 24. The key glycosylation between pentasaccharide donor **125** and pseudodisaccharide acceptor **126** under promotion by TMSOTf afforded the expected α-pseudoheptasaccharide **122** in 74% yield. The TBDPS group on O-6 of Man-III was removed with TBAF, making intermediate **142** ready for phosphorylation with compound **36** using the phosphoramidite method. *In situ* cleavage of the PA group at O-2 of Man-I and cyanoethoxyl groups with MeNH$_2$ generated alcohol **143**, which underwent installation of the second phosphoethanolamine group by reaction with **123**, again using phosphoramidite chemistry. Phospholipidation of GPI intermediate **144** was accomplished by 1-*O*-debenzoylation of the inositol, followed by reaction with phosphoramidite **124**. Concomitant *in situ* oxidation and aminolysis (to remove the final cyanoexothyl group) afforded the partially deprotected GPI derivative **146**. At this stage, the authors demonstrated the feature of orthogonal amino-group protection by subjecting compound **146** to Pd-catalyzed hydrogenolysis to generate GPI **147** (75%), which contained a single free amino group at the Man-III phosphoethanolamine component for future coupling with peptides/glycopeptides. In addition, intermediate

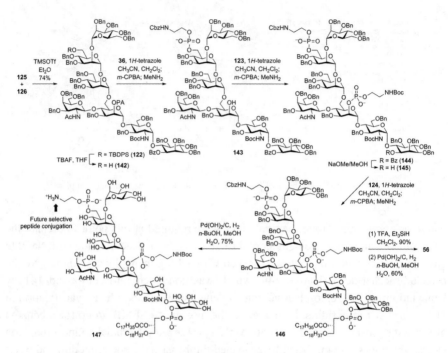

SCHEME 24. Completion of Thy-1 GPI anchor **56** by the Schmidt group (2003).

146 was subjected to a two-step global deprotection, including acidic and reductive conditions, resulting in formation of the rat brain Thy-1 GPI anchor **56** in 54% yield over the final two steps.

Schmidt and coworkers' 2003 synthesis of the rat brain Thy-1 GPI anchor, the second time they accomplished this feat, combined elements from their previous work in the field, including syntheses of GPI anchors (or related glycoconjugates) from yeast (1994), Thy-1 (1999), and *T. gondii* (2001). Thoughtfully designed building blocks were reused throughout these projects, a testament to their versatility. The syntheses of these GPIs relied on convergent assembly tactics in combination with expert use of glycosyl trichloroacetimidate and phosphoramidite chemistries for key bond-forming events. In the 2003 Thy-1 GPI synthesis, a strategy for orthogonal amino-group protection, which would permit future peptide/glycopeptide conjugation at Man-III, was demonstrated, although this feature was not applied. The Guo group was the first team to report the chemical synthesis of GPI-anchored peptides[30] and glycopeptides,[31] in 2003 and 2004, respectively (see Scheme 31), but these constructs did not contain GPI anchors of natural structures.

6. Synthesis of a Human Sperm CD52 Antigen GPI Anchor Containing an Acylated Inositol by the Guo Group (2003)

The second mammalian GPI anchor to be chemically synthesized was that of the human sperm CD52 antigen by the Guo group in 2003.[95] Related synthetic studies by the Guo group included the total synthesis of CD52 GPI-peptide/glycopeptide conjugates bearing GPI partial structures (see Scheme 31).[30,31] In addition to the intriguing and therapeutically relevant biological activities[96,97] of the sperm and lymphocyte CD52 antigens, these GPI-anchored glycopeptides exhibit a short peptide chain (12 amino acids) and interesting GPI structures,[98,99] making them an ideal model for synthetic studies on GPIs and GPI-anchored molecules. Notably, the 2003 report by Xue and Guo was the first synthesis of a GPI anchor of the CD52 antigen and also the first of a GPI containing an acylated inositol, which is a structural feature present in many GPIs.[1] Acylation at O-2 of the inositol is important during GPI biosynthesis, although it is often deleted after biosynthesis is accomplished, and it also confers resistance of GPIs to cleavage by bacterial PI phospholipase C.[2,100] To develop a synthetic strategy to access these GPIs for structural and biological studies, the inositol-acylated GPI **148** was targeted for synthesis.

A convergent strategy was used for synthesizing **148** (Scheme 25), whose fully protected precursor **149** would undergo late-stage installation of phosphoethanolamine by phosphoramidite **150**, which exhibited orthogonal protection of the amino group to permit future peptide/glycopeptide coupling. Uniquely, rather than perform phospholipidation at a late stage as in other reported GPI syntheses, this component was attached to the pseudodisaccharide fragment prior to the key glycosylation reaction. This decision was made because of the observation of an undesired cyclophosphitamidation reaction during the attempted phospholipidation of inositol-acylated pseudopentasaccharide **158**, which in fact generated compound **161** (Scheme 26). Because such reactivity was not observed by other research groups, or on pseudodisaccharide model systems,[101] it was proposed that the coexistence of the 2-O-acyl group on the inositol along with the trimannose moiety forced the GPI intermediate **159** into a conformation that favored the intramolecular redox cyclization reaction that was observed. Therefore, the key coupling to give **149** would take place between the trimannose thioglycoside **151** and phospholipidated pseudodisaccharide **152**. The former intermediate was built up from the monosaccharide mannose derivatives **153**, **6** and **154**, while the latter was synthesized from the inositol derivative **155**, the azido glucosyl fluoride **156**, and phosphoramidite **157**. Glycosyl halides were primarily used as sugar donors, and the phosphoramidite method was used for phosphotriester formation.

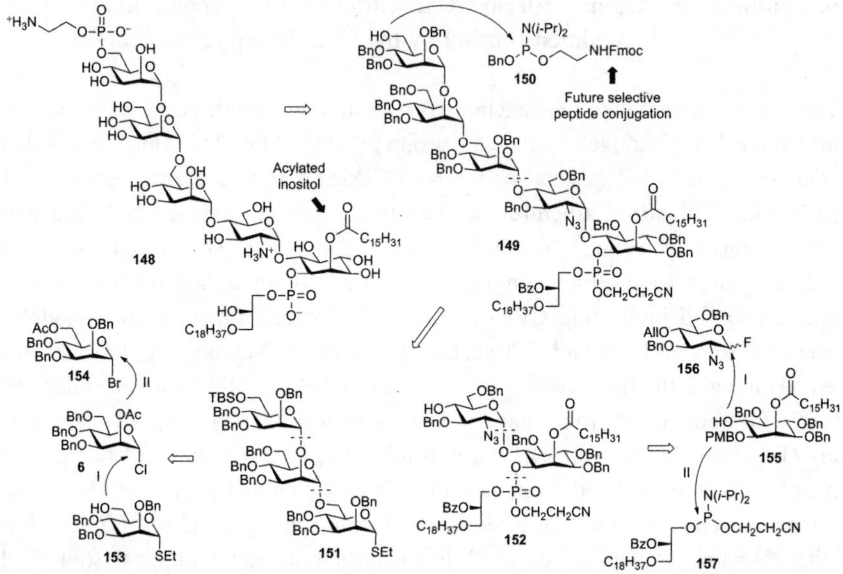

SCHEME 25. Retrosynthesis of GPI anchor **148** of the sperm CD52 antigen by the Guo group.

The synthesis of trimannose thioglycoside **151** was straightforward and relied on the orthogonal relationship of glycosyl halides and thioglycosides (Scheme 27). The thioglycoside (**153**) of Man-I was obtained from methyl α-D-mannopyranoside (**162**). Benzylation and sulfuric acid-promoted acetolysis of **162** gave diacetate **163**, which on BF$_3$-promoted glycosylation with ethanethiol and subsequent deacetylation at O-6 provided **153**. Koenigs–Knorr glycosylation of the Man-I acceptor **153** by the Man-II glycosyl chloride **6**[56] in the presence of AgOTf provided α-dimannoside **164** in 82% yield with complete stereoselectivity. Removal of the temporary stereodirecting 2-*O*-acetyl group on Man-II generated **165**, which was further glycosylated by glycosyl bromide **154** (also formed from **163** by treatment with HBr/AcOH), under the influence of AgOTf, to give the α-trimannoside product **166** in moderate (53%) yield. Finally, the 6-*O*-acetyl group on Man-III was exchanged for a TBS group to furnish **151**.

Motivated by the typical 50% material loss when resolving inositol enantiomers, Guo devised an inositol preparation based largely on tin-mediated chemistry for utilizing both enantiomers, namely (+)-**20** and (−)-**20**, to improve efficiency (Scheme 28). Compound (+)-**20** underwent stannylene acetal-directed allylation at O-6, leaving the O-1 position free for *p*-methoxybenzylation. Next, the less-stable

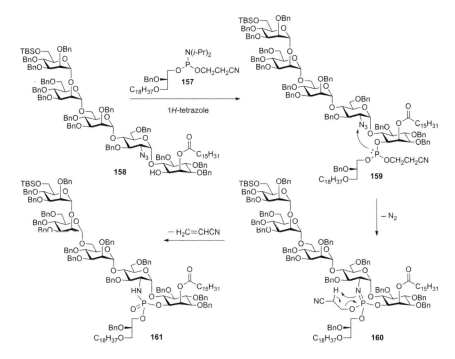

SCHEME 26. Unexpected cyclophosphitamidation reaction of the inositol-acylated GPI intermediate **159** (Guo group).

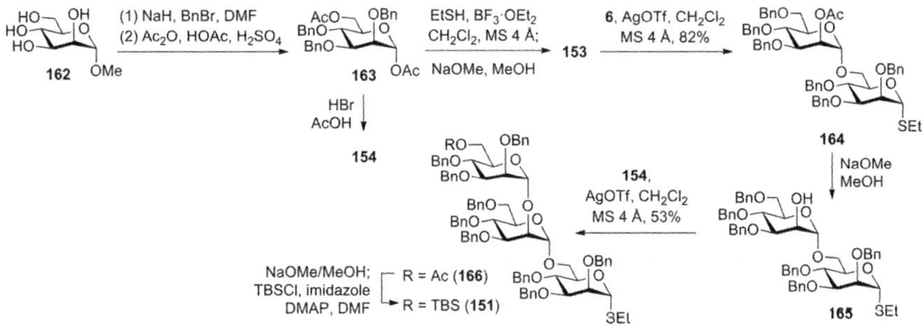

SCHEME 27. Synthesis of trimannose thioglycoside **151** by the Guo group.

trans-cyclohexylidene ring was selectively methanolyzed under acidic conditions, freeing the O-4 and O-5 positions for benzylation. The product from this sequence (**167**) was treated under acidic conditions for a prolonged period to cleave the more

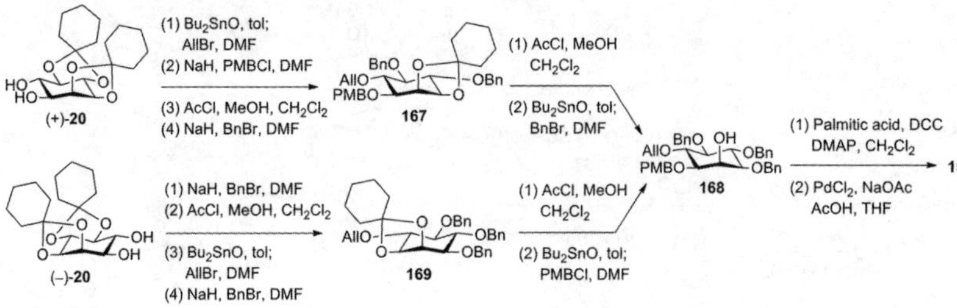

SCHEME 28. Synthesis of inositol derivative **155** by the Guo group.

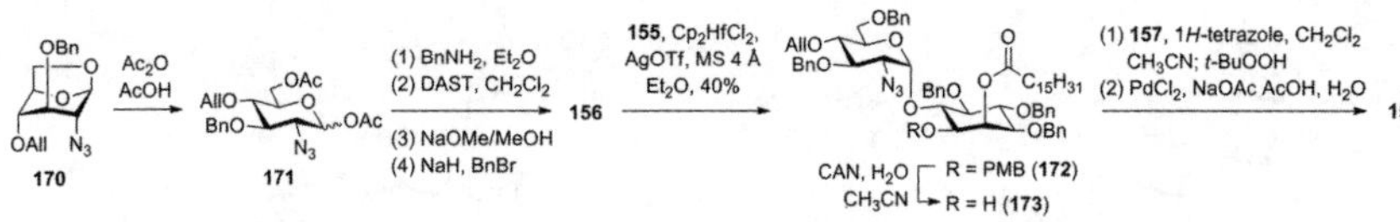

SCHEME 29. Synthesis of phospholipidated pseudodisaccharide **152** by the Guo group.

stable *cis*-cyclohexylidene ring, thus generating a diol whose O-3 position was selectively benzylated by using tin chemistry. This route gave the inositol derivative **168**, which was also accessed from the enantiomer (−)-**20** by using the opposite sequence. Dibenzylation of (−)-**20** was followed by removal of the *trans*-cyclohexylidene group, selective allylation at O-6, and benzylation at O-5. The product **169** converged to the common intermediate **168** by cleavage of the *cis*-cyclohexylidene substituent and selective *p*-methoxybenzylation at O-1. Transformation of **168** into the correct derivative involved esterification of the free axial alcohol with hexadecanoic (palmitic) acid in the presence of *N,N'*-dicyclohexylcarbodiimide (DCC) and 4-dimethylaminopyridine (DMAP), followed by Pd-mediated deallylation to give the 2-*O*-acylated inositol **155**.

Synthesis of the phospholipidated pseudodisaccharide **152** was completed as shown in Scheme 29. The azido glucosyl fluoride building block **156** was prepared from the known anhydro sugar **170**.[102] First, acetolysis of **170** gave the diacetate **171**, whose anomeric acetate was exchanged for fluoride through selective deacetylation and treatment with DAST. Deacetylation at O-6 followed by benzylation gave donor **156**, which underwent coupling with **155** in the presence of Cp$_2$HfCl$_2$/AgOTf.[41] An anomeric mixture was generated in 70% yield (α:β = 4:3), from which the desired α-pseudodisaccharide **172** was isolated in 40% yield. Oxidative hydrolysis of the PMB group gave compound **173**, which was then phosphitylated with phosphoramidite **157** in the presence of 1*H*-tetrazole. *In situ* oxidation occurred on treatment with *t*-BuOOH, and subsequent Pd-mediated deallylation afforded the pseudodisaccharide phosphotriester **152**. Notably, no migration of the 2-*O*-hexadecanoyl group on the inositol was observed during removal of the PMB group or phosphitylation.

Owing to the early installation of the alkylglycerophosphate moiety, the synthetic end-game was considerably shorter than GPI syntheses previously reported (Scheme 30). The key coupling of trimannose thioglycoside **151** and pseudodisaccharide acceptor **152** was carried out with NIS/TfOH in CH_2Cl_2, with the addition of Et_2O to improve α-stereoselectivity. Desilylation at O-6 of Man-III with boron trifluoride etherate in the same vessel afforded the desired pseudopentasaccharide intermediate **149** in 52% yield over two steps. Facile installation of the phosphoethanolamine group by using phosphoramidite chemistry generated the fully protected GPI derivative **174**. An orthogonal 9-fluorenylmethoxycarbonyl (Fmoc) group was used to block the phosphoethanolamine group, a feature designed to allow future coupling with peptides/glycopeptides (see Scheme 31 for related work). In this case, the Fmoc group was selectively removed with DBU in 87% yield, and subsequent hydrogenolytic removal of the remaining protecting groups provided the target GPI **148** in 84% yield.

Xue and Guo's first total synthesis of a GPI anchor of sperm CD52 antigen was founded on well-established synthetic methods, such as the Koenigs–Knorr and Suzuki glycosylations to form glycosidic bonds and phosphoramidite chemistry to form phosphotriester bonds. However, problems arising from the incorporation of a 2-*O*-acylated inositol group necessitated the development of a novel assembly strategy involving early installation of the phospholipid. Other noteworthy aspects of the synthesis were (i) an efficient preparation of inositol **168** that employed both enantiomers of the starting material and (ii) orthogonal protection of the Man-III

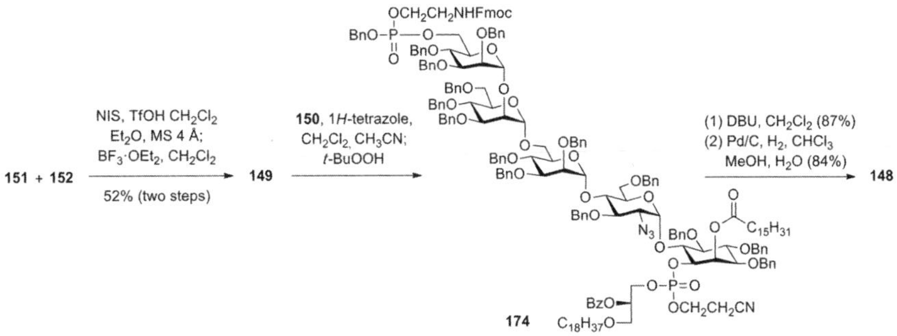

SCHEME 30. Completion of a GPI anchor of human CD52 bearing an acylated inositol (**148**) by the Guo group.

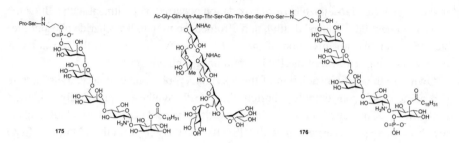

SCHEME 31. GPI-peptide/glycopeptide conjugates **175** and **176** chemically prepared by the Guo group.

phosphoethanolamine group for future peptide coupling. With regard to the latter feature, the Guo group published closely related chemical syntheses of GPI-peptide/glycopeptide conjugates **175** and **176** in 2003 and 2004 (Scheme 31).[30,31] Since they contained only GPI partial structures (without the glycerolipid moiety), they are not discussed in detail here. Current research efforts by the Guo laboratory concerning the synthesis of GPI-anchored peptides/proteins are focused on developing chemoenzymatic methods,[33–35] which likewise are not covered here.

7. Synthesis of a *Plasmodium falciparum* GPI Anchor Containing an Acylated Inositol by the Fraser-Reid Group (2004)

The Fraser-Reid group explored the consequences of incorporating a 2-*O*-acylated inositol into GPIs in parallel with, but independent of, the Guo group. A brief 2004 communication by Fraser-Reid group[103] describing the synthesis of a *P. falciparum* GPI "prototype" (bearing unnatural truncated lipids) was followed by a comprehensive report[104] that included a fully lipidated target GPI anchor. The broader implications of this work lie in the fact that some protozoan parasites, including the malaria-causing agent *P. falciparum*, use free GPI anchors and GPI-anchored proteins to modulate the immune response of the host. Recently, the development of synthetic GPI-based constructs as antiparastic vaccines has emerged as a promising area of research.[105]

Fraser-Reid's group set out to synthesize the GPI anchor **177** of *P. falciparum* bearing an acylated inositol (Scheme 32). In contrast to Guo's early-stage phospholipidation to avoid a side reaction, Fraser-Reid disconnected the target to pseudopentasaccharide **178**, which after introduction of phosphoethanolamine would undergo late-stage acylation at O-2 of the inositol and phospholipidation at O-1 via an

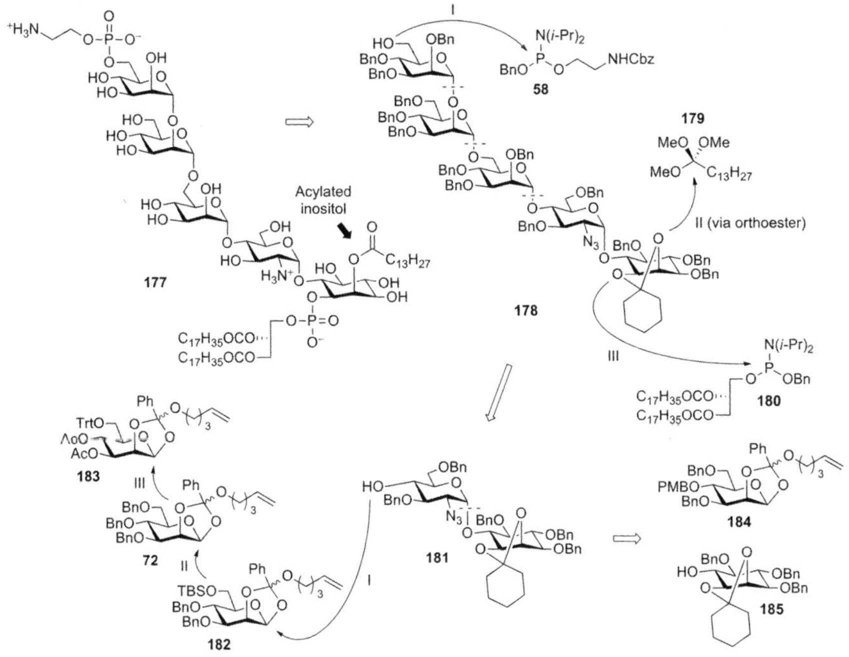

SCHEME 32. Retrosynthesis of the *P. falciparum* GPI anchor **177** by the Fraser-Reid group

intermediate orthoester. The preparation of **178** made use of a linear-assembly strategy involving consecutive mannosylations of pseudodisaccharide **181** with *n*-pentenyl orthoester (NPOE)[73] building blocks **182**, **72**, and **183**. A novel strategy was developed to enhance the α-stereoselectivity in the formation of pseudodisaccharide **181**. Rather than employing an azido glucosyl donor, which often gives low stereoselectivity, the authors used the mannosyl NPOE **184**, which enhanced stereoselective formation of the glycosidic bond while acting as a latent azido glucose unit via inversion at the 2-position with an azide nucleophile. The synthesis of **177** was thus centered on the versatility of NPOEs as building blocks.

The mannosyl NPOEs **182–184** were prepared from the readily available precursor **70**[68] (Scheme 33). The NPOE **182** of Man-I was obtained by silylation at O-6 followed by benzylation. The Man-III diacetate **183** was generated in a single flask from precursor **70** by sequential treatment with TrtCl and Ac$_2$O. The latent azido glucosyl building block **184**, at this stage in the form of a mannosyl NPOE, was synthesized by tin-mediated selective dibenzylation at O-3 and O-6, leaving the O-4 position available for *p*-methoxybenzylation.

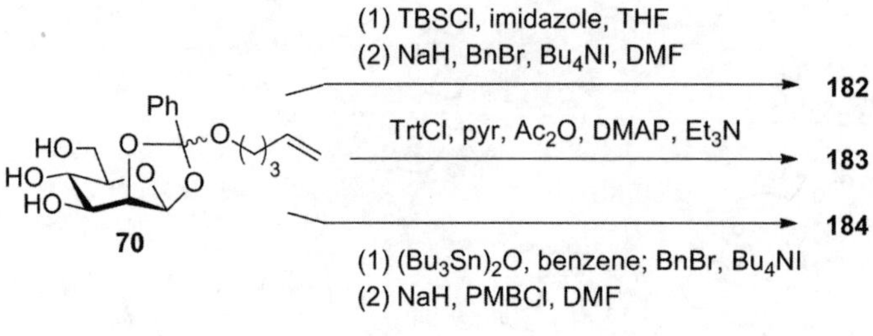

SCHEME 33. Synthesis of mannosyl NPOEs **182–184** by the Fraser-Reid group.

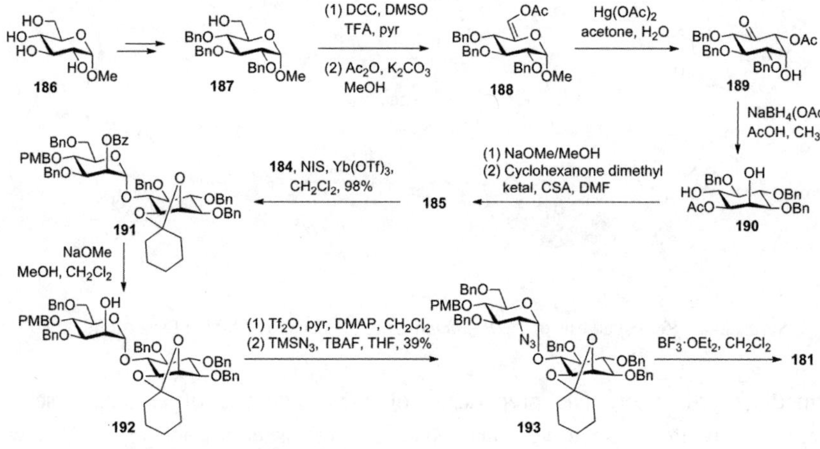

SCHEME 34. Synthesis of pseudodisaccharide **181** by the Fraser-Reid group.

Pseudodisaccharide **181** was prepared according to Scheme 34. *De novo* synthesis of the inositol building block **185** from methyl α-D-glucopyranoside (**186**) was performed by using a modified version of Bender's method.[106] After elaboration of the glucoside **186** to the inositol derivative **190**, which was highlighted by stereo-controlled Hg(II)-mediated Ferrier rearrangement and ketone reduction, conversion into compound **185** was accomplished by deacetylation and *cis*-cyclohexylidenation. Stereoselective glycosylation of the inositol acceptor **185** by the mannosyl NPOE **184** under activation by NIS/Yb(OTf)$_3$ in CH$_2$Cl$_2$ afforded the desired pseudodisaccharide **191** in an impressive 98% yield. Transformation into the desired azido glucose (2-equatorial) configuration was accomplished by debenzoylation, triflation at O-2, and azide displacement by DeShong's procedure (TMSN$_3$/TBAF),[107] although the desired product **193** was obtained in only 39% yield. Nevertheless, the concept introduced by Fraser-Reid is an attractive one, given first the notoriously low stereoselectivity of azido glucose–inositol couplings, and second, the prospect that azide displacement might be optimized. Finally, pseudodisaccharide acceptor **181** was obtained by treatment of **193** with boron trifluoride etherate to cleave the PMB group.

Stepwise elongation of the pseudodisaccharide by mannose building blocks took advantage of the propensity of NPOE donors to form α-configured products (Scheme 35). Thus, acceptor **181** reacted with the NPOE **182** of Man-I in the presence of NIS/BF$_3$·OEt$_2$, providing α-pseudotrisaccharide **194** in 79% yield. Protecting-group manipulations of the Man-I unit furnished compound **195**, which had a free O-6 position, and this was linked to Man-II **72** under the same activation conditions to form the α-pseudotetrasaccharide **196** in excellent yield (99%). Debenzoylation at O-2 of the Man-II component gave alcohol **197**, which was elongated with the NPOE **183** of Man-III, again activated by NIS/BF$_3$·OEt$_2$, to generate the all-α-linked pseudopentasaccharide **198** in 75% yield. The acyl protecting groups on Man-III were replaced by benzyl groups, and subsequent removal of the O-6 trityl group of Man-III by mild acid gave intermediate **178**, which was ready for the three phosphorylation reactions.

Completion of the GPI synthesis is shown in Scheme 36. Introduction of the phosphoethanolamine group to the Man-III O-6 position of pseudopentasaccharide **178** using the phosphoramidite method was successful. Cleavage of the 1,2-cyclohexylidene acetal group from the inositol was accomplished with CSA/ethylene glycol to give diol

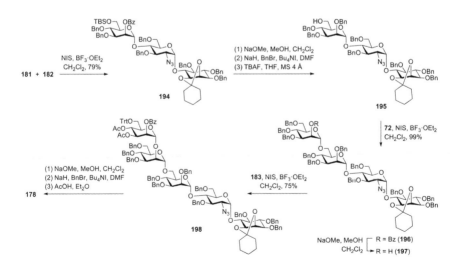

SCHEME 35. Construction of pseudopentasaccharide **178** by the Fraser-Reid group.

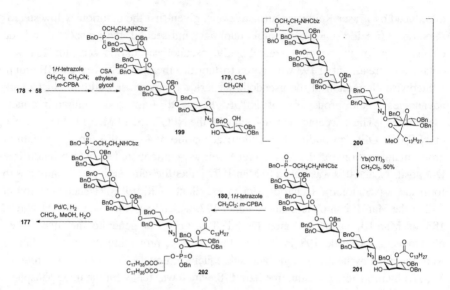

SCHEME 36. Completion of the target GPI **177** by the Fraser-Reid group.

199. Model studies by Fraser-Reid predicted that introduction of an acyl group at the axial O-2 position of **199** using manipulations of protecting groups would be precarious, and so an alternative approach involving intermediate cyclic orthoesters was pursued. The intermediate diol **199** was therefore allowed to react with the trimethyl orthoester **179** in the presence of CSA to give the intermediate GPI-orthoester **200**, which underwent rearrangement with Yb(OTf)$_3$ to provide a 2:1 mixture of regioisomers in favor of the desired 2-O-acylated product **201** (isolated in 50% yield). Despite Guo's observation that attempted late-stage phospholipidation was thwarted by the coexistence of 2-O-acylated inositol and the trimannose moiety (see Scheme 26), Fraser-Reid encountered no problems in converting compound **201** into **202** (60–70%) using phosphoramidite chemistry, suggesting that minor structural differences in the substrate and/or phosphoramidite reagent can impact the reaction outcome. Global deprotection by Pd-catalyzed hydrogenolysis, although challenging because of unanticipated solubility issues, was eventually completed to deliver the target GPI anchor **177**.

In summary, the Fraser-Reid group accomplished the second total synthesis of an inositol-acylated GPI anchor using a linear-assembly strategy in concert with NPOE chemistry. Four mannosyl NPOEs, all accessed from a single synthetic precursor (**70**), were used for stereoselective glycosidic bond formation, including the glucosamine–inositol linkage. In the latter instance, a mannose-configured NPOE donor was employed to facilitate the typically difficult α-glycosylation of inositol, and after serving this purpose was converted into an azido glucose moiety by invertive introduction of an azido group at the 2-position. Also contributing to the overall efficiency of the synthesis was the use of Bender's *de novo* inositol synthesis, which obviated the need for cumbersome enantiomeric resolution protocols. In the final synthetic sequence, an inositol 1,2-orthoester intermediate was used to attach regioselectively the correct (phospho)lipids to the GPI.

8. Synthesis of a *P. falciparum* GPI Anchor Containing an Acylated Inositol by the Seeberger Group (2005)

The Seeberger laboratory became involved in the study of GPI anchors in the early 2000s, when they used chemical synthesis to construct a GPI-based antitoxin vaccine against malaria, which was effective in a rodent model of the disease.[16] An intriguing aspect of this project was the application of automated solid-phase oligosaccharide synthesis for rapid generation of large quantities of GPI intermediates, which could be elaborated to conjugate vaccines.[17] These exciting developments offered further support to the notion that synthetic GPIs—indeed, carbohydrates in general—can be harnessed to develop novel therapeutic approaches.[105] Seeberger expanded on this early work in 2005 with the solution-phase total synthesis of a fully lipidated GPI anchor of *P. falciparum*.[108]

The inositol-acylated GPI anchor **203** was almost identical in structure to that synthesized by Fraser-Reid (compound **177**, see Scheme 32), but it contained the additional mannose unit present in natural GPIs from *P. falciparum* (Scheme 37). The target compound was accessed from the GPI pseudohexasaccharide **204**, which had orthogonal protecting groups, including PMB, allyl, and triisopropylsilyl (TIPS), for sequential installation of hexadecanoyl, phospholipid, and phosphoethanolamine groups. For the two phosphorylation events, the *H*-phosphonate method[44] was used. Intermediate **204** was formed convergently from the tetramannosyl trichloroacetimidate **207** and pseudodisaccharide alcohol **208**, which were both prepared in a straightforward manner using the building blocks **41**, **44**, and **209–211**. Schmidt's trichloroacetimidate method,[63] in combination with 2-*O*-acyl protecting groups, was used for the stereocontrolled formation of glycosidic bonds.

Synthesis of the tetramannose fragment **207** involved extension of the Man-I acceptor **209** using mannosyl trichloroacetimidate donors (Scheme 38). Compound **209** was glycosylated by imidate **41** in the presence of TMSOTf, generating dimannoside **212**, which then underwent selective deacetylation by treatment with AcCl/MeOH to give product **213** in 88% yield over two steps. This process was repeated

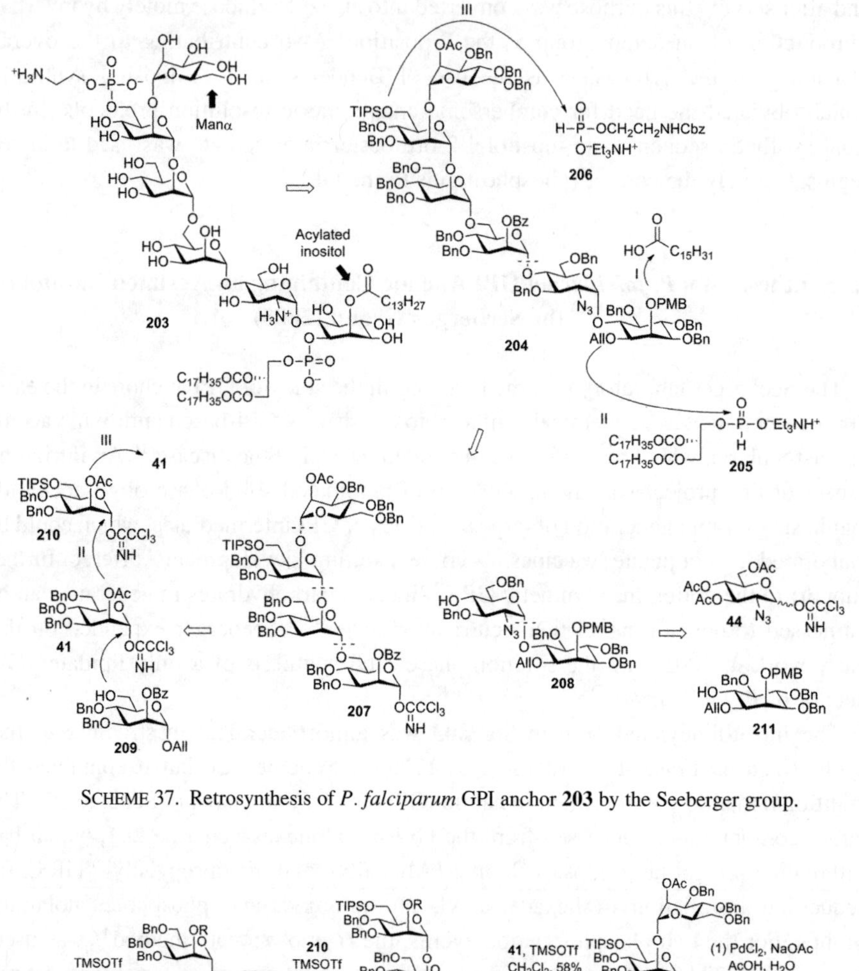

SCHEME 37. Retrosynthesis of *P. falciparum* GPI anchor **203** by the Seeberger group.

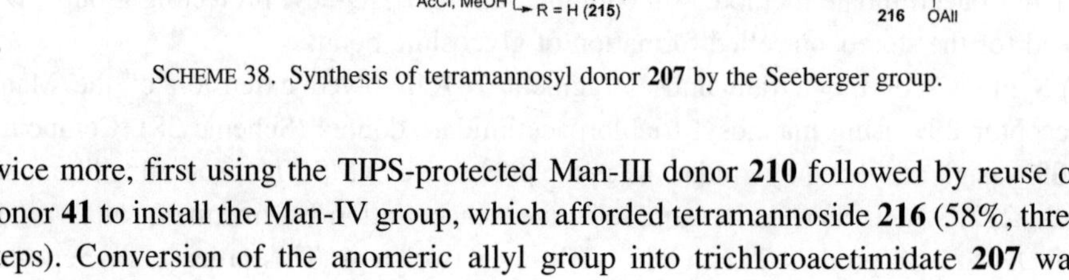

SCHEME 38. Synthesis of tetramannosyl donor **207** by the Seeberger group.

twice more, first using the TIPS-protected Man-III donor **210** followed by reuse of donor **41** to install the Man-IV group, which afforded tetramannoside **216** (58%, three steps). Conversion of the anomeric allyl group into trichloroacetimidate **207** was accomplished by Pd-mediated deallylation and subsequent treatment with trichloroacetonitrile and DBU.

The pseudodisaccharide fragment was efficiently prepared starting from the known inositol diol **217** generated from compound **190** by deacetylation/allylation (Scheme 39).[106,109] Selective protection at O-2 of **217** with a PMB group allowed for late-stage hexadecanoylation of this site. The union of the resulting product **211** and the azido glucosyl trichloroacetimidate **44**[63] under standard activation conditions yielded the α,β mixture **218** in 89% yield with good stereoselectivity (α:β = 4:1). This inseparable mixture was carried forward by deacetylation to give a triol, which underwent successive 4,6-*O*-benzylidenation, benzylation, and regioselective benzylidene cleavage to afford the α anomer **208**, which was separated from the corresponding β anomer at this stage.

The key glycosylation between tetramannosyl trichloroacetimidate **207** and pseudodisaccharide alcohol **208** was performed in the presence of TMSOTf, while the 2-*O*-benzoyl group of the donor ensured α-stereoselectivity through neighboring-group participation (Scheme 40). The reaction generated pseudohexasaccharide **204** in 94% yield, and the subsequent transformation of this to the intermediate **220** required deacylation followed by benzylation. The fully protected GPI intermediate **220** could be obtained in nearly gram quantities, an important accomplishment considering the large amount of material required for generating and testing antimalarial vaccine candidates.

The remaining steps centered on lipidation and phosphorylation events. After oxidative removal of the PMB group at O-2 of the inositol, the axial hydroxyl group thus exposed was esterified with hexadecanoic acid in the presence of DCC and DMAP to give **221**. Subsequent removal of the neighboring allyl group at O-1 of the inositol to furnish compound **222** was successfully performed by using an excess of $PdCl_2$ without acyl migration or decomposition. This outcome was in contrast to Fraser-Reid's previously reported difficulties in a nearly identical system. The phospholipidation step was performed by using *H*-phosphonate **205** in the presence of PivCl and pyridine, and this was followed by iodine-mediated oxidation, to give **223**

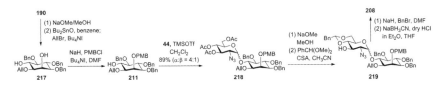

SCHEME 39. Synthesis of pseudodisaccharide acceptor **208** by the Seeberger group.

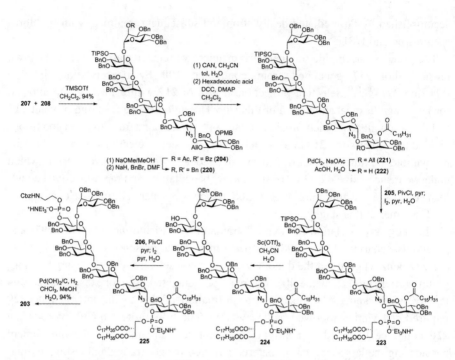

SCHEME 40. Completion of GPI anchor **203** of *P. falciparum* by the Seeberger group.

without any side reactions involving the azido group. After the 6-*O*-TIPS group on Man-III had been hydrolyzed in the presence of Sc(OTf)$_3$ to give compound **224**, the *H*-phosphonate method was again used to install the phosphoethanolamine group. Final deprotection of intermediate **225** was effected by Pd-catalyzed hydrogenolysis in a mixture of solvents, which provided the target GPI **203** in excellent (94%) yield.

Seeberger's synthesis of the *P. falciparum* GPI anchor **203** bearing an acylated inositol group featured a logical and straightforward convergent assembly strategy based on previous GPI syntheses. Proven methods were used for the formation of glycosidic and phosphodiester bonds, while difficulties encountered by the Guo and Fraser-Reid groups in the late-stage lipidation and phosphorylation steps were avoided. Impressively, gram-scale quantities of complex pseudohexasaccharide intermediates were obtained on scale-up, a critical point given the group's development of GPI-based antimalarial vaccines. Subsequent advances in this area by the Seeberger group include the 2008 development of synthetic GPI glycan arrays to characterize malaria-induced antibody responses,[110] which should aid in the design of improved vaccine candidates. Also reported by the Seeberger laboratory in 2008 was a project describing the semisynthesis of a GPI-anchored prion–protein construct, which hinged on the use of native chemical ligation to couple the GPI and protein fragments.[29] As these research efforts involved partial GPI structures (missing the phosphoglycerolipid), they are not covered in detail here, but the target GPI–prion conjugate **226** is shown in Scheme 41.

9. Synthesis of a *T. cruzi* GPI Anchor by the Vishwakarma Group (2005)

In 2000, Ferguson and coworkers discovered that GPIs from *T. cruzi*, the causative agent of Chagas' disease, have potent proinflammatory activity. This leads to various pathologies and, interestingly, was attributed to unsaturated fatty acids at the *sn*2 position of the GPI acylalkylglycerophosphate.[111] In 2005, the Vishwakarma group reported the first synthesis of a *T. cruzi* GPI anchor,[112] although the authors targeted a GPI containing saturated lipids that was more synthetically accessible and would be compatible with global benzyl-group protection (for the synthesis of *T. cruzi* GPIs bearing naturally occurring unsaturated lipids and a 2-aminoethyl phosphonate group, see Nikolaev's work in Section III.1).

Vishwakarma's chemical synthesis of the GPI anchor **227** adopted a convergent strategy resembling previous syntheses by other groups (Scheme 42). The key intermediate **228** was obtained from compounds **230** and **231**, the former of which was constructed in a [2+2] fashion using disaccharides **232** and **233** rather than the

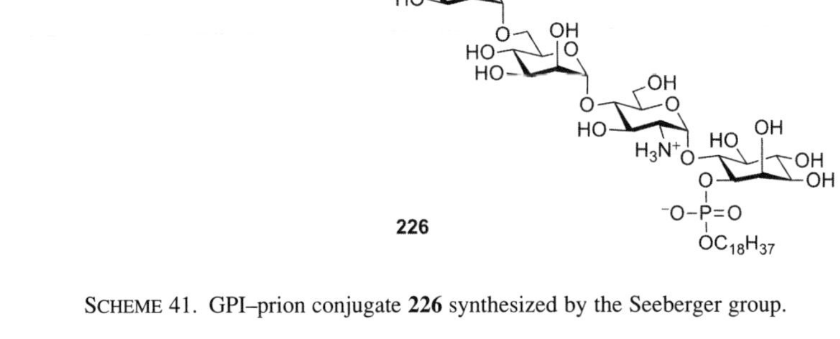

SCHEME 41. GPI–prion conjugate **226** synthesized by the Seeberger group.

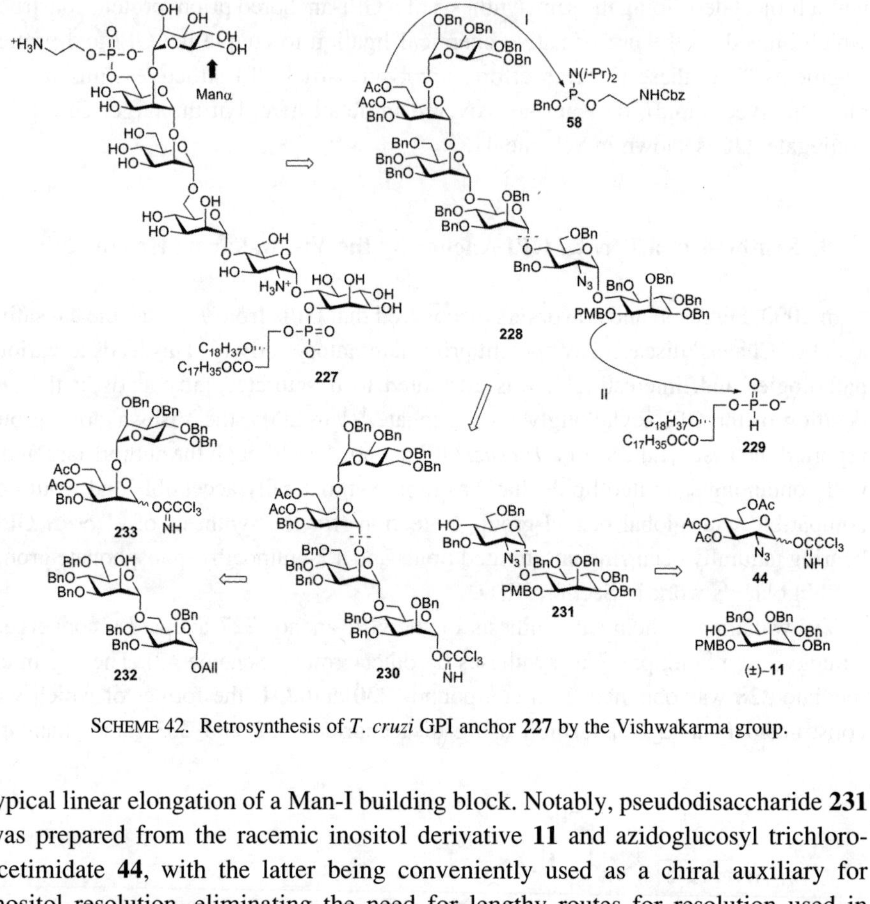

SCHEME 42. Retrosynthesis of *T. cruzi* GPI anchor **227** by the Vishwakarma group.

typical linear elongation of a Man-I building block. Notably, pseudodisaccharide **231** was prepared from the racemic inositol derivative **11** and azidoglucosyl trichloroacetimidate **44**, with the latter being conveniently used as a chiral auxiliary for inositol resolution, eliminating the need for lengthy routes for resolution used in previous GPI syntheses. Glycosyl NPOE[73] and trichloroacetimidate donors[63] were used for glycosylations, while both phosphoramidite[60] and *H*-phosphonate[44] methods were used for phosphorylations.

The [2+2] construction of tetramannosyl donor **230** from dimannosides **232** and **233** is shown in Scheme 43. Allyl α-D-mannopyranoside (**128**) was used as a common starting material for three of the four mannose building blocks. En route to **232**, compound **128** was advanced by tritylation at O-6, perbenzylation at the O-2, 3, and 4 positions, and detritylation to give **234**. Subsequent α-stereoselective glycosylation of **234** by the mannosyl NPOE **72** in the presence of NIS and TESOTf followed by debenzoylation furnished compound **232** (92%, two steps). The other dimannoside **233** was accessed from monomers **235** and **236**, each of which arose from allyl

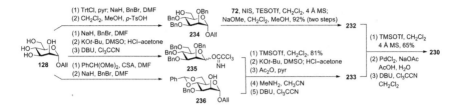

SCHEME 43. Synthesis of tetramannosyl donor **230** by the Vishwakarma group.

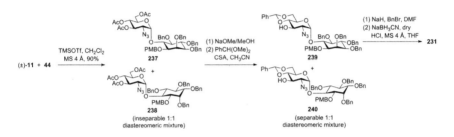

SCHEME 44. Synthesis of pseudodisaccharide **231** by the Vishwakarma group.

mannoside **128**, the first by successive benzylation, deallylation, and 1-*O*-trichloroacetimidation and the second by sequential 4,6-*O*-benzylidenation and selective benzylation at O-3. Compounds **235** and **236** were then joined via Schmidt glycosylation in 81% yield. A series of subsequent protecting-group manipulations were used to generate **233**, including concomitant deallylation/benzylidene hydrolysis on treatment with *t*-BuOK then HCl, peracetylation, selective anomeric deacetylation with MeNH$_2$, and finally treatment with DBU and trichloroacetonitrile. The [2+2] coupling of **232** and **233** under activation by TMSOTf proceeded in moderate (65%) yield, perhaps because of the lack of a participating 2-*O*-acyl group on the donor. Subsequent transformation into the tetramannosyl donor **230** was completed by Pd-mediated anomeric deallylation and treatment with DBU and trichloroacetonitrile.

In the preparation of pseudodisaccharide **231** (Scheme 44), the azidoglucosyl group was used as an efficient chiral auxiliary, and this eliminated the yield-depleting techniques for inositol resolution that are traditionally used. The racemic mixture of inositols (±)-**11** was treated with the azido glucosyl trichloroacetimidate **44** in the presence of TMSOTf to generate a 1:1 mixture of diastereomers **237** and **238**. The high α-stereoselectivity achieved using donor **44**, as previously reported by Schmidt and other groups, prevented the formation of a complex mixture of products. As **237** and **238** were inseparable at this stage, they were subjected to deacetylation followed by 4,6-*O*-benzylidenation, which produced the separable diastereomers **239** and **240**. The desired isomer **239** was carried forward to compound **231** by benzylation and regioselective benzylidene cleavage. This approach for resolving inositol enantiomers was efficient and should be useful in future GPI syntheses.

Vishwakarma's completion of the synthesis of GPI anchor **227** (Scheme 45) commenced with the coupling of **230** and **231**, promoted by TMSOTf, to afford the α-linked pseudohexasaccharide **228** in 70% yield. The product was designed to permit orthogonal protection of the O-4 and O-6 positions of Man-III to allow first the installation of phosphoethanolamine and second the synthesis of 4-deoxy Man-III GPI analogues for biological studies. In this synthesis, the Man-III moiety was subjected to deacetylation, 6-*O*-silylation, 4-*O*-benzylation, and finally desilylation. This lengthy sequence furnished intermediate **241**, which was then treated with

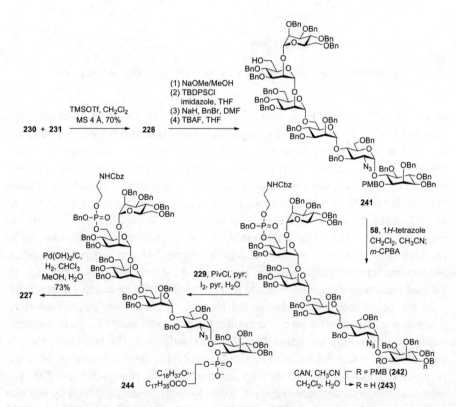

SCHEME 45. Completion of *T. cruzi* GPI anchor **227** by the Vishwakarma group.

phosphoramidite **58** in the presence of 1*H*-tetrazole, followed by *in situ* oxidation with *m*-CPBA, to install the phosphoethanolamine group. Phospholipidation of the inositol was accomplished by CAN-mediated oxidative cleavage of the PMB group, followed by coupling with building block **229** using the *H*-phosphonate method. Global deprotection by hydrogenolysis over Pd(OH)$_2$/C provided the target GPI **227** in 73% yield.

Vishwakarma and coworkers thus accomplished the first chemical synthesis of a *T. cruzi* GPI anchor. Overall, standard building blocks, assembly strategies, and carbohydrate synthetic methods were used to access the target. Of note was the efficient preparation of the pseudodisaccharide, which was highlighted by inositol resolution using the azido glucosyl group as a chiral auxiliary.

10. Synthesis of a Fully Phosphorylated and Lipidated Human Sperm CD52 Antigen GPI Anchor by the Guo group (2007)

In 2007, Wu and Guo reported the total synthesis of a fully phosphorylated and lipidated GPI anchor of the human sperm CD52 antigen,[113,114] which followed their 2003 synthesis of a related CD52 antigen GPI that contained an acylated inositol but lacked the phosphoethanolamine group at O-2 of Man-I of the natural product (see Section II.6). Thus, the target molecule **245**, bearing an additional phosphoethanolamine unit, was synthesized by a similar convergent strategy (Scheme 46). To install both phosphoethanolamine groups, a late-stage double phosphorylation using phosphoramidite **58** was performed on the intermediate GPI diol **246**, which in turn was accessed from trimannose thioglycoside **247** and phospholipidated pseudodisaccharide **248**. The latter compound was obtained by consecutive glycosylation, lipidation, and phospholipidation of inositol **249** by building blocks **104**, **250**, and **251**. For acylation of the inositol, the unsaturated hexadecane-9-enoyl group was used to act as either a latent aldehyde (on oxidative cleavage of the C—C bond) for conjugation chemistry or a latent hexadecanoyl group (on reductive global deprotection).

The synthesis of phospholipidated pseudodisaccharide **248** is shown in Scheme 47. The required inositol derivative **249** was readily accessed from the previously reported compound **168** by acetylation followed by deallylation. Reminiscent of a previous route by Ley (see Scheme 19), glycosylation of **249** by the azido glucosyl bromide **104** under Lemieux conditions[42] generated pseudodisaccharide **252** in moderate (56%) yield but with complete α-stereoselectivity. Deacetylation at O-2 of the inositol was followed by esterification with hexadec-9-enoic acid in the presence of DCC/DMAP to give **254**, which then underwent removal of the PMB group with

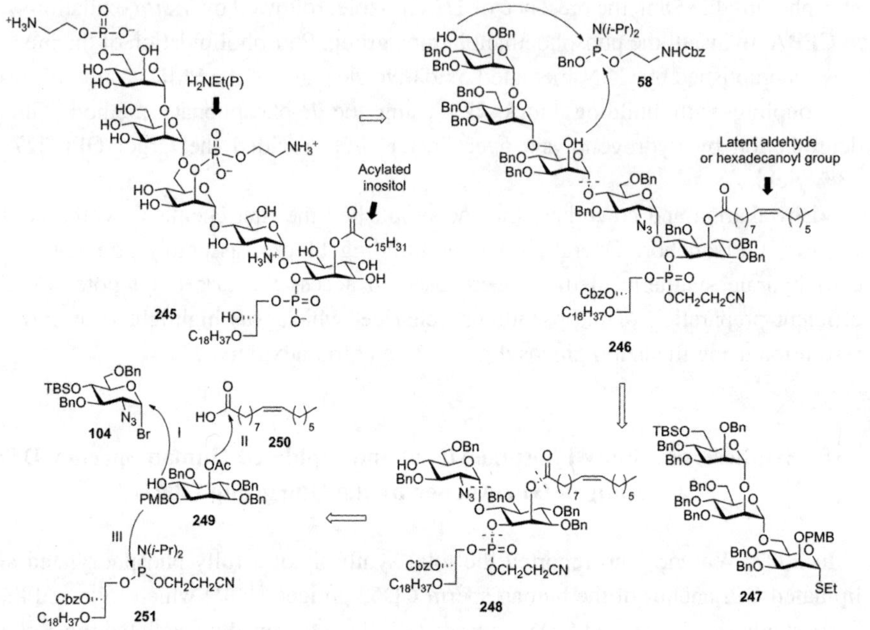

SCHEME 46. Retrosynthesis of fully phosphorylated and lipidated human sperm CD52 antigen GPI anchor **245** by the Guo group.

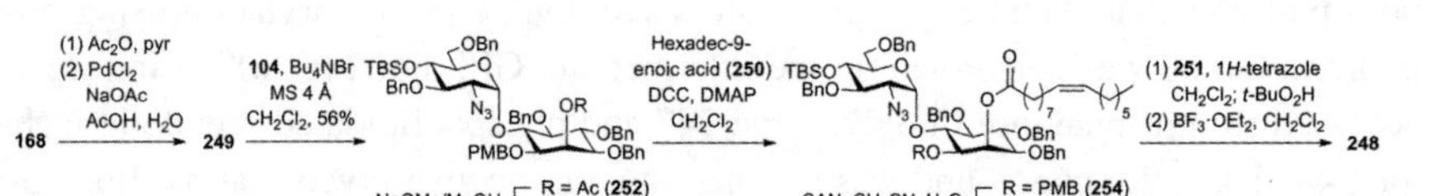

SCHEME 47. Synthesis of pseudodisaccharide **248** by the Guo group.

CAN. Phospholipidation of **255** using phosphoramidite **251** in the presence of 1*H*-tetrazole followed by *t*-BuOOH-mediated oxidation provided an intermediate phosphotriester, which was subjected to desilylation with boron trifluoride etherate to give **248**. Phospholipidation was performed at this stage rather than later, so as to prevent the cyclophosphitamidation side reaction previously discussed (see Scheme 26).

Completion of the target molecule is shown in Scheme 48. Prior to the key glycosylation, the trimannose thioglycoside **247** (synthesized according to Guo's previous route, see Scheme 27) was converted into trichloroacetimidate **256** via NIS-mediated hydrolysis of the thioacetal followed by reaction with DBU and

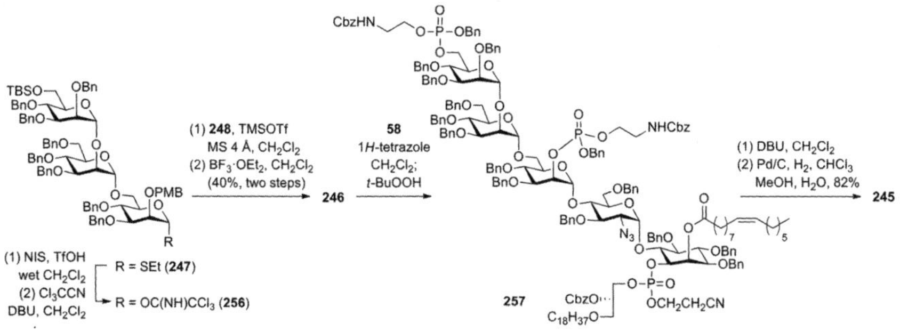

SCHEME 48. Completion of fully phosphorylated and lipidated human sperm CD52 antigen GPI anchor 245 by the Guo group.

trichloroacetonitrile. The union of trichloroacetimidate 256 and pseudodisaccharide 248 under Schmidt glycosylation conditions formed the desired intermediate, which was immediately treated with boron trifluoride etherate to remove simultaneously the TBS and PMB groups. This sequence provided the α-pseudopentasaccharide diol 246 in moderate (40%) yield over two steps. Double phosphorylation of 246 using phosphoramidite 58, a difficult task, proceeded in 50% yield to give the fully protected GPI 257. Global deprotection was accomplished by treatment with DBU to remove the cyanoethoxyl group, followed by debenzylation using Pd-catalyzed hydrogenolysis to afford the target GPI 245 in 82% yield.

In summary, the synthesis of 245, a fully phosphorylated and lipidated GPI, was guided by Guo's 2003 synthesis of a CD52 antigen GPI while using an orthogonal protection strategy that enabled late-stage double phosphorylation. The target molecule and related GPI derivatives were used to investigate their relative abilities to bind CAMP factor, a pore-forming toxin secreted by *Streptococcus agalacticae* that is thought to utilize GPI anchors as its binding partners.[114] Initial results suggested that both the intact GPI anchor 245 and its corresponding truncated phospholipidated pseudodisaccharide bound CAMP factor with similarly strong affinities. In general, such studies using well-characterized synthetic GPIs and GPI derivatives should prove instrumental in elucidating GPI structure–activity relationships in various contexts.

III. DIVERSITY-ORIENTED APPROACHES TO GPI SYNTHESIS

The so-called classic GPI syntheses thus far detailed here were accomplished by strategies having limited potential for structural diversity, and this can be attributed primarily to the global-protection tactics used. Traditionally, carbohydrate chemists have used benzyl groups for permanent protection of hydroxyl groups because of their ease of installation, stability under a broad range of reaction conditions, numerous choices for orthogonal protection, and mild conditions for global deprotection. However, the latter feature, deprotection via Pd-catalyzed hydrogenolysis, constitutes a considerable synthetic shortcoming given its incompatibility with many important functional groups. In this category are alkenes, alkynes, azides, thiols, thioethers, and other potentially reducible or catalyst-poisoning functionalities. Therefore, global protection by benzyl groups in GPI synthesis precludes the incorporation of biologically important unsaturated fatty acids and indispensable functional handles such as "click" chemistry, imaging, and affinity-purification tags. Furthermore, the natural diversity of GPIs arising from core structure modifications, such as additional glycans, phosphoethanolamine groups, and variation in lipids (see Fig. 1), has prompted the development of newer general methods for efficient synthesis of a range of GPIs from common building blocks. Research at the recent forefront of GPI synthesis has been focused on developing diversity-oriented synthetic strategies to address these drawbacks. Progress in this area by the Guo, Nikolaev, and Seeberger groups is discussed in this section.

1. Synthesis of *T. cruzi* GPI Anchors Containing Unsaturated Lipids by the Nikolaev Group (2006)

As already mentioned, Ferguson and co-workers identified GPIs from *T. cruzi* exhibiting potent proinflammatory activity that was thought to be associated with unsaturated fatty acids in the $sn2$ position of the GPI acylalkylglycerophosphate.[111] In an effort to establish rigorous structure–activity relationships in this system, the Nikolaev group set out to synthesize *T. cruzi* GPIs containing unsaturated fatty acids that are present in the natural structure.[115] The GPI anchors **258** and **259** (Scheme 49), bearing the potentially crucial $sn2$-oleoyl [(Z)-octadec-9-enoyl] and linoleoyl (*cis,cis*-octadec-9,12-dienoyl) groups, respectively, were targeted for chemical synthesis.

Nikolaev's 2006 publication was the first total synthesis of GPIs containing unsaturated lipids and accordingly required the development of a novel strategy to obviate the need for reductive conditions employed in global benzyl-group deprotection. Acyl-based permanent protection of hydroxyl groups in the form of benzoic esters was chosen, despite potential compatibility issues with the ester-linked lipid in the acylalkylglycerophosphate moiety. The authors' hypothesis was that, upon deprotection by mild base in a polar solvent, the nascent amphiphilic GPI intermediate may undergo micelle formation, thereby protecting the bond of the fatty ester during

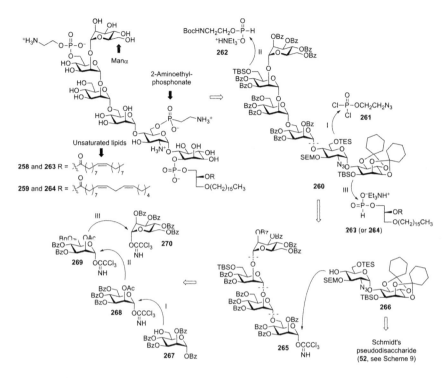

SCHEME 49. Retrosynthesis of *T. cruzi* GPIs **258** and **259** by the Nikolaev group.

selective removal of the benzoic esters. Further increasing the synthetic challenge, the target GPIs had a 2-aminoethylphosphonate group linked to O-6 of the glucosamine component, which is a parasite-specific modification of some GPIs.

The retrosynthesis of GPIs **258** and **259** is shown in Scheme 49. Both targets were accessed from the pseudohexasaccharide **260**, which was poised to undergo a triad of phosphorylation events with building blocks **261**, **262**, and **263/264**, to install the 2-aminoethylphosphonate, phosphoethanolamine, and phosphoglycerolipid components, respectively. Intermediate **260** was assembled convergently from the benzoyl-protected tetramannosyl donor **265** and a derivative of Schmidt's pseudodisaccharide **52**, which in turn were built up from their respective monomers. For all steps involving glycosidic bond formation, the Schmidt method[61] using glycosyl trichloroacetimidate donors was applied.

The tetramannose fragment **265** was prepared by the stepwise joining of mannose monosaccharides **267–270** (Scheme 50), a process entailing sequential stereoselective α-mannosylations, each aided by neighboring-group participation. The Man-I

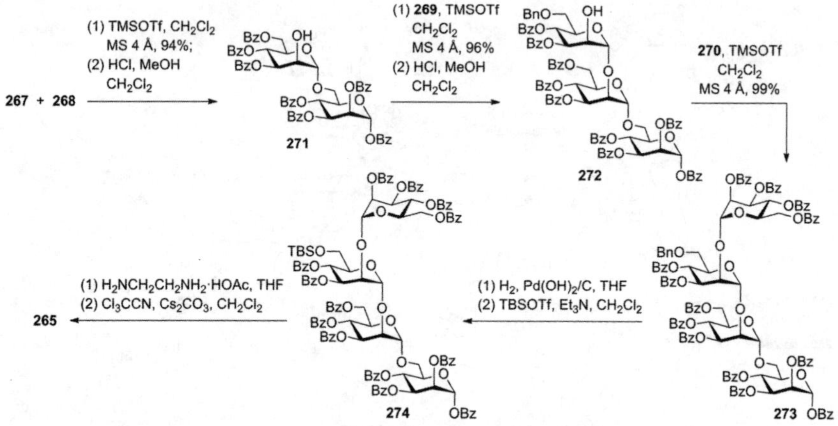

SCHEME 50. Synthesis of tetramannosyl donor **265** by the Nikolaev group.

component (**267**) was coupled with the Man-II component (**268**) in the presence of TMSOTf, providing an α-dimannoside in 94% yield that was subjected to selective deacetylation at O-2 of Man-II with HCl/MeOH to give compound **271**. The same two-step sequence was used to extend the glycan chain with the Man-III component (**269**, 96% yield) and expose the O-2 position of Man-III. Subsequent reaction of this site with the Man-IV component **270**, again promoted by TMSOTf, afforded fully α-configured tetramannose product **273** in nearly quantitative yield. The O-6 benzyl group on the Man-III precursor was then exchanged for a TBS group via Pd-catalyzed hydrogenolysis, followed by treatment with TBSOTf/Et$_3$N. Finally, the anomeric benzoyl group of the Man-I precursor was converted into a trichloroacetimidate by regioselective debenzoylation with ethylenediamine followed by treatment with Cl$_3$CCN/Cs$_2$CO$_3$.

For the pseudodisaccharide fragment, the authors converted known compound **52**, previously described by Schmidt (see Scheme 9 for its preparation),[58] into a suitably protected derivative containing only acid-labile groups whose late-stage removal would not interfere with the unsaturated lipid chains. However, rather than perform the separation of inositol diastereomers **43** and **51** as reported in the literature,[58,116] it was found that conducting the glycosylation directly on the diastereomeric mixture afforded readily separable pseudodisaccharide products **52** and **275** (Scheme 51), indicating that the azidoglucose component acted as a supplementary chiral auxiliary. Notably, this approach relied on the exceptional α-stereoselectivity observed in the glycosylation reaction previously fine tuned by Schmidt[58]; otherwise, a complex and possibly inseparable mixture of isomers would have been generated.

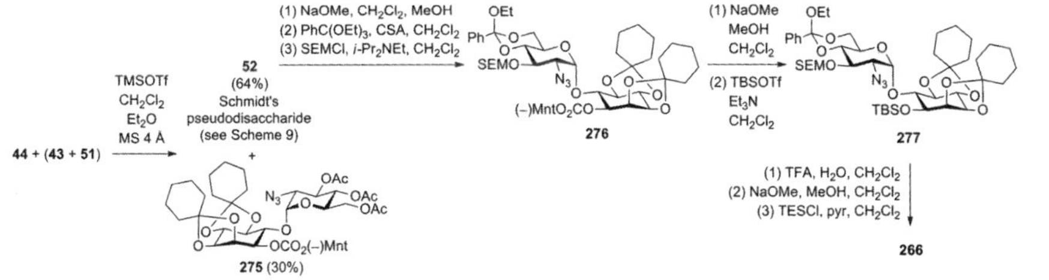

SCHEME 51. Synthesis of pseudodisaccharide **266** by the Nikolaev group.

From Schmidt's pseudodisaccharide **52**, several protecting-group manipulations were required to obtain **266**. First, the azidoglucose component was deacetylated and then treated with PhC(OEt)$_3$ under acidic conditions to form a 4,6-orthoester derivative. Subsequent protection of the free O-3 position with an acid-labile 2-trimethylsilylethoxymethyl (SEM) group, using SEMCl/i-Pr$_2$Net, gave compound **276**, which was subjected to cleavage of the inositol 1-O-(−)-menthylcarbonate group with NaOMe. Protection of the exposed hydroxyl group with a TBS ether was accomplished by treatment with TBSOTf/Et$_3$N to give the protected pseudodisaccharide **277**. The final sequence involved removal of the 4,6-orthoester from the azidoglucose, followed by selective triethylsilylation at O-6. This synthetic route from **52**, while somewhat lengthy, was very efficient, providing **266** in 50% yield over seven steps.

Elaboration of the tetramannose and pseudodisaccharide fragments to the target GPI anchors is shown in Scheme 52. The key glycosylation reaction between the tetramannosyl trichloroacetimidate **265** and pseudodisaccharide **266** proceeded smoothly, giving the fully protected α-pseudohexasaccharide **260** in 71% yield. At this stage, three phosphorylation events were required to decorate the GPI with the appropriate structures found in GPIs of *T. cruzi*. First, the unique 2-aminoethylphosphonate group was installed at O-6 of the azidoglucose in three steps, including selective cleavage of the TES ether using AcOH-buffered TBAF, 1H-tetrazole-promoted phosphorylation with 2-azidoethylphosphonodichloridate **261** (generated from 2-bromoethylphosphonate in two steps), and methanolysis of the intermediate chlorophosphonate to give **278**. Subsequent Staudinger reduction and treatment with Boc anhydride converted the azido groups into N-Boc-protected amino groups. The next transformation involved selective O-6 desilylation of the Man-III TBS group with triethylamine trihydrofluoride, which exposed this position for introduction of the phosphoethanolamine unit by using H-phosphonate derivative **262** in the presence of pivaloyl chloride, followed by *in situ* iodine-mediated oxidation to

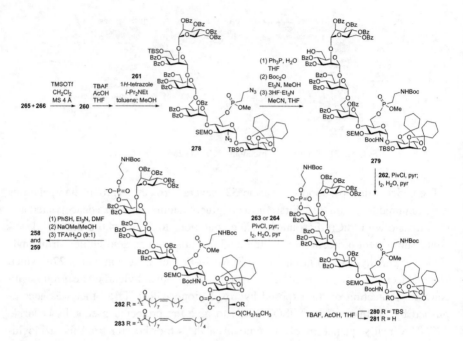

SCHEME 52. Completion of *T. cruzi* GPI anchors **258** and **259** by the Nikolaev group using acyl-based global protection.

phosphodiester **280**. Subsequent removal of the TBS from O-1 of the inositol under strong conditions, namely TBAF/AcOH at 55 °C, set the stage for installation of the unsaturated lipid-containing acylalkylglycerophosphate components. Again, the *H*-phosphonate method was successfully employed, resulting in the phospholipidation of **281** with precursors **263** or **264** to introduce oleic [(*Z*)-octadec-9-enoic] or linoleic (*cis,cis*-octadec-9,12-dienoyl) acids, respectively, into the GPI.

The global deprotection of fully protected GPI intermediates **282** and **283** required three reactions. First, demethylation of the 2-aminoethylphosphonate group was effected by using PhSH and Et₃N. The second and most critical deprotection step required base-promoted debenzoylation, a process that risked cleavage of the GPI fatty esters. Nikolaev used 0.05 M NaOMe in MeOH, which successfully delivered the products, albeit in moderate (38–40%) yields, most probably a result of poor selectivity. Finally, the purified intermediates were subjected to treatment with aqueous 90% TFA to hydrolyze the remaining acetal and *N*-Boc protecting groups, which proceeded with high efficiency to afford, after purification by reverse-phase chromatography, the target GPIs **258** and **259**.

Overall, use of benzoyl groups as an acyl-based global-protection strategy to access GPIs containing unsaturated lipids was successful, but compatibility issues with acylglycerophosphates render this approach a precarious one for synthesizing functionalized GPIs. Shortly after their 2006 publication, Nikolaev and co-workers adapted their synthetic strategy by using acetal- and silyl-based protection *in lieu* of acyl-based protection to address this issue.[117] In this work, a convergent [3+3] strategy was used to synthesize GPI **258** from the trimannose thioglycoside **284** and pseudotrisaccharide **285**, which were both permanently protected with acetal or silyl groups (Scheme 53). Briefly, these two components combined in the presence of methyl triflate and 2,6-di-*tert*-butyl-4-methylpyridine (DTBMP), with subsequent selective removal of the TES group using TBAF and AcOH to generate α-pseudohexasaccharide **286** in 75% yield. The 2-aminoethylphosphonate, phosphoethanolamine, and alkyloleoylglycerophosphate groups were installed as previously to arrive at the fully protected GPI **290**. Final deprotection was accomplished by a two-step sequence comprising desilylation with triethylamine trihydrofluoride and TFA-promoted acetal hydrolysis. The target oleoyl [(Z)-octadec-9-enoyl] GPI **258** was obtained in enhanced (67%) yield over two steps, although the desilylation step required 18 days to complete.

In summary, Nikolaev's innovative strategies for access to diversely structured GPIs were used for the synthesis of biologically important GPIs of *T. cruzi* containing unsaturated lipids. Both strategies were focused on using alternative protecting groups that would not require Pd-catalyzed hydrogenolysis for their removal. Although acyl-based protection probably led to compatibility issues with the acylalkylglycerophosphate during global deprotection with NaOMe, an improved strategy based on acetal and silyl protection was developed that featured a higher-yielding, but time-consuming, global deprotection. Synthetic GPIs **258** and **259** were subjected to a preliminary biological evaluation, which showed activity similar to that of the naturally occurring GPIs in a Toll-like receptor stimulation assay.[115] More detailed biological studies of these compounds are ongoing.

2. Synthesis of a GPI Anchor Containing Unsaturated Lipid Chains by the Guo Group (2010)

In an effort to develop a robust approach for the synthesis GPI anchors bearing unsaturated lipids and other important functionalities, the Guo group focused on a global hydroxyl-protection strategy based on the PMB group, which can be removed under relatively mild acidic conditions, such as 5–10% TFA, or oxidation, as with

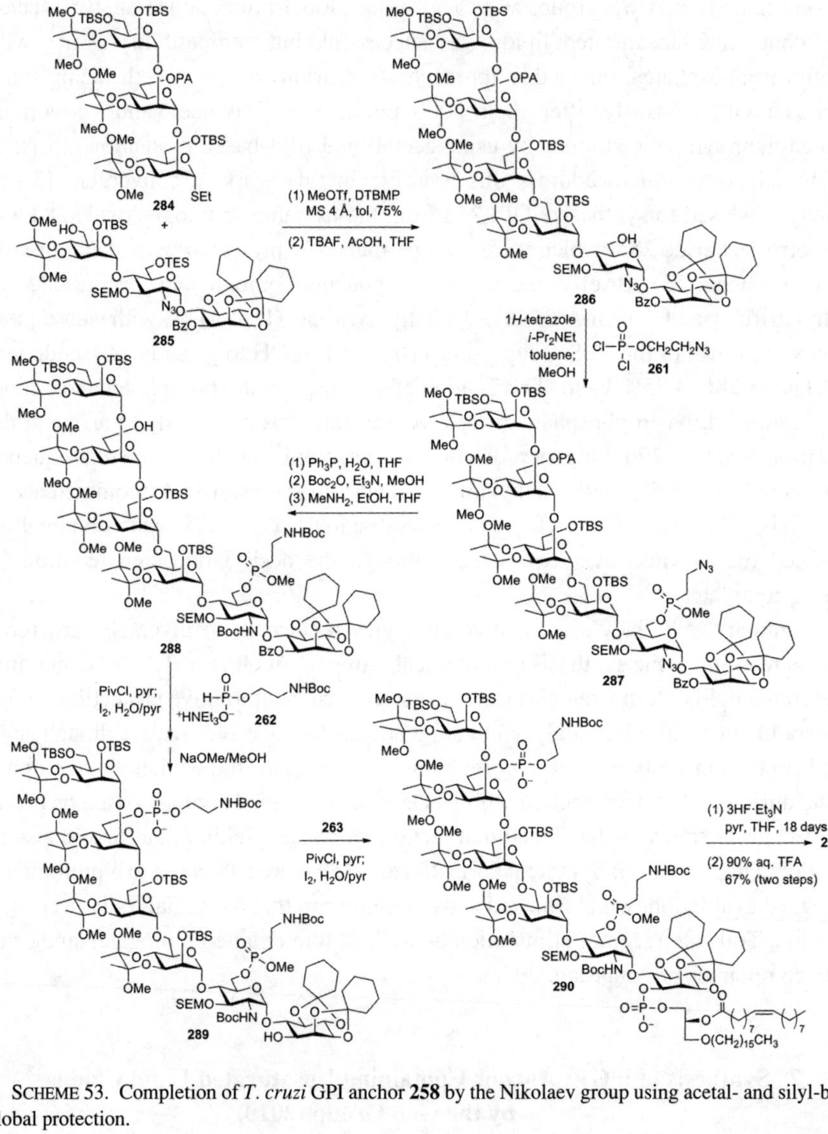

SCHEME 53. Completion of *T. cruzi* GPI anchor **258** by the Nikolaev group using acetal- and silyl-based global protection.

CAN or 2,3-dichloro-5,6-dicyano-1,4-benzoquinone (DDQ). It was expected that PMB protection would offer a major improvement in functional-group tolerance during the global deprotection, thus allowing access to highly complex and sensitively functionalized GPI anchors.

CHEMICAL SYNTHESIS OF GLYCOSYLPHOSPHATIDYLINOSITOL ANCHORS 195

In 2010, Swarts and Guo provided a proof of concept for the PMB-protection strategy by synthesizing a GPI anchor containing unsaturated lipids,[118] specifically a dioleoylglycerophosphate moiety. The retrosynthetic analysis, shown in Scheme 54, was guided by previous convergent syntheses from the Guo group. The target GPI anchor **291** was accessed from intermediate **292**, which was formed in a key glycosylation reaction between the trimannosyltrichloroacetimidate donor **294** and pseudodisaccharide **295**. The trimannose fragment, joined together from PMB-protected monomers **296–298**, contained orthogonal silyl ether protecting groups on the Man-I and Man-III components for late-stage phosphorylation with the phosphoethanolamine precursor **293**. The pseudodisaccharide fragment, already containing the dioleoylglycerophosphate group at O-1 of the inositol, was synthesized from the optically pure inositol **299**, the azido glucosyl trichloroacetimidate **300**, and the phospholipid precursor **301**.

The preparation of trimannoside **294** involved the α-stereoselective coupling of mannose building blocks **296–298**, which were prepared from a common intermediate, mannose pentaacetate **302** (Scheme 55). The Man-I building block, thioglycoside alcohol **296** bearing a free 6-OH group, was obtained from diol **303**[119] via sequential

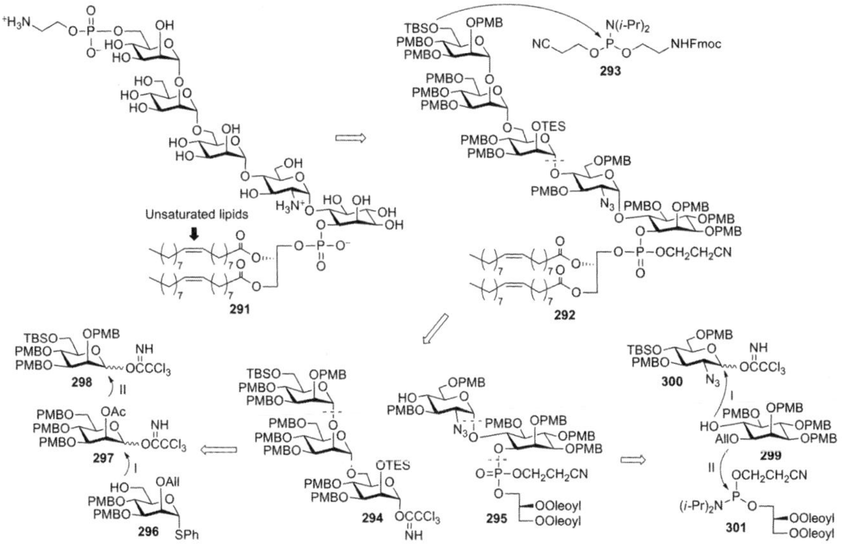

SCHEME 54. Retrosynthetic analysis of a GPI anchor bearing unsaturated lipid chains by the Guo group using the PMB-protection strategy.

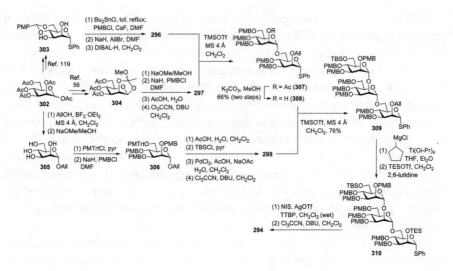

SCHEME 55. Synthesis of PMB-protected mannose derivatives and elaboration to the trimannosyl donor **294** by the Guo group.

stannylene acetal-directed regioselective *p*-methoxybenzylation at O-3, allylation at O-2, and regioselective opening of the *p*-methoxybenzylidene ring using DIBAL-H. The Man-II component **297** was prepared from **302** via orthoester **304**,[56] which after deacetylation and methoxybenzylation was treated with acetic acid to open the orthoester ring, providing a hemiacetal that was converted into the corresponding trichloroacetimidate. Synthesis of mannose derivative **298** started with a BF$_3$-promoted glycosidation of **302** with allyl alcohol. Subsequent deacetylation gave tetraol **305**, which then underwent differentiation of the 6-position by reaction with *p*-methoxytrityl (PMTrt) chloride and methoxybenzylation of the 2-, 3-, and 4-positions to generate **306**. After exchanging the PMTrt protecting group for a TBS group, Pd-catalyzed deallylation freed up the anomeric hydroxyl group for transformation into trichloroacetimidate **298** by treatment with trichloroacetonitrile and DBU. In general, the preparation of PMB-protected mannose building blocks was comparable to the corresponding benzyl-protected variants in terms of both strategy and efficiency.

With monosaccharides **296–298** in hand, the trimannosyl donor **294** was constructed by using the Schmidt glycosylation method.[61] The coupling of Man-I component **296** and Man-II component **297** proceeded well using catalytic TMSOTf in CH$_2$Cl$_2$, and complete α-stereoselectivity was imparted by the participating acetyl group at O-2. Deacetylation with K$_2$CO$_3$/MeOH exposed the O-2 position, resulting in the dimannoside alcohol **308** in 66% yield over two steps. Coupling of **308** with the

Man-III trichloroacetimidate **298**, also under Schmidt conditions, gave trimannoside **309** in a 5:1 α:β ratio, a figure that was improved to exclusively α (76% yield) upon changing the reaction solvent to diethyl ether. The trimannose thioglycoside **309** was also used as a divergence point for the preparation of other functionalized GPI anchors (see Section III.4). In this instance, compound **309** was elaborated to the trichloroacetimidate **294** in four steps, including titanium-mediated deallylation using Cha's protocol[120] and triethylsilylation to change the protecting group at O-2 of the Man-I component, followed by treatment with NIS/AgOTf/2,4,6-tri-*tert*-butylpyrimidine (TTBP) to hydrolyze the anomeric thioacetal, and final treatment with trichloroacetonitrile and DBU.

For the pseudodisaccharide fragment, the PMB-protected inositol derivative **299** was required (Scheme 56). The O-1 and O-6 positions of racemic diol (±)-**20** were simultaneously protected as allyl ethers, which were unaffected by the subsequent cyclohexylidene-group removal and *p*-methoxybenzylation reactions. Next, deallylation of the bis-allyl ether (±)-**311** using Cha's protocol[120] provided inositol derivative (±)-**312** in 85% yield. For differentiation of the PMB-protected diol (±)-**312** at O-1 and O-6, regioselective stannylene acetal-directed allylation was used to generate compound (±)-**299** in 72% yield. Enantiomeric resolution of racemate (±)-**299** was achieved by acylation of the free O-6 position with the chiral reagent (1*S*)-(−)-camphanic chloride to yield HPLC-separable diastereomers. Finally, the desired, purified diastereomer (−)-**313** was subjected to saponification to provide the enantiomerically pure inositol **299** in 44% yield over two steps (maximum yield 50%).

The synthesis of phospholipidated pseudodisaccharide **295** is depicted in Scheme 57. Preparation of the azido glucosyl donor **300** began with the transformation of compound **314**[121] into the allyl glycoside **315** via BF$_3$-promoted glycosidation with AllOH, deacetylation, and formation of the 4,6-*p*-methoxybenzylidene acetal. A PMB group was installed at the O-3 position, which was followed by regioselective

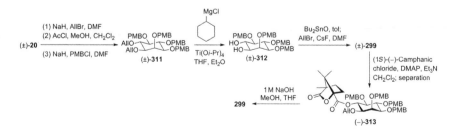

SCHEME 56. Synthesis of inositol derivative **299** by the Guo group.

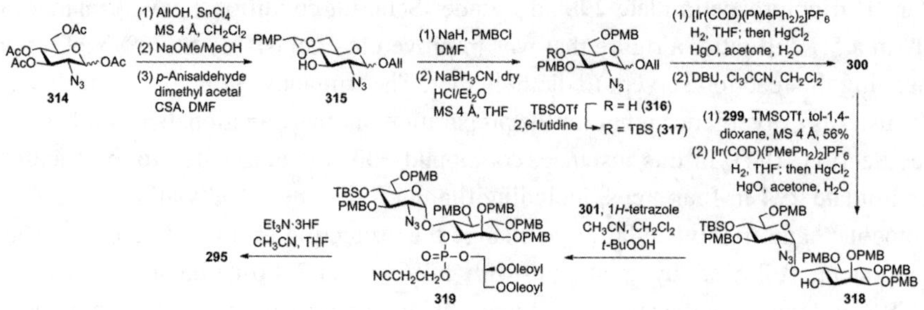

SCHEME 57. Synthesis of pseudodisaccharide **295** by the Guo group.

opening of the *p*-methoxybenzylidene ring using dry HCl and NaBH$_3$CN to provide compound **316**. After use of TBSOTf to protect the O-4 position to afford compound **317**, the anomeric allyl ether was removed by Ir(I)-catalyzed isomerization to the corresponding vinyl ether and subsequent hydrolysis by Hg(II).[122] The resulting hemiacetal was converted into trichloroacetimidate **300** by using trichloroacetonitrile and DBU.

Coupling of the inositol derivative **299** and azido glucosyl donor **300** was performed in toluene–1,4-dioxane in the presence of TMSOTf to generate the pseudodisaccharide intermediate in 80% yield as an inseparable α,β mixture (α:β = 2.3:1). Subsequent deallylation of the mixture using the Ir–Hg protocol produced the α-pseudodisaccharide **318**, which was readily separable from the accompanying β anomer. Phospholipidation of the inositol at O-1 using phosphoramidite **301**, followed by *in situ* oxidation with *t*-BuOOH, installed the dioleoylglycerophosphate moiety to produce compound **319**. Treatment of this intermediate with triethylamine trihydrofluoride cleaved the O-4 TBS group of the azido glucose derivative, which required several days for completion but cleanly produced the desired pseudodisaccharide **295**.

The final stage of the synthesis centered on union of the trimannose and pseudodisaccharide fragments (Scheme 58). The trimannose trichloroacetimidate **294** reacted smoothly with **295** in the presence of TMSOTf to afford the desired α-pseudopentasaccharide **292** in 64% yield. Overnight treatment of **292** with Et$_3$N·3HF removed the TBS and TES groups to provide diol **320**. Installation of the phosphoethanolamine unit by treatment with phosphoramidite **293** for a short period (1 h) was selective for the O-6 position of Man-III to give **321**. Alternatively, the diol could be used to access a fully phosphorylated GPI bearing two phosphoethanolamine groups. To obtain the target GPI anchor, compound **291**, a three-step, one-flask protocol was developed to efficiently remove all of the protecting groups from **321** in under 4 h: first, Zn-mediated reduction of the azide; second, removal of the base-

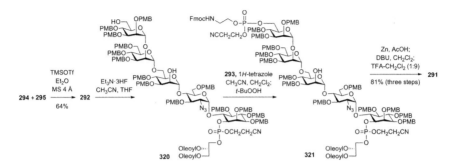

SCHEME 58. Completion of a GPI anchor bearing unsaturated lipid chains (**291**) using the PMB-protection strategy by the Guo group.

labile Fmoc and cyanoethoxyl groups with DBU; and third, hydrolysis of all PMB ethers with 10% TFA. The target GPI **291** was obtained in 81% yield over the final three steps.

This proof-of-concept synthesis established the PMB-protection strategy as a general approach to the synthesis of functionalized GPIs and potentially other complex functionalized glycoconjugates. Overall, the preparation of PMB-protected monosaccharide and inositol building blocks was conducted efficiently, and the performance of these intermediates in Schmidt glycosylation reactions was comparable to their benzyl-protected counterparts. The highlight of this convergent synthesis was the three-step, one-flask reaction that afforded the target GPI **291** in 81% yield in under 4 h. In the next two sections, use of the PMB-protection strategy by the Guo group for access to a complex naturally occurring GPI anchor, as well as "clickable" GPI anchors, is discussed.

3. Synthesis of a Human Lymphocyte CD52 Antigen GPI Anchor Containing a Polyunsaturated Arachidonoyl Lipid by the Guo Group (2011)

The structural diversity of GPI lipids comes about by the "lipid-remodeling" process of GPI-anchored proteins.[14,123–125] GPI biosynthesis begins from cellular PI, which in mammalian systems typically contains a polyunsaturated arachidonoyl lipid (20:4) at the *sn*2 position. The arachidonoyl [(5Z,8Z,11Z,14Z)-icosa-5,8,11,14-tetraenoyl] group can be replaced by other lipids through a series of transformations in the Golgi apparatus, and as a result, most mature mammalian GPI-anchored proteins contain only saturated fatty acids. However, several GPI-anchored proteins bearing

2-arachidonoyl PI have been identified, including a major form of the human lymphocyte CD52 antigen.[98,126] The effects of unremodeled lipids on the function of lymphocyte CD52 and other GPI-anchored proteins have not been thoroughly explored, although in general it is expected that the structure of GPI lipids has important implications on the localization and function of GPI-APs. For example, a recent study has demonstrated that saturated fatty acids resulting from lipid remodeling are required for the association of mammalian GPI-APs with lipid rafts.[127]

To develop a strategy for accessing GPIs relevant to studying the effects of lipid remodeling and to continue synthetic studies of GPI anchors and the human CD52 antigen (see Sections II.6 and II.10), the Guo group targeted the lymphocyte CD52 GPI anchor **322** (Scheme 59) for total synthesis.[128] The target molecule contained modifications of the GPI core glycan present in the human lymphocyte CD52 antigen, including additional phosphoethanolamine and mannose groups linked to the O-2 positions of Man-I and Man-III, respectively. Most importantly, compound **322** contained an octadecanoyl-arachidonoyl-glycerophosphate component, which is present in the lymphocyte CD52 antigen and other GPI-anchored proteins that do not undergo lipid remodeling. To achieve the synthetically challenging incorporation of a

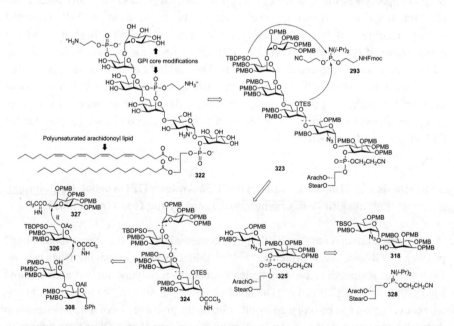

SCHEME 59. Retrosynthetic analysis of human lymphocyte CD52 antigen GPI anchor **322** containing a polyunsaturated arachidonoyl lipid by the Guo group.

highly sensitive polyunsaturated arachidonoyl lipid in GPI **322**, the PMB-protection strategy was used.

A convergent approach was used for the construction of GPI **322** that relied on a key glycosylation between PMB-protected coupling partners, namely the tetramannosyl donor **324** and the pseudodisaccharide acceptor **325** (Scheme 59). Silyl ether protecting groups at the O-2 position of Man-I and O-6 position of Man-III would be selectively removed at a late stage to enable double phosphorylation of these sites with phosphoramidite **293**. The glycosyl donor **324** would arise from sequential α-mannosylations of disaccharide **308** using monomers **326** and **327**. Acceptor **324** could be accessed through phospholipidation of compound **318** with phosphoramidite **328**, which would attach the arachidonoyl-containing phosphoglycerolipid. Guided by previous success, the Schmidt method[71] was used to couple stereoselectively all PMB-protected glycosylation partners. Some PMB-protected building blocks (**308**, **318**) from Guo's 2010 synthesis of a GPI anchor containing unsaturated lipids were also used in this synthesis, showing their versatility.

The preparation of PMB-protected tetramannosyl donor **324** is shown in Scheme 60. To obtain the Man-III building block **326**, the orthoester **304** was subjected to consecutive deacetylation, silylation at O-6, *p*-methoxybenzylation at O-3 and O-4, acid-promoted regioselective opening of the orthoester ring, and finally treatment of the resulting hemiacetal with trichloroacetonitrile and DBU. The resultant compound **326** was then treated with **308** in the presence of TMSOTf to furnish the α-trimannoside **329** in 72% yield. Deacetylation of **329** gave the acceptor **330**, which was glycosylated by the Man-IV donor **327** to afford tetramannoside **331** in 76% yield, although stereoselectivity for this reaction was low (α:β = 3:2). At this stage, the allyl group at O-2 of Man-I was exchanged for a TES group using Cha's deallylation protocol,[120] followed by treatment with TESOTf/2,6-lutidine to furnish compound **332**. Transformation of the anomeric thioacetal of **332** to the

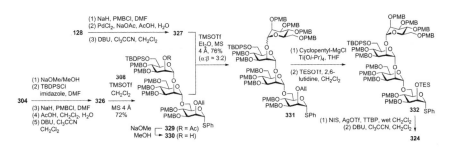

SCHEME 60. Synthesis of PMB-protected tetramannosyl donor **324** by the Guo group.

trichloroacetimidate **324** was accomplished by treatment with NIS/AgOTf/TTBP followed by trichloroacetonitrile and DBU.

Given the documented sensitivity of arachidonoyl-containing compounds to air and light (leading to oxidation of the alkene) and heat and base (leading to rearrangements),[129–131] care was taken to avoid these conditions in the preparation of phosphoramidite **328** and throughout the completion of the synthesis (Scheme 61). Phospholipidation of the afore-described PMB-protected pseudodisaccharide **318** was performed by using the arachidonoyl-containing phosphoramidite **328** in the presence of 1*H*-tetrazole. *In situ* oxidation with *t*-BuOOH, followed by removal of the TBS group with triethylamine trihydrofluoride, provided compound **325**. The key glycosylation between acceptor **325** and the tetramannosyl donor **324** was performed under conventional Schmidt conditions, leading to the α-pseudohexasaccharide **323** in moderate (39%) yield. Subsequent desilylation to remove the Man-I TES and Man-III TBDPS groups gave diol **333**, which underwent double phosphorylation with phosphoramidite **293** to provide the fully PMB-protected GPI **334**. Application of the previously described three-step, one-flask global-deprotection protocol involving consecutive treatments with Zn/AcOH, DBU, and 10% TFA delivered the target GPI anchor **322** in 90% yield over the final three steps (under 3 h of reaction time).

Construction of the lymphocyte CD52 GPI anchor **322** constituted the first total synthesis of a GPI anchor bearing unremodeled fatty acids, namely a polyunsaturated arachidonoyl lipid at the *sn*2 position of the PI moiety. Inclusion of the oxidation- and reduction-sensitive arachidonoyl lipid was made possible by utilizing global protection by the PMB group, which featured a mild and rapid three-step, one-flask deprotection protocol that gave the target GPI anchor in 90% yield in under 3 h. Furthermore, two additional structural modifications in the GPI core present in the

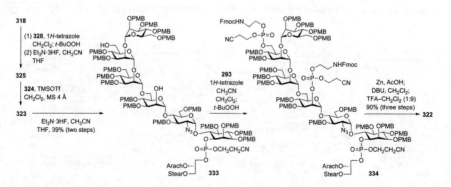

SCHEME 61. Completion of the human lymphocyte CD52 antigen GPI anchor **322** by the Guo group.

lymphocyte CD52 antigen, specifically additional phosphoethanolamine and mannose units, were incorporated into the target molecule, which further demonstrated the applicability of the PMB-protection strategy to the efficient synthesis of highly complex and sensitively functionalized GPIs.

4. Synthesis of "Clickable" GPI Anchors by the Guo Group (2011)

Given the rapid influx of techniques in chemical biology that use "click" chemistry for chemoselective conjugation, particularly the powerful [3 + 2] cycloaddition reactions between azides and alkynes/cyclooctynes,[132–135] the arming of GPIs with "click" tags provides a platform for developing novel approaches for exploring their biological properties and therapeutic applications (Fig. 2). For example, "clickable" GPIs could be deployed in studies involving molecular imaging, profiling of GPI-anchored proteins ("GPIomics"), and biophysical characterization of GPIs and GPI-anchored molecules on the surfaces of live cells. However, "click" chemistry tags and other useful functionalities, including many imaging and affinity-purification tags, are often incompatible with traditional protecting-group methodologies used in carbohydrate synthesis.

In 2011, Swarts and Guo used the PMB-protection strategy to synthesize "clickable" GPI anchors.[136] Retrosynthetic analysis (Scheme 62) suggested a convergent–divergent strategy that would optimize efficiency while enabling access to a diverse set of GPIs. Flexibility was emphasized in the synthetic design to

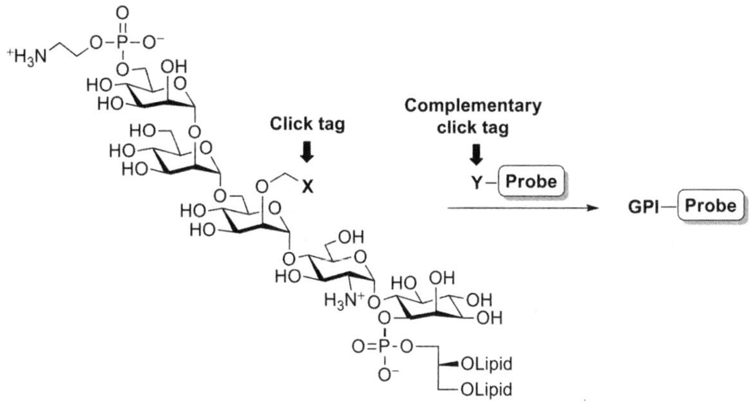

FIG. 2. Probing GPI anchors and GPI-anchored molecules with "click" chemistry.

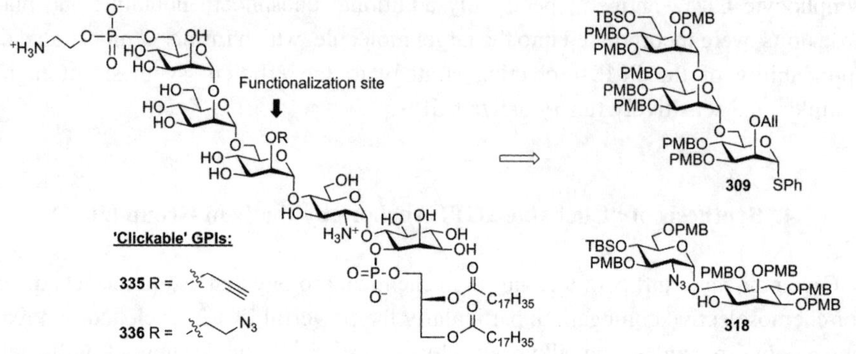

SCHEME 62. Convergent–divergent strategy for the synthesis of "clickable" GPI anchors by the Guo group.

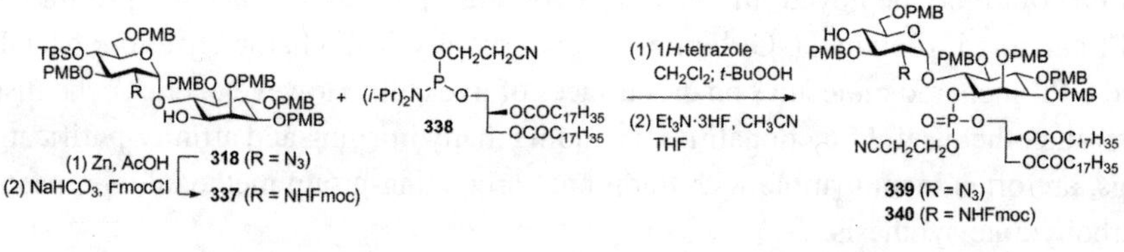

SCHEME 63. Synthesis of pseudodisaccharides **339** and **340** by the Guo group.

accommodate the structural heterogeneity of naturally occurring GPIs, particularly in lipid chains, as well the ability to functionalize GPIs with "click" tags for biological studies. Accordingly, the afore-described trimannose thioglycoside **309** and pseudodisaccharide **318** were used as adaptable, PMB-protected intermediates that would allow for a choice of phosphoglycerolipid structure and functional tag in the target GPI. Using these intermediates, two generations of "clickable" GPIs were synthesized. In this article, synthesis of the second-generation Alkynyl-GPI **335** and Azido-GPI **336**, as well as subsequent chemoselective labeling of these GPIs via "click" chemistry, is discussed.

For both of the target GPI anchors **335** and **336**, a distearoylglycerophosphate group was chosen as the phospholipid (Scheme 63). Toward the synthesis of Alkynyl-GPI **335**, the pseudodisaccharide **318** was phospholipidated by using phosphoramidite **338** under conventional conditions, followed by treatment with triethylamine trihydrofluoride to remove the TBS group, which gave the pseudodisaccharide **339**. En route to Azido-GPI **336**, the azido group of pseudodisaccharide **318** was replaced by an Fmoc group to give compound **337**, which subsequently underwent phospholipidation–desilylation in the same manner to furnish the Fmoc-protected pseudodisaccharide **340**.

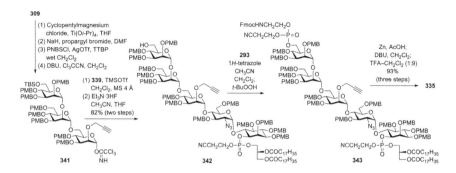

SCHEME 64. Completion of "clickable" Alkynyl-GPI **335** by the Guo group.

Completion of the synthesis of Alkynyl-GPI **335** from trimannoside **309** and pseudodisaccharide **339** is shown in Scheme 64. First, the terminal alkyne "click" tag was installed in the trimannose fragment by consecutive Cha deallylation[120] and propargylation of the O-2 position of Man-I. Following hydrolysis of the anomeric thioacetal using *p*-nitrobenzenesulfenyl chloride (PNBSCl) and AgOTf/TTBP[137] in wet CH_2Cl_2, the intermediate hemiacetal was converted into trichloroacetimidate **341** by treatment with DBU and trichloroacetonitrile. The key glycosylation reaction of **341** with pseudodisaccharide **339**, promoted by TMSOTf, furnished an intermediate that was immediately subjected to removal of the TBS group on O-6 of Man-III with triethylamine trihydrofluoride to give **342**. This alkyne-modified pseudopentasaccharide was isolated in good yield (82%) over two steps, and no β anomer was observed. Installation of the phosphoethanolamine group into Man-III using phosphoramidite **293** proceeded well and gave compound **343**, which was finally subjected to the aforementioned three-step, one-flask global deprotection to provide the target Alkynyl-GPI **335** in excellent yield (93%, three steps) in under 3 h.

The synthesis of Azido-GPI **336**, shown in Scheme 65, followed a similar route. First, the O-2 allyl group on Man-I of compound **309** was exchanged for a 2-azidoethyl group in five steps, including Cha deallylation,[120] base-promoted alkylation using *tert*-butyl bromoacetate, ester reduction with $LiAlH_4$, mesylation of the primary hydroxyl group, and finally displacement of the mesylate with NaN_3.[138] Transformation into the trichloroacetimidate **344** was accomplished by treatment with NIS/AgOTf/TTBP in wet CH_2Cl_2, followed by DBU and trichloroacetonitrile. The TMSOTf-promoted coupling of **344** with *N*-Fmoc-protected pseudodisaccharide **340** gave the desired α-pseudopentasaccharide intermediate, which was desilylated with triethylamine trihydrofluoride to generate compound **345** in moderate yield (30%, 88% based on recovered **340**). The remaining steps involved installation of phosphoethanolamine at Man-III as already described to give **346**, followed by global

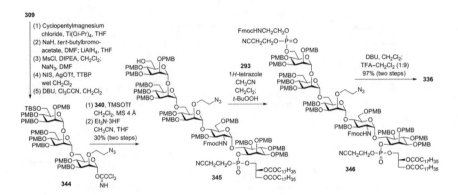

SCHEME 65. Completion of "clickable" Azido-GPI **336** by the Guo group.

deprotection, which proceeded in two steps (DBU and then 10% TFA) to afford Azido-GPI **336** in 97% yield over the final two steps.

To establish Alkynyl-GPI **335** and Azido-GPI **336** as "clickable" tools for the biological study of GPI anchors, they were coupled via "click" chemistry to imaging and affinity probes (Scheme 66). Alkynyl-GPI **335** was conjugated with Azide-Fluor 488 (**347**), a commercially available regioisomeric mixture, via Cu(I)-catalyzed [3 + 2] cycloaddition under standard conditions.[139,140] This Cu-catalyzed azide–alkyne cycloaddition (CuAAC) reaction led to complete conversion of the starting material into the GPI-Fluor conjugate **348**, a promising tool for visualizing GPI anchors by fluorescence microscopy in cellular contexts. Azido-GPI **336** was effectively coupled via Cu-free, strain-promoted azide–cyclooctyne [3 + 2] cycloaddition (Cu-free "click") to BARAC-biotin (**349**), a biarylazacyclooctynone-based affinity probe developed by Jewett et al.[141] This step effected virtually quantitative conversion of the starting material into GPI-Biotin **350**. Conjugation of GPIs to affinity tags such as biotin is a potentially useful strategy for pull-down experiments to probe cell-surface, GPI-anchored proteomics.

Alkynyl-GPI **335** and Azido-GPI **336**, which were not accessible by traditional tactics for carbohydrate protection, were synthesized using Guo's PMB-protection strategy. These "clickable" GPIs were demonstrated to undergo efficient coupling reactions via both CuAAC and Cu-free "click" reactions and therefore constitute versatile tools for the development of chemical approaches to studying the biology of GPI anchors. In addition, this work further validated the PMB-protection strategy as an alternative global-protection option, most usefully when sensitively functionalized GPI anchors are targeted for synthesis.

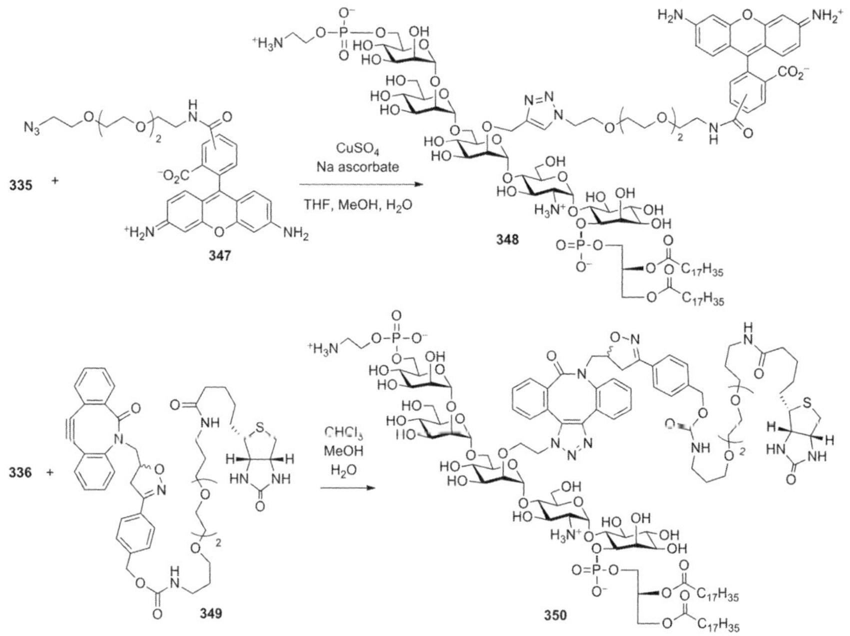

SCHEME 66. "Click" reactions of Alkynyl-GPI **335** and Azido-GPI **336** by the Guo group.

5. General Synthetic Strategy for Branched GPI Anchors by the Seeberger Group (2011)

In addition to methods focused on expanding functional group tolerance, such as the efforts by Nikolaev and Guo already discussed, the development of strategies that address structural diversity in terms of branching sugars and other substituents is also an important goal. As already discussed (see Fig. 1 in the introduction), the conserved core structure of GPI anchors is often decorated in a species- and cell-type dependent manner, with additional sugars, phosphoethanolamine groups, and lipid chains at various positions. In 2011, the Seeberger group developed a general method for the synthesis of GPI anchors containing, in principle, any type of structural modification that is compatible with global benzyl-group protection.[142] A general retrosynthetic analysis for this strategy is shown in Scheme 67. It was expected that diverse target GPI anchors could be accessed from orthogonally protected GPI intermediates containing first of all the desired branching sugars at Man-I and Man-III and, second, orthogonal protecting groups at sites for late-stage phospholipidation, O-2 acylation

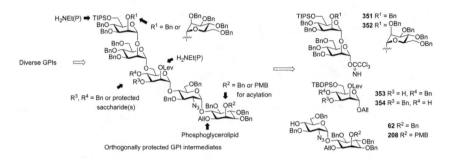

SCHEME 67. General retrosynthesis for branched GPI anchors by the Seeberger group.

of the inositol, and selective installation of phosphoethanolamine. These orthogonally protected GPI intermediates could be prepared from the interchangeable mannose building blocks **351–354** and pseudodisaccharides **62** and **208**.

To demonstrate the applicability of this general method, Seeberger conducted the total synthesis of *T. gondii* GPI anchor **355**, which contains a core modification of an α-Glc-(1→4)-β-Gal-(1→4) disaccharide at Man-I. As shown in Scheme 68, the target GPI (**355**) was first disconnected to the orthogonally protected GPI intermediate **356**, which was properly set up to undergo late-stage introduction of the phospholipid and phosphoethanolamine groups by the *H*-phosphonate method.[44] The preparation of **356** centered on Man-I building block **354**, which possessed the appropriate array of orthogonal protecting groups to allow the following [3+2+2] assembly sequence: first, elongation at the O-3 position with donors **357** and **358** to give a trisaccharide; next, [3+2] coupling with donor **351**; followed by subsequent [5+2] coupling with pseudodisaccharide **62**; and, finally, optional installation of a phosphoethanolamine group at the O-2 position of Man-I (not performed in this synthesis). For the glycosylation steps, a combination of glycosyl phosphates[143] and trichloroacetimidates[71] was used as donors.

Preparation of the central Man-I derivative **354** and its elaboration to trisaccharide fragment **365** is shown in Scheme 69. En route to compound **354**, the previously reported allyl mannoside **359**[144] was protected at O-4 by the oxidatively cleavable 2-naphthylmethyl (NAP) group, which was followed by acid-catalyzed deacetonation and selective tin-mediated benzylation at O-3 to give compound **360**. Subsequent conversion into **354** was accomplished by *N,N'*-diisopropylcarbodiimide (DIC)/ DMAP-promoted levulinoylation (Lev) at O-2, followed by oxidative cleavage of the NAP group at O-4 with DDQ. Alternatively, this sequence could be modified to expose the O-3 position to obtain the Man-I derivative **353**, which would be useful for the synthesis of GPIs of *T. brucei* containing additional sugars at this location.

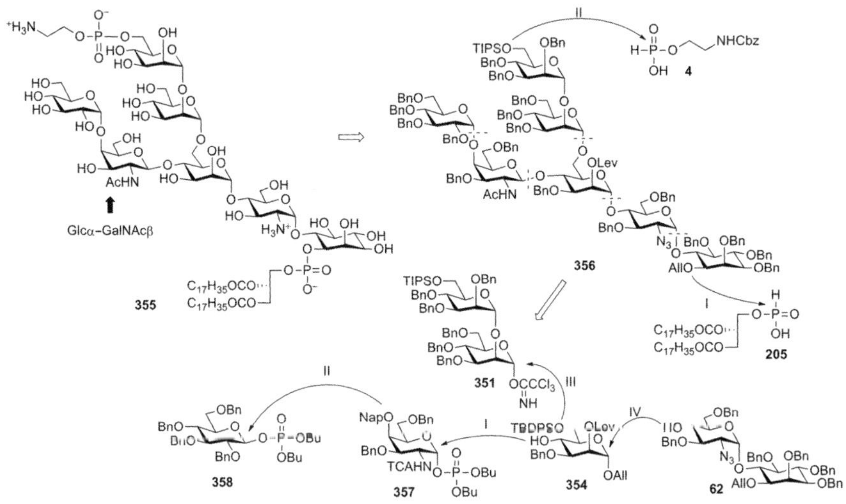

SCHEME 68. Retrosynthetic analysis of *T. gondii* GPI anchor **355**, using the general method for branched GPI synthesis by the Seeberger group.

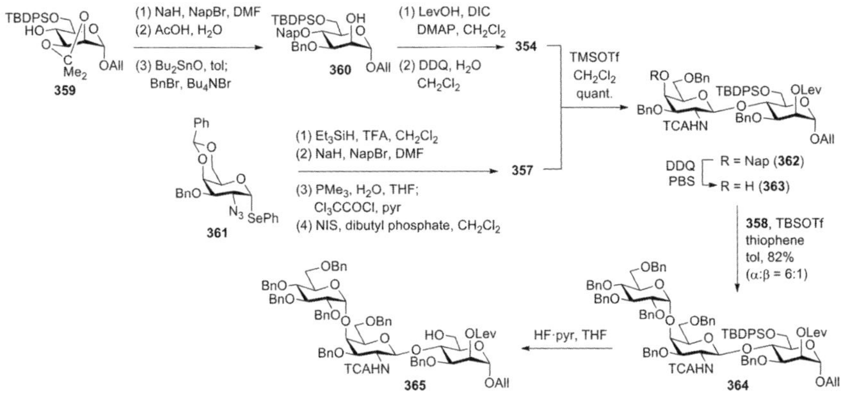

SCHEME 69. Synthesis of central trisaccharide **365** by the Seeberger group.

The galactosyl phosphate derivative **357**, the coupling partner of Man-I derivative **354**, was synthesized from the previously reported 2-azido selenoglycoside **361**[145] in four steps, including regioselective benzylidene cleavage, naphthylmethylation at O-4, Staudinger reduction followed by *N*-trichloroacetylation, and finally formation of the glycosyl phosphate by treatment with NIS/dibutyl phosphate. The Man-I acceptor

354 and galactosyl phosphate donor 357 combined in the presence of TMSOTf to give the β-disaccharide 362, which was treated with DDQ to remove the NAP group, furnishing compound 363 quantitatively over two steps. Subsequent glycosylation of 363 with the glucosyl phosphate donor 358 in the presence of TBSOTf and thiophene (a reported α-enhancing additive)[146] afforded 364 (82% yield, α:β = 6:1), which underwent desilylation by HF·pyridine to generate the central trisaccharide 365.

Scheme 70 depicts the [3+2+2] assembly of building blocks 365, 351, and 62 to generate the orthogonally protected pseudoheptasaccharide intermediate 356. TBSOTf and added thiophene were again used to promoted α-stereoselective glycosylation between the dimannosyl donor 351 and the acceptor 365, resulting in the formation of pentasaccharide 366 in 69% yield. Reductive conversion of the trichloroacetamido group to an acetamido group gave compound 367, which was converted into the trichloroacetimidate donor 368 by deallylation using the Ir–Hg protocol[122] and subsequent treatment with trichloroacetonitrile and DBU. Reaction of this donor with the pseudodisaccharide 62 in the presence of catalytic TMSOTf forged the key glycosidic bond, delivering compound 356 in 71% yield. The participating Lev group at O-2 ensured α-stereoselectivity in this crucial coupling reaction.

At this stage, compound 356 was subjected to Ir–Hg deallylation to expose O-1 of the inositol, which was phospholipidated by using *H*-phosphonate 205 and iodine-mediated oxidation to give intermediate 369 (Scheme 71). Acid-promoted removal of the TIPS group at O-6 of Man-III prepared the intermediate for the second phosphorylation, which was accomplished by using the *H*-phosphonate 4 to afford 370. Hydrazinolysis of the O-2 Lev group on Man-I deprotected the 2-position of Man-I, which had the potential to undergo further phosphorylation at this site if desired. In this synthesis, intermediate 370 was taken forward to global deprotection by hydrogenolysis to provide the target GPI 355 in 62% yield.

In this work, the Seeberger group synthesized the *T. gondii* GPI anchor 355, which carried a branching disaccharide at Man-I, using a [3+2+2] assembly strategy. The most important aspect of this strategy is its potential applicability to the synthesis of

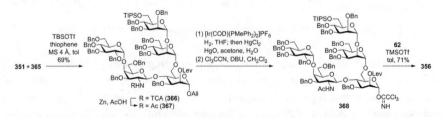

SCHEME 70. Synthesis of orthogonally protected GPI intermediate 356 by the Seeberger group.

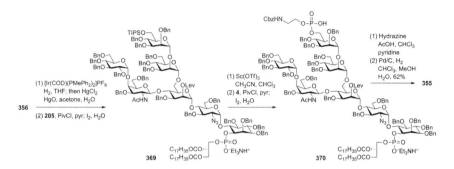

SCHEME 71. Completion of synthesis of the *T. gondii* GPI anchor **355**, using the general method for branched GPI synthesis by the Seeberger group.

virtually any type of GPI anchor bearing common modifications, including acyl groups on O-2 of the inositol, phosphoethanolamine on O-2 of Man-I, sugars on O-3 or O-4 of Man-I, and a sugar on O-2 of Man-III. This flexibility is a result of carefully designed building blocks, in particular, the central Man-I components **353** and **354**, and a well-optimized assembly sequence. In addition, synthetic **355** was used in combination with Seeberger's GPI microarray technology[109] to confirm that it was indeed the immunogenic "low-molecular-weight antigen" of *T. gondii* responsible for antibody production in toxoplasmosis patients.

IV. CONCLUSIONS AND OUTLOOK

Starting with Ogawa's seminal work in the early 1990s, the field of chemical synthesis of GPI anchors has enjoyed marked success in melding the knowledge and methods of carbohydrate, inositol, lipid, and phosphate chemistry into a reliable framework for constructing these complex natural products. Numerous naturally occurring GPI anchors from protozoan parasites, yeast, and mammals have been synthesized, and through this, many new synthetic methods have been developed and tested. Currently, the approach to GPI synthesis has shifted toward developing novel diversity-oriented strategies that have enabled access to new classes of molecules, including GPI anchors bearing the biologically important unsaturated lipids, indispensable functionalities such as "click" chemistry tags, and highly branched and modified core structures.

In the future, a combination of such diversity-oriented strategies, with the goal of a truly general method for the synthesis of any type of target GPI anchor, would be a highly valuable contribution. Another goal of great importance is improvement of efficiency by limiting the number of steps in a synthesis, which, taking into account preparation of the monomers, can become prohibitively large. In this context, solid-phase synthesis, one-flask oligosaccharide synthesis, and reactivity tuning of glycosyl acceptors and donors will all contribute toward meeting this goal. Synthetic GPIs and GPI derivatives have the potential to play vital roles in deducing structure–activity relationships of these molecules in various contexts, exploring the scope of GPI anchorage as a posttranslational modification, probing therapeutic applications, and in general gaining insight into their fundamental biological functions.

Acknowledgments

The authors thank the National Institutes of Health (NIH, R01GM090270) and National Science Foundation (NSF, CHE-1053848, 0715275, and 0407144) for financial support.

References

1. M. A. J. Ferguson and A. F. Williams, Cell-surface anchoring of proteins via glycosyl-phosphatidylinositol structures, *Annu. Rev. Biochem.*, 57 (1988) 121–138.
2. P. T. Englund, The structure and biosynthesis of glycosyl phosphatidylinositol protein anchors, *Annu. Rev. Biochem.*, 62 (1993) 121–138.
3. H. Ikezawa, Glycosylphosphatidylinositol (GPI)-anchored proteins, *Biol. Pharm. Bull.*, 25 (2002) 409–417.
4. M. G. Paulick and C. R. Bertozzi, The glycosylphosphatidylinositol anchor: A complex membrane-anchoring structure for proteins, *Biochemistry*, 47 (2008) 6991–7000.
5. M. A. Ferguson, The structure, biosynthesis and functions of glycosylphosphatidylinositol anchors, and the contributions of trypanosome research, *J. Cell Sci.*, 112 (1999) 2799–2809.
6. B. Eisenhaber, P. Bork, and F. Eisenhaber, Post-translational GPI lipid anchor modification of proteins in kingdoms of life: Analysis of protein sequence data from complete genomes, *Protein Eng.*, 14 (2001) 17–25.
7. H. T. He, J. Finne, and C. Goridis, Biosynthesis, membrane association, and release of N-CAM-120, a phosphatidylinositol-linked form of the neural cell adhesion molecule, *J. Cell Biol.*, 105 (1987) 2489–2500.
8. P. J. Robinson, M. Millrain, J. Antoniou, E. Simpson, and A. L. Mellor, A glycophospholipid anchor is required for Qa-2-mediated T cell activation, *Nature*, 342 (1989) 85–87.
9. D. D. Eardley and M. E. Koshland, Glycosylphosphatidylinositol: A candidate system for interleukin-2 signal transduction, *Science*, 251 (1991) 78–81.
10. G. A. M. Cross, Cellular and genetic aspects of antigenic variation in Trypanosomes, *Annu. Rev. Immunol.*, 8 (1990) 83–110.
11. T. Kinoshita, N. Inoue, and J. Takeda, Defective glycosyl phosphatidylinositol anchor synthesis and paroxysmal nocturnal hemoglobinuria, in J. D. Frank, (Ed.), *Advances in Immunology*, Vol. 60, Academic Press, New York, 1995, pp. 57–103.

12. K. Hwa, Glycosyl phosphatidylinositol-linked glycoconjugates: Structure, biosynthesis, and function, *Adv. Exp. Med. Biol.*, 491 (2001) 207–214.
13. M. A. J. Ferguson, S. W. Homans, R. A. Dwek, and T. W. Rademacher, Glycosyl-phosphatidylinositol moiety that anchors *Trypanosoma brucei* variant surface glycoprotein to the membrane, *Science*, 239 (1988) 753–759.
14. M. Fujita and Y. Jigami, Lipid remodeling of GPI-anchored proteins and its function, *Biochim. Biophys. Acta*, 1780 (2008) 410–420.
15. Y. Varma and T. Hendrickson, Methods to study GPI anchoring of proteins, *ChemBioChem*, 11 (2010) 623–636.
16. L. Schofield, M. C. Hewitt, K. Evans, M.-A. Siomos, and P. H. Seeberger, Synthetic GPI as a candidate anti-toxic vaccine in a model of malaria, *Nature*, 418 (2002) 785–789.
17. M. C. Hewitt, D. A. Snyder, and P. H. Seeberger, Rapid synthesis of a glycosylphosphatidylinositol-based malaria vaccine using automated solid-phase oligosaccharide synthesis, *J. Am. Chem. Soc.*, 124 (2002) 13434–13436.
18. M. G. Paulick, M. B. Forstner, J. T. Groves, and C. R. Bertozzi, A chemical approach to unraveling the biological function of the glycosylphosphatidylinositol anchor, *Proc. Natl. Acad. Sci. U.S.A.*, 104 (2007) 20332–20337.
19. E. S. Nakayasu, D. V. Yashunsky, L. L. Nohara, A. C. T. Torrecilhas, A. V. Nikolaev, and I. C. Almeida, GPIomics: Global analysis of glycosylphosphatidylinositol-anchored molecules of *Trypanosoma cruzi*, *Mol. Syst. Biol.*, 5 (2009) 261.
20. R. Gigg and J. Gigg, Synthesis of glycosylphosphatidylinositol anchors, in D. G. Large and C. D. Warren, (Eds.) *Glycopeptides and Related Compounds: Synthesis, Analysis, and Applications*, Marcel Dekker, Inc., New York, 1997, pp. 327–392.
21. Z. Guo and L. Bishop, Chemical synthesis of GPIs and GPI-anchored glycopeptides, *Eur. J. Org. Chem.* (2004) 3585–3596.
22. A. V. Nikolaev and N. Al-Maharik, Synthetic glycosylphosphatidylinositol (GPI) anchors: How these complex molecules have been made, *Nat. Prod. Rep.*, 28 (2011) 970–1020.
23. M. Martín-Lomas, N. Khiar, S. García, J.-L. Koessler, P. M. Nieto, and T. W. Rademacher, Inositolphosphoglycan mediators structurally related to glycosyl phosphatidylinositol anchors: Synthesis, structure and biological activity, *Chem. Eur. J.*, 6 (2000) 3608–3621.
24. K. Pekari, D. Tailler, R. Weingart, and R. R. Schmidt, Synthesis of the fully phosphorylated GPI anchor pseudohexasaccharide of *Toxoplasma gondii*, *J. Org. Chem.*, 66 (2001) 7432–7442.
25. K. Ruda, J. Lindberg, P. J. Garegg, S. Oscarson, and P. Konradsson, Synthesis of the Leishmania LPG core heptasaccharyl *myo*-inositol, *J. Am. Chem. Soc.*, 122 (2000) 11067–11072.
26. M. Hederos and P. Konradsson, Synthesis of the *Trypanosoma cruzi* LPPG heptasaccharyl *myo*-inositol, *J. Am. Chem. Soc.*, 128 (2006) 3414–3419.
27. P. H. Seeberger, R. L. Soucy, Y.-U. Kwon, D. A. Snyder, and T. Kanemitsu, A convergent, versatile route to two synthetic conjugate anti-toxin malaria vaccines, *Chem. Commun.* (2004) 1706–1707.
28. Y.-U. Kwon, R. L. Soucy, D. A. Snyder, and P. H. Seeberger, Assembly of a series of malarial glycosylphosphatidylinositol anchor oligosaccharides, *Chem. Eur. J.*, 11 (2005) 2493–2504.
29. C. Becker, X. Liu, D. Olschewski, R. Castelli, R. Seidel, and P. Seeberger, Semisynthesis of a glycosylphosphatidylinositol-anchored prion protein, *Angew. Chem. Int. Ed. Engl.*, 47 (2008) 8215–8219.
30. J. Xue, N. Shao, and Z. Guo, First total synthesis of a GPI-anchored peptide, *J. Org. Chem.*, 68 (2003) 4020–4029.
31. N. Shao, J. Xue, and Z. Guo, Chemical Synthesis of a skeleton structure of sperm CD52—A GPI-anchored glycopeptide, *Angew. Chem. Int. Ed. Engl.*, 43 (2004) 1569–1573.
32. M. G. Paulick, A. R. Wise, M. B. Forstner, J. T. Groves, and C. R. Bertozzi, Synthetic analogues of glycosylphosphatidylinositol-anchored proteins and their behavior in supported lipid bilayers, *J. Am. Chem. Soc.*, 129 (2007) 11543–11550.

33. X. Guo, Q. Wang, B. M. Swarts, and Z. Guo, Sortase-catalyzed peptide–glycosylphosphatidylinositol analogue ligation, *J. Am. Chem. Soc.*, 131 (2009) 9878–9879.
34. Z. Wu, X. Guo, Q. Wang, B. M. Swarts, and Z. Guo, Sortase A-catalyzed transpeptidation of glycosylphosphatidylinositol derivatives for chemoenzymatic synthesis of GPI-anchored proteins, *J. Am. Chem. Soc.*, 132 (2010) 1567–1571.
35. Z. Wu, X. Guo, and Z. Guo, Chemoenzymatic synthesis of glycosylphosphatidylinositol-anchored glycopeptides, *Chem. Commun.*, 46 (2010) 5773–5774.
36. C. Murakata and T. Ogawa, A total synthesis of GPI anchor of *Trypanosoma brucei*, *Tetrahedron Lett.*, 32 (1991) 671–674.
37. C. Murakata and T. Ogawa, Stereoselective total synthesis of the glycosyl phosphatidylinositol (GPI) anchor of *Trypanosoma brucei*, *Carbohydr. Res.*, 235 (1992) 95–114.
38. D. R. Mootoo, P. Konradsson, and B. Fraser-Reid, n-Pentenyl glycosides facilitate a stereoselective synthesis of the pentasaccharide core of the protein membrane anchor found in *Trypanosoma brucei*, *J. Am. Chem. Soc.*, 111 (1989) 8540–8542.
39. C. Murakata and T. Ogawa, Synthetic study on glycophosphatidyl inositol (GPI) anchor of *Trypanosoma brucei*: Glycoheptaosyl core, *Tetrahdron Lett.*, 31 (1990) 2439–2442.
40. R. Verduyn, C. J. J. Elie, C. E. Dreef, G. A. van der Marel, and J. H. van Boom, Stereospecific synthesis of partially protected 2-azido-2-deoxy-D-glucosyl D-*myo*-inositol: Precursor of a potential insulin mimetic and membrane protein anchoring site, *Recl. Trav. Chim. Pays-Bas*, 109 (1990) 591–593.
41. T. Matsumoto, H. Maeta, K. Suzuki, and G.-i. Tsuchihashi, New glycosidation reaction 1: Combinational use of Cp_2ZrCl_2-$AgClO_4$ for activation of glycosyl fluorides and application to highly β-selective glycosidation of D-mycinose, *Tetrahedron Lett.*, 29 (1988) 3567–3570.
42. R. U. Lemieux, K. B. Hendriks, R. V. Stick, and K. James, Halide ion catalyzed glycosidation reactions. Syntheses of α-linked disaccharides, *J. Am. Chem. Soc.*, 97 (1975) 4056–4062.
43. W. Koenigs and E. Knorr, Über Enige Derivate des Traubenzuckers und der Galactose, *Ber. Dtsch. Chem. Ges.*, 34 (1901) 957–981.
44. I. Lindh and J. Stawinski, A general method for the synthesis of glycerophospholipids and their analogs via H-phosphonate intermediates, *J. Org. Chem.*, 54 (1989) 1338–1342.
45. K. Koike, M. Sugimoto, S. Sato, Y. Ito, Y. Nakahara, and T. Ogawa, Total synthesis of globotriaosyl-E and Z-ceramides and isoglobotriaosyl-E-ceramide, *Carbohydr. Res.*, 163 (1987) 189–208.
46. T. Ogawa and M. Matsui, Regioselective enhancement of the nucleophilicity of hydroxyl groups through trialkylstannylation: A route to partial alkylation of polyhydroxy compounds, *Carbohydr. Res.*, 62 (1978) C1–C4.
47. W. Kinzy and R. R. Schmidt, Glycosylimidate, 16. Synthese des Trisaccharids aus der "Repeating Unit" des Kapselpolysaccharids von *Neisseria meningitidis* (Serogruppe L), *Liebigs Ann. Chem.*, 1985 (1985) 1537–1545.
48. M. Ek, P. J. Garegg, H. Hultberg, and S. Oscarson, Reductive ring openings of carbohydrate benzylidene acetals using borane-trimethylamine and aluminium chloride. Regioselectivity and solvent dependance, *J. Carbohydr. Chem.*, 2 (1983) 305–311.
49. P. J. Garegg, T. Iverson, R. Johansson, and B. Lindberg, Synthesis of some mono-*O*-benzyl- and penta-*O*-methyl-*myo*-inositols, *Carbohydr. Res.*, 130 (1984) 322–326.
50. D. C. Billington, R. Baker, J. J. Kulagowski, and I. M. Mawer, Synthesis of *myo*-inositol 1-phosphate and 4-phosphate, and of their individual enantiomers, *J. Chem. Soc. Chem. Commun.* (1987) 314–316.
51. C. E. Dreef, M. Douwes, C. J. J. Elie, G. A. van der Marel, and J. H. van Boom, Application of the bifunctional phosphonylating agent bis[6-(trifluoromethyl)benzotriazol-1-yl] methylphosphonate towards the preparation of isosteric D-myo-inositol phospholipid and phosphate analogues, *Synthesis*, 1991 (1991) 443–447.

52. S. Cottaz, J. S. Brimacombe, and M. A. J. Ferguson, Parasite glycoconjugates. Part 1. The synthesis of some early and related intermediates in the biosynthetic pathway of glycosyl-phosphatidylinositol membrane anchors, *J. Chem. Soc. Perkin Trans. 1* (1993) 2945–2951.
53. T. Ogawa, Haworth Memorial Lecture. Experiments directed towards glycoconjugate synthesis, *Chem. Soc. Rev.*, 23 (1994) 397–407.
54. S. Sato, M. Mori, Y. Ito, and T. Ogawa, An efficient approach to O-glycosides through $CuBr_2$-Bu_4NBr mediated activation of glycosides, *Carbohydr. Res.*, 155 (1986) C6–C10.
55. A. Demchenko, T. Stauch, and G.-J. Boons, Solvent and other effects on the stereoselectivity of thioglycoside glycosidations, *Synlett*, 1997 (1997) 818–820.
56. P. J. Garegg and L. Maron, Improved synthesis of 3,4,6-tri-*O*-benzyl-α-D-mannopyranosides, *Acta Chem. Scand. Ser. B*, B33 (1979) 39–41.
57. T. Ogawa and K. Sasajima, Synthesis of a model of an inner chain of cell-wall proteoheteroglycan isolated from *Piricularia oryzae*: Branched D-mannopentaosides, *Carbohydr. Res.*, 93 (1981) 67–81.
58. T. G. Mayer, B. Kratzer, and R. R. Schmidt, Synthesis of a GPI anchor of yeast (*Saccharomyces cerevisae*), *Angew. Chem. Int. Ed. Engl.*, 33 (1994) 2177–2181.
59. T. G. Mayer and R. R. Schmidt, Glycosyl phosphatidylinositol (GPI) anchor synthesis based on versatile building blocks—Total synthesis of a GPI anchor of yeast, *Eur. J. Org. Chem.* (1999) 1153–1165.
60. W. Bannwarth and A. Trzeciak, A simple and effective chemical phosphorylation procedure for biomolecules, *Helv. Chim. Acta*, 70 (1987) 175–186.
61. R. R. Schmidt and J. Michel, Facile synthesis of alpha- and beta-O-glycosyl imidates; preparation of glycosides and disaccharides, *Angew. Chem. Int. Ed. Engl.*, 19 (1980) 731–732.
62. B. Kratzer, T. G. Mayer, and R. R. Schmidt, Synthesis of D-*erythro*-sphingomyelin and of D-erythroceramide-1-phosphoinositol, *Tetrahedron Lett.*, 34 (1993) 6881–6884.
63. G. Grundler and R. R. Schmidt, Glycosylimidate, 13. Anwendung des Trichloracetimidat-verfahrens auf 2-Azidoglucose- und 2-Azidogalactose-Derivate, *Liebigs Ann. Chem.* (1984) 1826–1847.
64. R. R. Schmidt and T. Maier, New synthesis of D-*ribo*- and L-*lyxo*-phytosphingosine: Transformation into the corresponding lactosyl-ceramides, *Carbohydr. Res.*, 174 (1988) 169–179.
65. U. E. Udodong, R. Madsen, C. Roberts, and B. Fraser-Reid, A ready, convergent synthesis of the heptasaccharide GPI membrane anchor of rat brain Thy-1 glycoprotein, *J. Am. Chem. Soc.*, 115 (1993) 7886–7887.
66. A. S. Campbell and B. Fraser-Reid, Support studies for installing the phosphodiester residues of the Thy-1 glycoprotein membrane anchor, *Bioorg. Med. Chem.*, 2 (1994) 1209–1219.
67. C. Roberts, C. L. May, and B. Fraser-Reid, Streamlining the *n*-pentenyl glycoside approach to the trimannoside component of the Thy-1 membrane anchor, *Carbohydr. Lett.*, 1 (1994) 89–93.
68. C. Roberts, R. Madsen, and B. Fraser-Reid, Studies related to synthesis of glycophosphatidylinositol membrane-bound protein anchors. 5. n-Pentenyl ortho esters for mannan components, *J. Am. Chem. Soc.*, 117 (1995) 1546–1553.
69. R. Madsen, U. E. Udodong, C. Roberts, D. R. Mootoo, P. Konradsson, and B. Fraser-Reid, Studies related to synthesis of glycophosphatidylinositol membrane-bound protein anchors. 6. Convergent assembly of subunits, *J. Am. Chem. Soc.*, 117 (1995) 1554–1565.
70. A. S. Campbell and B. Fraser-Reid, First synthesis of a fully phosphorylated GPI membrane anchor: Rat brain Thy-1, *J. Am. Chem. Soc.*, 117 (1995) 10387–10388.
71. G. Grundler and R. R. Schmidt, Glycosylimidate, 13. Anwendung des Trichloroacetimidat-verfahrens auf 2-Azidoglucose- und 2-Azidogalactose-Derivate, *Leibigs Ann. Chem.* (1984) 1826–1847.
72. D. R. Mootoo, V. Date, and B. Fraser-Reid, *n*-Pentenyl glycosides permit the chemospecific liberation of the anomeric center, *J. Am. Chem. Soc.*, 110 (1988) 2662–2663.
73. M. Mach, U. Schlueter, F. Mathew, B. Fraser-Reid, and K. C. Hazen, Comparing n-pentenyl orthoesters and n-pentenyl glycosides as alternative glycosyl donors, *Tetrahedron*, 58 (2002) 7345–7354.

74. B. Fraser-Reid, U. E. Udodong, Z. Wu, H. Ottosson, J. R. Merritt, C. S. Rao, C. Roberts, and R. Madsen, *n*-Pentenyl glycosides in organic chemistry: A contemporary example of serendipity, *Synlett*, 1992 (1992) 927–942.
75. H. Paulsen and A. Bünsch, Synthese der Pentasaccharid-kette des Forsmann-antigens, *Carbohydr. Res.*, 100 (1982) 143–167.
76. J. P. Vacca, S. J. DeSolms, and J. R. Huff, The total synthesis of D- and L-*myo*-inositol 1,4,5-trisphosphate, *J. Am. Chem. Soc.*, 109 (1987) 3478–3479.
77. H. Hori, Y. Nishida, H. Ohrui, and H. Meguro, Regioselective de-O-benzylation with Lewis acids, *J. Org. Chem.*, 54 (1989) 1346–1353.
78. D. K. Baeschlin, A. R. Chaperon, L. G. Green, M. G. Hahn, S. J. Ince, and S. V. Ley, 1,2-Diacetals in synthesis: Total synthesis of a glycosylphosphatidylinositol anchor of *Trypanosoma brucei*, *Chem. Eur. J.*, 6 (2000) 172–186.
79. D. K. Baeschlin, A. R. Chaperon, V. Charbonneau, L. G. Green, S. V. Ley, U. Lücking, and E. Walther, Rapid assembly of oligosaccharides: Total synthesis of a glycosylphosphatidylinositol anchor of *Trypanosoma brucei*, *Angew. Chem. Int. Ed.*, 110 (1998) 3609–3614.
80. S. V. Ley, R. Leslie, P. D. Tiffin, and M. Woods, Dispiroketals in synthesis (part 2): A new group for the selective protection of diequatorial vicinal diols in carbohydrates, *Tetrahedron Lett.*, 33 (1992) 4767–4770.
81. G.-J. Boons, P. Grice, R. Leslie, S. V. Ley, and L. L. Yeung, Dispiroketals in synthesis (part 5): A new opportunity for oligosaccharide synthesis using differentially activated glycosyl donors and acceptors, *Tetrahedron Lett.*, 34 (1993) 8523–8526.
82. S. V. Ley, S. Mio, and B. Meseguer, Dispiroketals in synthesis (part 22): Use of chiral 2,2′-bis(phenylthiomethyl)dihydropyrans as new protecting and resolving agents for 1,2-diols, *Synlett*, 1996 (1996) 791–792.
83. D. R. Mootoo, P. Konradsson, U. Udodong, and B. Fraser-Reid, Armed and disarmed *n*-pentenyl glycosides in saccharide couplings leading to oligosaccharides, *J. Am. Chem. Soc.*, 110 (1988) 5583–5584.
84. S. Mehta and B. Mario Pinto, Phenylselenoglycosides as novel, versatile glycosyl donors. Selective activation over thioglycosides, *Tetrahedron Lett.*, 32 (1991) 4435–4438.
85. S. V. Ley, G.-J. Boons, R. Leslie, M. Woods, and D. M. Hollinshead, Dispiroketals in synthesis (part 3): Selective protection of diequatorial vicinal diols in carbohydrates, *Synthesis*, 1993 (1993) 689–692.
86. A. Mallet, J.-M. Mallet, and P. Sinaÿ, The use of selenophenyl galactopyranosides for the synthesis of α and β-(1→4)-C-disaccharides, *Tetrahedron: Asymmetry*, 5 (1994) 2593–2608.
87. P. Grice, S. V. Ley, J. Pietruszka, H. M. I. Osborn, H. W. M. Priepke, and S. L. Warriner, A new strategy for oligosaccharide assembly exploiting cyclohexane-1,2-diacetal methodology: An efficient synthesis of a high mannose type nonasaccharide, *Chem. Eur. J.*, 3 (1997) 431–440.
88. L. Hans, Synthesis of a tri- and a hepta-saccharide which contain α-L-fucopyranosyl groups and are part of the complex type of carbohydrate moiety of glycoproteins, *Carbohydr. Res.*, 139 (1985) 105–113.
89. Y. Watanabe, M. Mitani, T. Morita, and S. Ozaki, Highly efficient protection by the tetraisopropyldisiloxane-1,3-diyl group in the synthesis of *myo*-inositol phosphates as inositol 1,3,4,6-tetrakisphosphate, *J. Chem. Soc. Chem. Commun.* (1989) 482–483.
90. K. C. Nicolaou, T. J. Caulfield, H. Kataoka, and N. A. Stylianides, Total synthesis of the tumor-associated Lex family of glycosphingolipids, *J. Am. Chem. Soc.*, 112 (1990) 3693–3695.
91. A. Vasella, C. Witzig, J.-L. Chiara, and M. Martin-Lomas, Convenient synthesis of 2-azido-2-deoxy-aldoses by diazo transfer, *Helv. Chim. Acta*, 74 (1991) 2073–2077.
92. D. Tailler, V. Ferrières, K. Pekari, and R. R. Schmidt, Synthesis of the glycosyl phosphatidyl inositol anchor of rat brain Thy-1, *Tetrahedron Lett.*, 40 (1999) 679–682.

93. K. Pekari and R. R. Schmidt, A variable concept for the preparation of branched glycosyl phosphatidyl inositol anchors, *J. Org. Chem.*, 68 (2003) 1295–1308.
94. M. Goebel, H.-G. Nothofer, G. Roß, and I. Ugi, A facile synthesis of per-O-alkylated glycono-δ-lactones from per-O-alkylated glycopyranosides and a novel ring contraction for pyranoses, *Tetrahedron*, 53 (1997) 3123–3134.
95. J. Xue and Z. Guo, Convergent synthesis of a GPI containing an acylated inositol, *J. Am. Chem. Soc.*, 125 (2003) 16334–16339.
96. Y. Tsuji, H. Clausen, E. Nudelman, T. Kaizu, S.-I. Hakomori, and S. Isojima, Human sperm carbohydrate antigens defined by an antisperm human monoclonal antibody derived from an infertile woman bearing antisperm antibodies in her serum, *J. Exp. Med.*, 168 (1988) 343–356.
97. J. G. Gribben and M. Hallek, Rediscovering alemtuzumab: Current and emerging therapeutic roles, *Br. J. Haematol.*, 144 (2009) 818–831.
98. A. Treumann, M. R. Lifely, P. Schneider, and M. A. J. Ferguson, Primary structure of CD52, *J. Biol. Chem.*, 270 (1995) 6088–6099.
99. S. Schröter, P. Derr, H. S. Conradt, M. Nimtz, G. Hale, and C. Kirchhoff, Male-specific modification of human CD52, *J. Biol. Chem.*, 274 (1999) 29862–29873.
100. T. Kinoshita and N. Inoue, Dissecting and manipulating the pathway for glycosylphos-phatidylinositol-anchor biosynthesis, *Curr. Opin. Chem. Biol.*, 4 (2000) 632–638.
101. J. Xue and Z. Guo, Convergent synthesis of an inner core GPI of sperm CD52, *Bioorg. Med. Chem. Lett.*, 12 (2002) 2015–2018.
102. J. Xue and Z. Guo, A facile synthesis of Cerny epoxides and selectively blocked derivatives of 2-azido-2-deoxy-β-D-glucopyranose, *Tetrahedron Lett.*, 42 (2001) 6487–6489.
103. J. Lu, K. N. Jayaprakash, and B. Fraser-Reid, First synthesis of a malarial prototype: A fully lipidated and phosphorylated GPI membrane anchor, *Tetrahedron Lett.*, 45 (2004) 879–882.
104. J. Lu, K. N. Jayaprakash, U. Schlueter, and B. Fraser-Reid, Synthesis of a malaria candidate glycosylphosphatidylinositol (GPI) structure: A strategy for fully inositol acylated and phosphorylated GPIs, *J. Am. Chem. Soc.*, 126 (2004) 7540–7547.
105. N. Azzouz, F. Kamena, and P. H. Seeberger, Synthetic glycosylphosphatidylinositol as tools for glycoparasitology research, *OMICS*, 14 (2010) 445–454.
106. S. L. Bender and R. J. Budhu, Biomimetic synthesis of enantiomerically pure D-*myo*-inositol derivatives, *J. Am. Chem. Soc.*, 113 (1991) 9883–9885.
107. E. D. Soli, A. S. Manoso, M. C. Patterson, P. DeShong, D. A. Favor, R. Hirschmann, and A. B. Smith, Azide and cyanide displacements via hypervalent silicate intermediates, *J. Org. Chem.*, 64 (1999) 3171–3177.
108. X. Liu, Y.-U. Kwon, and P. H. Seeberger, Convergent synthesis of a fully lipidated glycosylphosphatidylinositol anchor of Plasmodium falciparum, *J. Am. Chem. Soc.*, 127 (2005) 5004–5005.
109. Z. J. Jia, L. Olsson, and B. Fraser-Reid, Ready routes to key *myo*-inositol component of GPIs employing microbial arene oxidation or Ferrier reaction, *J. Chem. Soc. Perkin Trans. 1* (1998) 631–632.
110. F. Kamena, M. Tamborrini, X. Liu, Y.-U. Kwon, F. Thompson, G. Pluschke, and P. H. Seeberger, Synthetic GPI array to study antitoxic malaria response, *Nat. Chem. Biol.*, 4 (2008) 238–240.
111. I. C. Almeida, M. M. Camargo, D. O. Procópio, L. S. Silva, A. Mehlert, L. R. Travassos, R. T. Gazzinelli, and A. J. Ferguson, Highly purified glycosylphosphatidylinositols from *Trypanosoma cruzi* are potent proinflammatory agents, *EMBO J.*, 19 (2000) 1476.
112. A. Ali, D. C. Gowda, and R. A. Vishwakarma, A new approach to construct full-length glycosylphosphatidylinositols of parasitic protozoa and [4-deoxy-Man-III]-GPI analogues, *Chem. Commun.* (2005) 519–521.
113. X. Wu and Z. Guo, Convergent synthesis of a fully phosphorylated GPI anchor of the CD52 antigen, *Org. Lett.*, 9 (2007) 4311–4313.

114. X. Wu, Z. Shen, X. Zeng, S. Lang, M. Palmer, and Z. Guo, Synthesis and biological evaluation of sperm CD52 GPI anchor and related derivatives as binding receptors of pore-forming CAMP factor, *Carbohydr. Res.*, 343 (2008) 1718–1729.
115. D. V. Yashunsky, V. S. Borodkin, M. A. J. Ferguson, and A. V. Nikolaev, The chemical synthesis of bioactive glycosylphosphatidylinositols from *Trypanosoma cruzi* containing an unsaturated fatty acid in the lipid, *Angew. Chem. Int. Ed. Engl.*, 45 (2006) 468–474.
116. T. G. Mayer and R. R. Schmidt, Glycosyl imidates, 78. An efficient synthesis of galactinol and isogalacatinol, *Liebigs Ann.* (1997) 859–863.
117. D. V. Yashunsky, V. S. Borodkin, P. G. McGivern, M. A. J. Ferguson, and A. V. Nikolaev, The chemical synthesis of glycosylphosphatidylinositol anchors from *Trypanosoma cruzi* Trypomastigote mucins, *Frontiers in Modern Carbohydrate Chemistry*, AmericanChemical Society, Vol. 960, pp. 285–306.
118. B. M. Swarts and Z. Guo, Synthesis of a glycosylphosphatidylinositol anchor bearing unsaturated lipid chains, *J. Am. Chem. Soc.*, 132 (2010) 6648–6650.
119. D. Crich and A. Banerjee, Stereocontrolled synthesis of the D- and L-*glycero*-β-D-*manno*-heptopyranosides and their 6-deoxy analogues. Synthesis of methyl α-L-rhamno-pyranosyl-(1-3)-D-*glycero*-β-D-*manno*-heptopyranosyl-(1-3)-6-deoxy-*glycero*-β-D-*manno*-heptopyranosyl-(1-4)-β-L-rhamnopyranosyl, a tetrasaccharide subunit of the lipopolysaccharide from *Plesimonas shigelloides*, *J. Am. Chem. Soc.*, 128 (2006) 8078–8086.
120. J. Lee and J. K. Cha, Selective cleavage of allyl ethers, *Tetrahedron Lett.*, 37 (1996) 3663–3666.
121. P. B. Alper, S.-C. Hung, and C.-H. Wong, Metal catalyzed diazo transfer for the synthesis of azides from amines, *Tetrahedron Lett.*, 37 (1996) 6029–6032.
122. J. J. Oltvoort, C. A. A. van Boeckel, J. H. De Koning, and J. H. van Boom, Use of the cationic iridium complex 1,5-cyclooctadiene-bis[methyldiphenylphosphine]-iridium hexafluorophosphate in carbohydrate chemistry: Smooth isomerization of allyl ethers to 1-propenyl ethers, *Synthesis* (1981) 305–308.
123. W. J. Masterson, J. Raper, T. L. Doering, G. W. Hart, and P. T. Englund, Fatty acid remodeling: A novel reaction sequence in the biosynthesis of trypanosome glycosyl phosphatidylinositol membrane anchors, *Cell*, 62 (1990) 73–80.
124. F. Reggiori, E. Canivenc-Gansel, and A. Conzelmann, Lipid remodeling leads to the introduction and exchange of defined ceramides on GPI proteins in the ER and Golgi of *Saccharomyces cerevisiae*, *EMBO J.*, 16 (1997) 3506–3518.
125. J. E. Ralton and M. J. McConville, Delineation of three pathways of glycosylphosphatidylinositol biosynthesis in *Leishmania mexicana*: Precursors from different pathways are assembled on distinct pools of phosphatidylinositol and undergo fatty acid remodeling, *J. Biol. Chem.*, 273 (1998) 4245–4257.
126. W. L. Roberts, J. J. Myher, A. Kuksis, M. G. Low, and T. L. Rosenberry, Lipid analysis of the glycoinositol phospholipid membrane anchor of human erythrocyte acetylcholinesterase. Palmitoylation of inositol results in resistance to phosphatidylinositol-specific phospholipase C, *J. Biol. Chem.*, 263 (1988) 18766–18775.
127. Y. Maeda, Y. Tashima, T. Houjou, M. Fujita, T. Yoko-o, Y. Jigami, R. Taguchi, and T. Kinoshita, Fatty acid remodeling of GPI-anchored proteins is required for their raft association, *Mol. Biol. Cell*, 18 (2007) 1497–1506.
128. S. Burgula, B. M. Swarts, and Z. Guo, Total synthesis of a glycosylphosphatidylinositol anchor of the human lymphocyte CD52 antigen, *Chem. Eur. J.*, 18 (2012) 1194–1201.
129. L. Han and R. K. Razdan, Total synthesis of 2-Arachidonylglycerol (2-Ara-Gl), *Tetrahedron Lett.*, 40 (1999) 1631–1634.
130. H. H. Seltzman, D. N. Fleming, G. D. Hawkins, and F. I. Carroll, Facile synthesis and stabilization of 2-arachidonylglycerol via its 1,3-phenylboronate ester, *Tetrahedron Lett.*, 41 (2000) 3589–3592.

131. A. Cartoni, A. Margonelli, G. Angelini, A. Finazzi-Agrò, and M. Maccarrone, Simplified chemical and radiochemical synthesis of 2-arachidonoyl-glycerol, an endogenous ligand of cannabinoid receptors, *Tetrahedron Lett.*, 45 (2004) 2723–2726.
132. S. R. Hanson, W. A. Greenberg, and C.-H. Wong, Probing glycans with the copper(I)-catalyzed [3+2] azide–alkyne cycloaddition, *QSAR Comb. Sci.*, 26 (2007) 1243–1252.
133. E. M. Sletten and C. R. Bertozzi, Bioorthogonal chemistry: Fishing for selectivity in a sea of functionality, *Angew. Chem. Int. Ed.*, 48 (2009) 6974–6998.
134. J. C. Jewett and C. R. Bertozzi, Cu-free click cycloaddition reactions in chemical biology, *Chem. Soc. Rev.*, 39 (2010) 1272–1279.
135. E. Lallana, R. Riguera, and E. Fernandez-Megia, Reliable and efficient procedures for the conjugation of biomolecules through Huisgen azide–alkyne cycloadditions, *Angew. Chem. Int. Ed.*, 50 (2011) 8794–8804.
136. B. M. Swarts and Z. Guo, Chemical synthesis and functionalization of clickable glycosylphosphatidylinositol anchors, *Chem. Sci.*, 2 (2011) 2342–2352.
137. D. Crich, F. Cai, and F. Yang, A stable, commercially available sulfenyl chloride for the activation of thioglycosides in conjunction with silver trifluoromethanesulfonate, *Carbohydr. Res.*, 343 (2008) 1858–1862.
138. S. Hagihara, A. Miyazaki, I. Matsuo, A. Tatami, T. Suzuki, and Y. Ito, Fluorescently labeled inhibitor for profiling cytoplasmic peptide:N-glycanase, *Glycobiology*, 17 (2007) 1070–1076.
139. V. V. Rostovtsev, L. G. Green, V. V. Fokin, and K. B. Sharpless, A stepwise Huisgen cycloaddition process: Copper(I)-catalyzed regioselective "ligation" of azides and terminal alkynes, *Angew. Chem. Int. Ed. Engl.*, 41 (2002) 2596–2599.
140. C. W. Tornøe, C. Christensen, and M. Meldal, Peptidotriazoles on solid phase: [1,2,3]-Triazoles by regiospecific copper(I)-catalyzed 1,3-dipolar cycloadditions of terminal alkynes to azides, *J. Org. Chem.*, 67 (2002) 3057–3064.
141. J. C. Jewett, E. M. Sletten, and C. R. Bertozzi, Rapid Cu-free click chemistry with readily synthesized biarylazacyclooctynones, *J. Am. Chem. Soc.*, 132 (2010) 3688–3690.
142. Y.-H. Tsai, S. Götze, N. Azzouz, H. S. Hahm, P. H. Seeberger, and D. Varon Silva, A general method for synthesis of GPI anchors illustrated by the total synthesis of the low-molecular-weight antigen from *Toxoplasma gondii*, *Angew. Chem. Int. Ed. Engl.*, 50 (2011) 9961–9964.
143. O. J. Plante, E. R. Palmacci, R. B. Andrade, and P. H. Seeberger, Oligosaccharide synthesis with glycosyl phosphate and dithiophosphate triesters as glycosylating agents, *J. Am. Chem. Soc.*, 123 (2001) 9545–9554.
144. P. Cmoch and Z. Pakulski, Comparative investigations on the regioselective mannosylation of 2,3,4-triols of mannose, *Tetrahedron: Asymmetry*, 19 (2008) 1494–1503.
145. S. Czernecki and E. Ayadi, Preparation of diversely protected 2-azido-2-deoxyglycopyranoses from glycals, *Can. J. Chem.*, 73 (1995) 343–350.
146. J. Park, S. Kawatkar, J.-H. Kim, and G.-J. Boons, Stereoselective glycosylations of 2-azido-2-deoxyglucosides using intermediate sulfonium ions, *Org. Lett.*, 9 (2007) 1959–1962.

EFFECT OF PROTEIN DYNAMICS AND SOLVENT IN LIGAND RECOGNITION BY PROMISCUOUS AMINOGLYCOSIDE-MODIFYING ENZYMES

Engin H. Serpersu and Adrianne L. Norris

Department of Biochemistry, Cellular and Molecular Biology, The University of Tennessee, Knoxville, Tennessee, USA

I. Introduction	222
II. Aminoglycoside Antibiotics	222
III. Aminoglycoside-Modifying Enzymes	224
IV. Thermodynamic Properties of Enzyme–AG Complexes	225
1. Enthalpy, Entropy, and Gibbs Energy Changes for AG Binding	225
2. Proton Linkage	229
3. Heat-Capacity Changes	231
4. Solvent Effects	234
V. Protein Dynamics in Substrate Recognition and Substrate Promiscuity of AGMEs	237
VI. Conclusions and Future Considerations	241
Acknowledgments	243
References	243

Abbreviations

2-DOS, 2-deoxystreptamine; AAC(2′)-Ic, aminoglycoside acetyltransferase(2′)-Ic; AAC(3)-IIIb, aminoglycoside acetyltransferase(3)-IIIb; AAC(6′)-Iy, aminoglycoside acetyltransferase(6′)-Iy; AG, aminoglycoside; AGME, aminoglycoside-modifying enzyme; AMP, adenosine 5′-phosphate; ANT(2″)-Ia, aminoglycoside nucleotidyltransferase (2″)-Ia; APH(3′)-IIIa, aminoglycoside phosphotransferase(3′)-IIIa; CoASH, coenzyme A; GNAT, general control of amino acid synthesis protein 5-related N-acetyltransferase; HSQC, heteronuclear single quantum coherence; ITC, isothermal titration calorimetry; NMR, nuclear magnetic resonance; RNA, ribonucleic acid; ΔASA, solvent-accessible surface area

I. INTRODUCTION

Aminoglycoside antibiotics are a large group of aminocyclitols that are used clinically to treat serious infections. They are among the earliest antibiotics to be used in clinical practice. Streptomycin was the second antibiotic to be discovered after penicillin, and it was used to cure infectious diseases, in particular, tuberculosis.[1] Aminoglycosides bind to the 16S RNA subunit in the bacterial ribosome and interfere with protein synthesis, eventually causing cell death.[2,3] Today, however, their efficacy is threatened, as with all other antibiotics, by an ever-increasing incidence of bacterial strains resistant to their action. The major mode of bacterial resistance to aminoglycoside antibiotics is enzymatic modification of the antibiotic by N-acetyl-, O-nucleotidyl-, and O-phosphotransferases.[4-6] Crystallographic, solution, and computational studies have shown structural features of various RNA–AG complexes that shed light on how modification of these compounds may disrupt their interaction with RNA.[7-13] A number of reviews have been published on various kinetic, biochemical, and structural aspects of the interaction of aminoglycoside antibiotics (AGs) with aminoglycoside-modifying enzymes (AGMEs) and nucleic acids.[4,6,14-20] In this article, we survey dynamic and thermodynamic aspects of AG–enzyme interactions and discuss dynamic properties and unusual effects of solvent in the formation of enzyme–AG complexes and their implications on substrate recognition and promiscuity of these enzymes. Even though some of these enzymes can catalyze side reactions, the term "promiscuity" is used here to define the ability of these enzymes to modify a broad range of AGs.

II. AMINOGLYCOSIDE ANTIBIOTICS

Representative structures of the two major families of aminoglycosides, namely the kanamycins and neomycins, are shown in Fig. 1. The primed ring is 6-amino-6-deoxy-D-glucose in kanamycin A and 2,6-diamino-2,6-dideoxy-D-glucose in neomycin B (henceforth neomycin). The unprimed ring in kanamycin A and neomycin is 2-deoxystreptamine (2-DOS). The double-primed ring is 3-amino-3-deoxy-D-glucose in kanamycin A and D-ribose in neomycin. The fourth ring in neomycin is identified as the triple-primed ring and is 2,6-diamino-2,6-dideoxy-D-glucose. Conformational aspects of enzyme-bound aminoglycosides have been reviewed earlier[21] and will not be repeated here. Suffice it to indicate that the most significant structural feature of

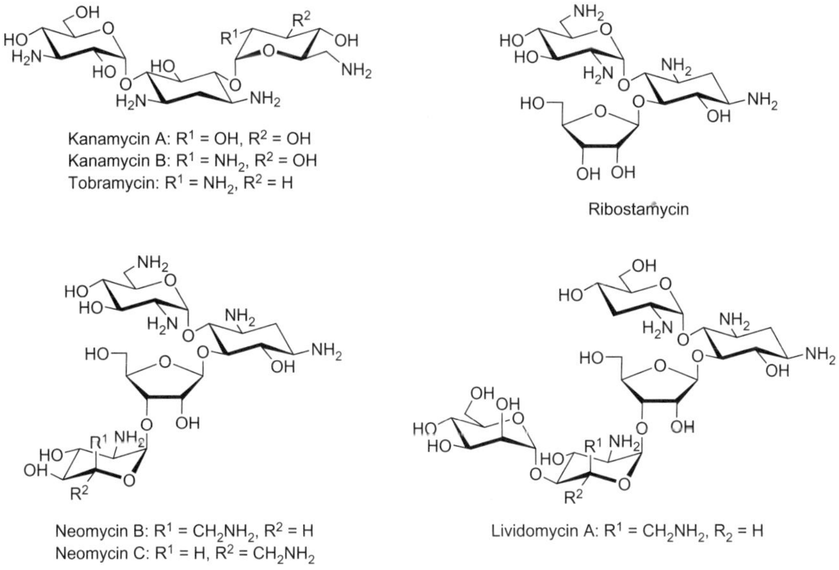

FIG. 1. Representative aminoglycoside structures.

enzyme-bound aminoglycosides is the remarkable similarity in the conformations of the primed and unprimed rings of AGs.

As determined by NMR studies using four different enzymes and a number of structurally diverse AGs, the enzyme-bound conformations of AGs showed that these two rings are superposable, regardless of the enzyme or the antibiotic.[22–25] Structures of AGs, also determined by NMR in complexation with 16S RNA, their natural target[8,11] and by X-ray crystallography in complex with the enzyme APH(3′)-IIIa],[26] also superimpose in the same manner. Figure 2 shows stereo pairs of overlaid conformations of several aminoglycosides bound to four different enzymes (yellow), 16S RNA (blue) determined by NMR and bound to APH(3′)-IIIa (red) taken from the crystal structures of APH(3′)-IIIa. Later, another crystallographic study performed with another aminoglycoside phosphotransferase, APH(3′)-IIa, also showed that the primed and unprimed rings of AGs bound to this enzyme superimpose well with those determined for APH(3′)-IIIa.[27] It is very clear that, while the primed and unprimed rings superimpose remarkably well, the rest of the molecules show a great diversity in their spatial orientation and cover a $\sim 180°$ span, even in the active site of the same enzyme.[21] Thus, the primed and unprimed rings of aminoglycosides appear to form the major recognition unit by these enzymes. In fact, this type of analysis allowed us

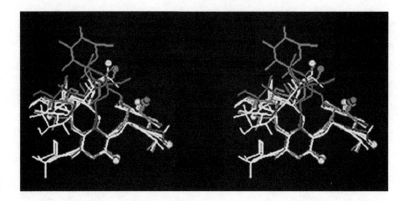

FIG. 2. Stereoview of enzyme/RNA-bound conformations of aminoglycoside antibiotics. In yellow are enzyme-bound AG structures determined with four different enzymes, while blue structures are RNA-bound gentamicin[11] and paromomycin[8] (for simplicity only three rings are shown), all derived via NMR. Red structures are APH(3′)-IIIa-bound kanamycin A and neomycin as determined by X-ray crystallography.[26] All structures are overlaid at the A and B rings. (Reprinted with permission from Ref. 25. Copyright 2002, American Chemical Society.) (See Color Insert.)

to suggest that the inability of the aminoglycoside nucleotidyltransferase (2″)-Ia [ANT(2″)-Ia] to modify neomycins is because the site of modification on neomycins by this enzyme, the 2″-site, moves ∼4.5 Å away from the attacking nucleophile, which would prevent the direct transfer of AMP to AG.[28]

III. Aminoglycoside-Modifying Enzymes

Currently, there are more than 50 enzymes known that can modify aminoglycoside antibiotics and render them ineffective against infectious bacteria. They show different levels of substrate promiscuity, ranging from the ability to modify more than a dozen AGs versus just one or two. However, they do show specificity with respect to the site of modification on the AGs. This specificity is incorporated into their nomenclature. For example, aminoglycoside phosphotransferase(3′) denotes an enzyme that phosphorylates the 3-OH group in the primed ring. This in itself, however, still does not specify a unique enzyme until the gene encoding the protein is also included, as in the aminoglycoside phosphotransferase(3′)-IIIa [APH(3′)-IIIa]. This is because multiple enzymes can catalyze the same modification at a given site on AGs. For example, there are more than a dozen enzymes, with varying degrees of homology, that acetylate the N-3 atom of AGs. While some of them are highly promiscuous, such as the aminoglycoside acetyltransferase(3)-IIIb [AAC(3)-IIIb] that acetylates more than 10 structurally different AGs, others such as the aminoglycoside acetyltransferase(3)-Ib can acetylate only gentamicin and fortimycin.[4] Figure 3 shows sequence alignments using a Clustal2.1 algorithm of (a) three proteins that can bind several of the same AGs but catalyze different modifications at different sites and (b) three proteins that catalyze the same reaction at the same site but have little substrate overlap.

Crystal structures of several AGMEs with and/or without bound substrates are also available,[26,27,29–38] many of which have been studied by using kinetic, biochemical, and biophysical techniques.[39–58] However, the molecular basis of differing substrate selectivity by these enzymes remains unexplained. Global parameters determined from thermodynamic or kinetic studies do not address such issues. Static structures provide guidance but fall short in predicting dynamic and thermodynamic aspects of enzyme–ligand complexes with AGMEs. In this article, we address these issues by attempting to describe molecular properties behind the AG selection of these enzymes through a combination of global properties of enzyme–AG complexes with site-specific data.

The superimposability of the primed and 2-DOS rings of all enzyme-bound conformations of AGs described in the previous section, along with the observation that the most buried parts of enzyme-bound AGs are the primed and 2-DOS rings (as demonstrated from the crystal structure of AAC(2′)-Ic[33] with three different AGs), may lead to a conclusion that all or most AGMEs that can modify these antibiotics may have active sites having similar structures and properties. However, in reality, it appears that the dynamic properties of both the ligands and the enzymes play a significant role in the formation of enzyme–AG complexes. These aspects are the main subject of this survey. In the following sections, we discuss the identification of factors that contribute in rendering the formation of enzyme–AG complexes thermodynamically favorable and the roles of these complexes in the process. First, we summarize global thermodynamic properties of various enzyme–AG complexes and then describe identification of the types of molecular interactions that are major contributors to these global properties, along with the sites that are responsible for these interactions.

IV. THERMODYNAMIC PROPERTIES OF ENZYME–AG COMPLEXES

1. Enthalpy, Entropy, and Gibbs Energy Changes for AG Binding

Thermodynamic data for AG binding to AGMEs are available for four enzymes, the aminoglycoside phosphotransferase(3′)-IIIa [APH(3′)-IIIa],[59–62] the aminoglycoside acetyltransferase(3)-IIIb [AAC(3)-IIIb],[63–65] the aminoglycoside nucleotidyltransferase

A

Clustal2.1 multiple sequence alignment

```
[AAC_3_-IIIb]    MTSATASFATRTSLAADLAALGLAWGDAIMVHAAVSRVGRLLDGPDTIIAALRDTVGPGG  60
[ANT_2"_]        MACYDCFFVQSMPRASKQQAR-----------YAVGRCLMLWSSNDVTQQGSRPKTKLG-  48
[APH_3'_-IIIa]   MAKMRISPELKKLIEKYRCVKDTEG-------MSPAKVYKLVGENENLYLKMTDSRYKG-  52
                 *:             .        : .:  * . :  .  *

[AAC_3_-IIIb]    TVLAYADWEARYEDLVDDAGRVPPEWREHVPPFDPQRSRAIRDNGVLPEFLRTTPGTLRS 120
[ANT_2"_]        -RMDTIQVTLIHKILAAADERNLPLWIGGGWAIDARLGRVTRKH---DDIDLTFPGERR- 103
[APH_3'_-IIIa]   ---TTYDVEREKDMMLWLEGKLPVPKVLHFERHDGWSNLLMSEA----DGVLCSEEYEDE 105
                    :   .  :     :         *   . .  .

[AAC_3_-IIIb]    GNPGASLVALGAKAEWFTADHPLDYGYGEGSPLAKLVEAGGKVLMLGAPLDTLTLLHHAE 180
[ANT_2"_]        GELEAIVEMLGGR-----VMEELDYGF--------LAEIGDELLDCEPAWWADEAYEIAE 150
[APH_3'_-IIIa]   QSPEKIIELYAECIRLFHSIDISDCPY--------TNSLDSRLAELDYLLNNDLADVDCE 157
                                     .  *  :       . ...:            .*

[AAC_3_-IIIb]    HLADIPGKRIKRIEVPFATPTGTQWRMIEEFDTGDPIVAGLAEDYFAGIVTEFLASGQGR 240
[ANT_2"_]        APQGSCPEAAEGVIAGRPVRCNSWEAIIWDYFYYADEVP--FVDWPTKHIESYRLACTS- 207
[APH_3'_-IIIa]   NWEEDTPFKDPRELYDFLKTEKPEEELVFSHGDLGDSNIFVKDGKVSGFIDLGRSGRADK 217
                   .          ::  ..              .    .  :  :  .    .

[AAC_3_-IIIb]    QGLIGAAPSVLVDAAAITAFGVTWLEKRFGTPSP------------- 274
[ANT_2"_]        ---LGAEKVEVLRAAFRSRYAA------------------------ 226
[APH_3'_-IIIa]   WYDIAFCVRSIREDIGEEQYVELFFDLLGIKPDWEKIKYYILLDELF 264
                    :.       :          :
```

B

```
[AAC_3_-IIa]     -------MHTQKAITEALQKLGVQSGDLLMVHASLKSIGPVEGGAETVVAALRSAVGPTG  53
[AAC_3_-IIIb]    MTSATASFATRTSLAADLAALGLAWGDAIMVHAAVSRVGRLLDGPDTIIAALRDTVGPGG  60
[AAC_3_-Ib]      ------MLWSSNDVTQQGSRPKTKLGGSMSIIATVK-IGPDE------ISAMRAVLDLFG  47
                        :  : .      *  : :.*  ::  :*          ::*:* .:.  *

[AAC_3_-IIa]     TVMGYASWDRSPYEETLNGA-RLDDNARRTWPPFDPATAGTYRGFGLLNQFLVQAPGARR 112
[AAC_3_-IIIb]    TVLAYADWEAR-YEDLVDDAGRVPPEWREHVPPFDPQRSRAIRDNGVLPEFLRTTPGTLR 119
[AAC_3_-Ib]      -------------------------KEFEDIPTYS----------------------- 57
                                          :  . *.:.

[AAC_3_-IIa]     SAHPDASMVAVGPLAETLTEPHELGHALGEGSPNERFVRLGGKALLLGAPLNSVTALHYA 172
[AAC_3_-IIIb]    SGNPGASLVALGAKAEWFTADHPLDYGYGEGSPLAKLVEAGGKVLMLGAPLDTLTLLHHA 179
[AAC_3_-Ib]      DRQPTNEYLANLLHSETFIALAAFDRGTAIG---------GLAAYVLPKFEQARSEIYIY 108
                 . :*  . :*   :*  :  :...*       *  .:*    :: : ::

[AAC_3_-IIa]     EAVADIPNKRWVTYEMPMPGRDGEVAWKTASDYDSNGILDCFAIEGKQDAVETIANAYVK 232
[AAC_3_-IIIb]    EHLADIPGKRIKRIEVPFATPTG-TQWRMIEEFDTG---DPIVAGLAEDYFAGIVTEFLA 235
[AAC_3_-Ib]      DLAVASSHRRLG----------------------------VATALISHLKRVAVELGA 138
                 :  . ..:*

[AAC_3_-IIa]     LGRHREGVVGFAQCYLFDAQDIVTFGVTYLEKHFGTTPIVPAHEAIERSCEPSG 286
[AAC_3_-IIIb]    SGQGRQGLIGAAPSVLVDAAAITAFGVTWLEKRFGTPSP--------------- 274
[AAC_3_-Ib]      YVIYVQADYGDDPAVALYTKLGVREDVMHFDIDPRTAT---------------- 176
                 :.  *     . .:    . .*  ::    *..
```

FIG. 3. Sequence alignments of (A) three promiscuous proteins having overlapping AG profiles: APH (3′)-IIIa, AAC(3)-IIIb, and ANT(2″)-Ia; (B) three proteins having high (AAC-IIIb), medium (AAC-IIa), and low (AAC-Ib) substrate promiscuity.

(2″)-Ia [ANT(2″)-Ia],[28,66,67] and the aminoglycoside acetyltransferase(6′)-Iy [AAC (6′)-Iy].[68] These studies showed that binding of AGs to all four enzymes is enthalpically favored and entropically disfavored. The only exceptions are with acetyltransferases, in which only kanamycin A with AAC(3)-IIIb and amikacin and netilmicin with AAC(6′)-Iy have barely positive $T\Delta S$ values. We should note that data acquired with AAC(6′)-Iy are reported as the observed enthalpy (ΔH_{obs}), whereas the others are reported as the intrinsic enthalpy (ΔH_{int}). As discussed later, ΔH_{obs} includes the contribution from buffer in the form of heat of ionization, and this is strongly dependent on the buffer used.

The binding enthalpy of AGs to AGMEs varies widely in an antibiotic- and enzyme-dependent manner. Differences as large as 40 kcal/mol can be observed for binding of different AGs to the same enzyme [APH(3′)-IIIa].[61] Similarly, binding of the same AG to different enzymes also shows large variations in binding enthalpy. As a consequence of enthalpy–entropy compensation, the entropic contribution ($T\Delta S$) to formation of the complex also shows large, antibiotic-dependent variations. In contrast to these observations, ΔG for binding to all enzymes for all AGs varies within 2–3 kcal/mol over the range between −6 and −8.5 kcal/mol. An example of data from isothermal titration calorimetry is shown in Fig. 4 to demonstrate the differences between a weak and strong binding of AGs to APH(3′)-IIIa.[61]

Binding enthalpies that are significantly different were observed, even for structurally very similar aminoglycosides, and these results allowed identification of sites in AGs that make the most significant contacts with enzymes. For example, the presence of —OH in kanamycin A versus —NH$_2$ in kanamycin B at the 2′-site constitutes the only difference between these two AGs. Their binding enthalpies to all four enzymes show differences ranging from 3.4–3.6 kcal/mol [AAC(6′)-Iy and APH(3′)-IIIa] to 6.9 kcal/mol [AAC(3)-IIIb]. Additional evidence consistent with these data comes from kinetic experiments on the metal-ion dependence of substrate preference with two AGMEs. For APH(3′)-IIIa, AGs with 2′-NH$_2$ are preferred substrates with MnATP over those with 2′-OH, and this is reversed with MgATP.[69] Exactly, the opposite is true for ANT(2″)-Ia, where the presence of —OH at the 2′-site makes AGs better substrates with MnATP.[66] It is noteworthy that several crystal structures of aminoglycoside-phospho and -acetyltransferases show that 2′-site of AGs does not interact with enzymes.[26,27,33] Therefore, it is difficult to rationalize these data based on static structures of AGMEs. The same conclusion was derived from the crystallographic studies with AAC(2′)-Ic in which the AG-dependent variations in k_{cat} values could not be rationalized from the static structures of the enzyme, as determined with three different AGs.[33] These observations clearly indicate that dynamic features of these enzymes play a significant role in ligand binding and catalysis.

Small differences at other sites in the AG structure also have an impact on the binding enthalpy in an enzyme-dependent manner. To this end, —OH versus —NH$_2$ at the 6′-site (paromomycin and neomycin, respectively) show very large differences

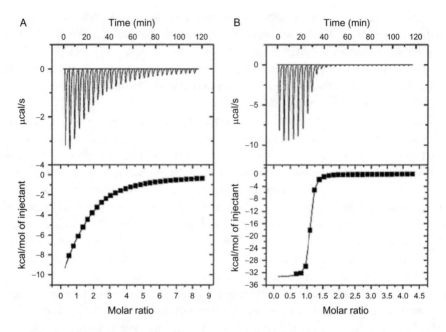

FIG. 4. Typical isotherms of weak (left) and tight (right) binding of AGs to an AGME. The upper panels show the raw data (thermal power). Time integration of the thermal power yields the heat of injection, which is plotted against the molar ratio of ligand to enzyme in the lower panels. Solid lines in the bottom panels constitute the least-squares fitting of the data to a one-site binding model.(Reprinted with permission from Ref. 61. Copyright 2004, American Chemical Society.)

in binding enthalpy ($\Delta\Delta H$) with APH(3′)-IIIa (14.8 kcal/mol) and AAC(6′)-Iy (7.2 kcal/mol), but less than 1 kcal/mol differences with AAC(3)-IIIb (0.3 kcal/mol) and ANT(2″)-Ia (0.9 kcal/mol). Similarly, the 3′-site (where —OH in kanamycin B becomes —H in tobramycin) was most significant for APH(3′)-IIIa (5.7 kcal/mol difference in ΔH_{int} between these two AGs), which is not surprising since this site is the site of modification by this enzyme. However, a 2.1 kcal/mol difference was also detected for AAC(3)-IIIb, which modifies a site on a different ring. On the other hand, AAC(6′)-Iy, an enzyme that modifies a site in the same ring, shows a difference of only 0.9 kcal/mol. As indicated earlier, entropic compensation narrows down the differences to be within 2–3 kcal/mol in the value of ΔG. There was no difference in binding enthalpies of kanamycin B and tobramycin to ANT(2″)-Ia; this enzyme modifies the site most distant from the 3′-OH group. In this case however, the difference in the entropic contribution results in a 1 kcal/mol difference in ΔG for binding of these two AGs to ANT(2″)-Ia.

The binding enthalpy of AGs to AGMEs in the presence of a co-substrate (or co-substrate analogue) shows interesting variations. Counterintuitively, the binding enthalpy of AGs to APH(3′)-IIIa becomes dramatically less favored when CaATP, a competitive inhibitor with respect to the substrate MgATP, is present.[61] The increase in enthalpy (becoming less negative) is between 11 and 20 kcal/mol, and this can be explained only by the surprising observations made in H/D exchange experiments that are described later. ANT(2″)-Ia separates kanamycins from neomycins inasmuch as the binding enthalpy of AGs becomes less favorable for kanamycins and more favorable for neomycins.[28] Interestingly, although neomycins cannot be adenylated by this enzyme, they do bind with high affinity to the protein. In contrast, the binding enthalpy of AGs to AAC(3)-IIIb becomes more favorable for all AGs when the coenzyme A (CoASH) is present.[63] This also holds true for AAC(6′)-Iy, although only one AG, lividomycin, was tested with this enzyme.[68]

2. Proton Linkage

Binding of all AGs to AGMEs is accompanied by a net change in the protonation state of functional groups in the ligand and the enzyme, indicating shifts in pK_as of these groups upon the formation of enzyme–AG complexes.[28,59,61,63] Similarly, shifts in pK_as were also observed in AG–nucleic acid interactions.[70,71] Changes in pK_as cause a net release or uptake of protons, and this triggers a response from the buffer. Therefore, ΔH_{obs} will be dependent on the heat of ionization of the buffer used in binding studies. Determination of intrinsic enthalpy requires the use of several buffers having different heats of ionization and the data to be analyzed by using the equation $\Delta H_{obs} = (\Delta n)(\Delta H_{ion}) + \Delta H_{int}$. This equation is the simplified form of the equation $\Delta H_{obs} = \Delta H_{int} + \Delta n[\alpha \Delta H_{ion} + (1-\alpha)\Delta H_{enz}] + \Delta H_{bind}$.[72,73] The ΔH_{obs} value denotes the observed binding enthalpy of formation of a complex in a given buffer, where ΔH_{ion} describes the heat of ionization of the buffer. The Δn term is the net number of protons transferred as a result of ligand binding, and ΔH_{int} is the intrinsic enthalpy of binding. The term $\Delta n[\alpha \Delta H_{ion} + (1-\alpha)\Delta H_{enz}]$ constitutes the heat of ionization of groups from the ionization of buffer and the protein to maintain pH, where α represents the fraction of protonation contributed by the buffer.[72,73] In addition, ΔH_{bind} constitutes the heat of binding of buffer to the enzyme. In the presence of a high salt concentration (i.e., 100 mM NaCl), ΔH_{bind} is assumed to be zero. Thus, a plot of ΔH_{obs} versus ΔH_{ion} yields a straight line with a slope of Δn and intercept of ΔH_{int}. A representative plot for different complexes of AAC(3)-IIIb is shown in Fig. 5.

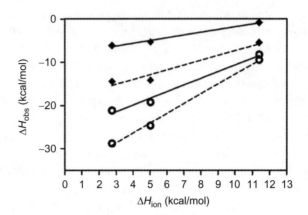

FIG. 5. Dependence of ΔH_{obs} on ΔH_{ion} for the binding of kanamycin B (solid lines) and paromomycin (dashed lines) in the presence (○) and the absence (♦) of CoASH.(Reprinted with permission from Ref. 63. Copyright 2010 American Chemical Society.)

Data available for APH(3′)-IIIa, AAC(3)-IIIb, and ANT(2″)-Ia show that there is no unique pattern for a particular enzyme or AG. The Δn term is different for each AG with a given enzyme, and binding of any AG to different enzymes also yields different Δn values, and these are further altered in the presence of the co-substrate. In general, there is always a net proton uptake when AGs bind to AGMEs. There are only a few exceptions, such as the binding of ribostamycin and amikacin to APH(3′)-IIIa, which occurs with a slight net release of protons, but is reversed in the presence of the metal–ATP co-substrate.[61] In the case of AAC(3)-IIIb, the only exception is formation of the AAC(3)-IIIb–kanamycin A complex, which yields $\Delta n \approx 0$. Formation of the ternary AAC(3)-IIIb–CoASH–kanamycin A complex, however, proceeds with a $\Delta n = 1.3$, denoting a net uptake of protons.[63]

We emphasize that caution should be exercised in the interpretation of the observed Δn values. The ΔH_{int} value determined in this manner still includes the heat of ionization of groups from both the protein and the ligand that are contributing to Δn, and this may represent the true ΔH_{int} value only when $\Delta n = 0$. Even if this condition is met, it still cannot be used as evidence against pK_a shifts in different functional groups because Δn represents the net proton balance. An example of this situation was observed in the binding of neomycin to APH(3′)-IIIa. At pH 6.7, Δn was determined to be ~ 0. However, pK_as of the amine groups of neomycin, as determined by NMR, indicated that the net proton uptake for the ligand alone should have been ~ 0.7 under these conditions. This observation clearly indicates that a net release of protons from the functional groups of the enzyme coincidentally matched the net uptake of the ligand, yielding a net value of zero for Δn.[59] In fact, computational studies have shown that a large number of ionizable groups alter their pK_as significantly when AGs bind to APH(3′)-IIIa in which ∼50% of the residues have ionizable groups (Serpersu and Ullmann, unpublished data). As already mentioned, the binding of kanamycin A to AAC(3)-IIIb proceeds with apparent $\Delta n \approx 0$. However, it is difficult to imagine that no changes will occur in pK_as of any one of the four amine groups of kanamycin A, as well as other functional groups of the enzyme, upon binding. We consider that this represents another coincidental cancellation of protonation and deprotonations under the experimental conditions, as observed with the binding of neomycin to APH(3′)-IIIa. Therefore, attempts to dissect Δn, based on ITC data alone, are likely to be misleading.

3. Heat-Capacity Changes

The most detailed thermodynamic data are available only for two AGMEs, APH (3′)-IIIa and AAC(3)-IIIb. The rest of this article is devoted mostly to the description and discussion of unusual thermodynamic properties of these two enzymes, both of which display very high substrate promiscuity and can modify more than ten structurally diverse AGs. Their dynamic and thermodynamic properties may hold keys to their ability to modify a broad range of AGs. The results may lead to an understanding of the molecular principles underlying the substrate promiscuity of these two enzymes that catalyze different reactions to modify different rings on AGs and show a large overlap in their substrate spectrum, despite the very small sequence homology (<5%) in their primary structure (Fig. 3).

The change in the heat capacity of proteins attributable to a ligand-binding event is affected by several factors, such as hydrophobic interactions, electrostatic charges, hydrogen bonds, intramolecular vibrations, and conformational entropy.[74] Changes in interactions of hydrophobic groups with solvent and alterations of low-frequency vibrational modes of the protein (i.e., stiffening or loosening of the protein) are among the most significant of these effects. Determination of the enthalpy of ligand binding as a function of temperature for a narrow temperature range usually yields a straight line. The change in heat capacity (ΔC_p) can be obtained from the slope of this line.

Although such data are available only for the binding of AGs to APH(3′)-IIIa[60] and AAC(3)-IIIb,[63] extremely surprising results highlighted some of the key thermodynamic factors that govern the substrate recognition and discrimination by these two highly promiscuous enzymes. These observations may reflect some of the general molecular properties of ligand recognition, not only by promiscuous AGMEs but also in the broader context of ligand recognition in general.

There are significant differences in the binding of kanamycin A and neomycin to APH(3′)-IIIa. The ΔC_p is temperature independent for the binding of neomycin to APH(3′)-IIIa, whereas strong dependence is observed for the binding of kanamycin A (Fig. 6). These observations were the first to demonstrate a difference in the interaction of this enzyme between kanamycins and neomycins, as previous kinetic and binding studies did not reveal any parameter that would distinguish interactions of these two classes of AGs with APH(3′)-IIIa. The dependence of enthalpy on temperature is unusually strong for both AGs and yields ΔC_p values of -1.6 kcal/mol/deg for neomycin and ~ -0.7 kcal/mol/deg below 30 °C and ~ -3.8 kcal/mol/deg above this temperature for kanamycin A.[60] These values are much higher than those determined for common carbohydrate–protein interactions, which are usually within the range of -0.1 to -0.5 kcal/mol/deg.[75,76] Neither the unusually negative values of ΔC_p nor the large difference in its values between the binding of kanamycin A and neomycin to APH(3′)-IIIa can be explained by changes in the solvent-accessible surface area (ΔASA) alone. Large conformational changes and/or significantly altered dynamic properties of the protein, such as changes in low-frequency vibrational modes, may have significant impact in the formation of these complexes.

A crystal structure of the enzyme in a binary complex with any AG is not available to help explain these differences. However, the crystal structures of the apoenzyme, nucleotide–enzyme, and enzyme–nucleotide–AG complexes show that no significant domain movements occur upon formation of these complexes.[26] Figure 7 shows superimposed structures of the apo-APH(3′)-IIIa,[31] APH(3′)-IIIa–MgAMPPNP,[36]

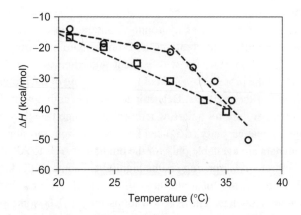

FIG. 6. Change in heat capacity upon interaction of APH(3′)-IIIa with neomycin (□) and kanamycin A (○). Reprinted with permission from Ref. 61. Copyright 2004, American Chemical Society.

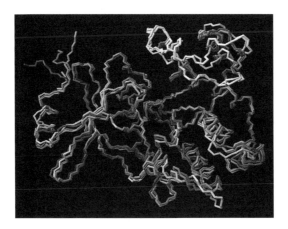

FIG. 7. Superimposed structures of APH(3′)-IIIa in apo (yellow), nucleotide (blue), and nucleotide–AG complexes with neomycin (green) and kanamycin (red). (See Color Insert.)

APH(3′)-IIIa–MgADP–kanamycin A,[26] and APH(3′)-IIIa–MgADP–neomycin[26] complexes. The major region showing a significant variation is the loop formed by residues 147–170 just above the AG-binding site, which is highlighted in white in all of the structures. As indicated by the authors, the caveat is that APH(3′)-IIIa is a ∼30 kDa monomeric enzyme having a strong tendency to form dimers via two intermolecular disulfide bridges. One of these is Cys-156, located in the middle of the loop covering the AG-binding site. In crystal structures of most complexes of this enzyme, the dimer has Cys-156 forming a disulfide bond with Cys-19 of the other monomer in the same unit cell. In other structures, the disulfide bond is with a monomer in the adjacent unit cell. It is not known to what extent these variations may affect the observed differences in the orientation of this loop. In solution, however, formation of the dimer is observed to have a profound effect on the binding affinity of AGs to this enzyme, despite a separation distance of ∼20 Å, as determined from crystallographic studies, between the two active sites. The binding affinity of neomycin to each monomer of the dimer differs by approximately three orders of magnitude.[59] In any case, the lack of domain movements observed in the static structures of the enzyme suggests that changes in dynamics of the protein must be one of the major contributors to ΔC_p, and this conclusion is also supported by NMR data, as described later in Section V.

Aminoglycoside-dependent differences in ΔC_p are also observed for AAC(3)-IIIb in an even more dramatic manner than that observed with APH(3′)-IIIa. For this enzyme, the binding of neomycin causes a decrease in heat capacity, while binding of

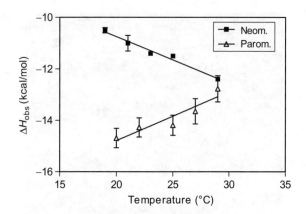

FIG. 8. Changes in heat capacity in the association of AAC(3)-IIIb to neomycin (■) and paromomycin (▲).(Adapted with permission from Ref. 64. Copyright 2011 American Chemical Society.)

paromomycin, a structurally similar AG, causes an increase in the heat capacity[64] (Fig. 8). As discussed later in Section V, these opposite trends are attributed to different conformations of a large loop that is at the AG-binding site of the enzyme. There is only one other case where binding of ligands to a protein causes the heat capacity to change in opposite directions.[77] In that instance, however, the ligands are structurally very different.

4. Solvent Effects

Solvent effects are one of the major contributors to ΔC_p, and the surprisingly large and unexpected AG-dependent differences observed in this parameter suggest that solvent effects may be significant for AG–enzyme interactions. Osmotic-stress experiments, performed by isothermal titration calorimetry (ITC), showed that there is a significant difference in solvent displacement between the binding of kanamycin A and neomycin to ANT(2″)-Ia. While the binding of kanamycin A expelled ~20 water molecules, there was no displacement with neomycin.[28] The crystal structure of AAC (2′)-Ic showed that the binding of tobramycin displaces six to eight water molecules.[33]

The binding enthalpy of kanamycins and neomycins to APH(3′)-IIIa showed a very unusual pattern that had never been observed before. In this situation, the $\Delta\Delta H$ ($\Delta H_{H_2O} - \Delta H_{D_2O}$) values between kanamycins and neomycins have opposite signs, such that kanamycins bind with a more favorable enthalpy in H_2O than in D_2O, whereas neomycins bind with more favorable enthalpy in D_2O than in H_2O (Fig. 9).[60]

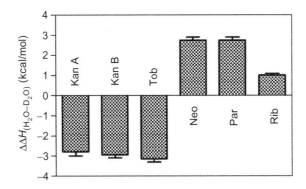

FIG. 9. Solvent-dependent differences in binding enthalpies of kanamycins and neomycins to APH(3′)-IIIa.(Reprinted with permission from Ref. 60. Copyright 2008 American Chemical Society.)

These data suggest that APH(3′)-IIIa uses solvent rearrangement to differentiate between the neomycin-class aminoglycosides and the kanamycin class. This factor may be one of the major contributors to the large differences observed in ΔC_p. It may be recalled that the 2-deoxystreptamine and primed rings of all enzyme-bound aminoglycosides are well superimposed, but differences arise in the rest of the antibiotic structure. Elastic network analysis demonstrated that there are two stretches of backbone residues located in two different domains of APH(3′)-IIIa that show differences in their correlated motions with kanamycin A and neomycin B (yellow segments in Fig. 10).[78]

The nonsuperimposable rings of neomycin and kanamycin A are oriented differently with respect to these two segments, suggesting that this interface is a potential region of solvent rearrangement that is responsible for the class-specific differences observed between kanamycins and neomycins in both $\Delta\Delta H$ and ΔC_p. This segment is highlighted in Fig. 10, which shows the superimposed crystal structures of APH(3′)-IIIa–MgADP–kanamycin A and APH(3′)-IIIa–MgADP–neomycin complexes. Again, these are suggestive of a significant role of protein dynamics and solvent interactions in ligand recognition by this enzyme. This is because, as shown in Fig. 10, both kanamycin A- and neomycin-bound structures of the enzyme, as determined by X-ray crystallography, are superimposable and do not show large differences in solvent-exposed surfaces between the two complexes that would yield clues for explaining such a dramatic solvent effect.

For the formation of enzyme–AG complexes of AAC(3)-IIIb, the solvent effect was again significant, and in some sense even more surprising than that observed with APH(3′)-IIIa. In this instance, a dramatic difference in $\Delta\Delta H$ values was observed for

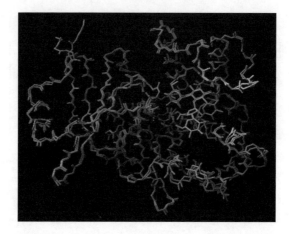

FIG. 10. Superimposed structures of APH(3′)-IIIa–MgADP–neomycin and APH(3′)-IIIa–MgADP–kanamycin A. Neomycin (green) and kanamycin (purple) are on the right side. The purple molecule on the left is MgADP. Yellow highlights those residues showing AG-dependent differences in their correlated motions. (See Color Insert.)

the binding of neomycin and paromomycin to AAC(3)-IIIb in H_2O and D_2O. These two AGs are structurally almost identical, having just a single change at the 6′-site (which is —OH in paromomycin and —NH_2 in neomycin), and yet the binding of neomycin to AAC(3)-IIIb is enthalpically more favored in D_2O than in H_2O, while the binding is exactly opposite for paromomycin.[64] This behavior is, however, only true for temperatures below 27 °C, and this is another very surprising and a first-time observation. These data show that the major contributors to the ΔC_p of the interaction of AAC(3)-IIIb with ligands are temperature and antibiotic dependent. When neomycin binds to AAC(3)-IIIb, the $\Delta\Delta H(\Delta H_{H_2O}-\Delta H_{D_2O})$ shows a strong dependence on temperature such that, at low temperatures, changes in low-frequency vibrational modes on the protein dominate the ΔC_p, but as the temperature increases, the balance between vibrational and solvent contributions shifts increasingly more toward the latter (Fig. 11). However, with paromomycin, there is no change in the $\Delta\Delta H$ value with temperature. This suggests that either solvent effects dominate at all temperatures, or effects from multiple contributions oppose each other and coincidentally cancel out, leaving $\Delta\Delta H$ unchanged. In either case, it is clear that AAC(3)-IIIb responds very differently to neomycin and paromomycin, and solvent reorganization plays a significant role in the AG-recognition process. The unusual interplay of contributions of changes in low-frequency vibrational modes of the protein (namely, stiffening or loosening of the enzyme), and solvent effects and their temperature and

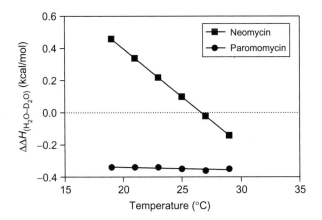

FIG. 11. Temperature dependency of $\Delta\Delta H$ for association of neomycin (■) and paromomycin (●) with AAC(3)-IIIb.(Adapted with permission from Ref. 64. Copyright 2011 American Chemical Society.)

AG dependence on the formation of enzyme–AG complexes, may not be specific to AAC(3)-IIIb because, to the best of our knowledge, there are no such data available for any other enzyme. These observations clearly indicate that dynamic features of the AGMEs play a very important role in rendering the binding of AGs to the enzyme thermodynamically favorable under different conditions. Observations such as these also show that solvent not only plays a significant role in substrate recognition but also has differential effects with each AG. Finally, it is very unlikely that the unusual solvent effects observed with different AGMEs that catalyze different reactions are coincidental. These properties are likely to apply to other AGMEs, or even other promiscuous enzymes in general.

V. Protein Dynamics in Substrate Recognition and Substrate Promiscuity of AGMEs

From fluid-breathing motions to conformational changes, structural dynamics are a property of proteins that is often important, if not critical, to their function. As many AGMEs bind to and/or modify a large number of structurally diverse AGs, it follows that dynamics could play a role in their promiscuous nature. For APH(3′)-IIIa, crystal structures of the apoenzyme as compared with the nucleotide or nucleotide/antibiotic complexes reveal that the protein is well structured in all features except that a 23-residue loop is altered in response to binding to the antibiotic (Fig. 7). Thus it was initially proposed that this enzyme's promiscuity results from an open antibiotic-binding site. Nuclear magnetic resonance experiments with uniformly ^{15}N-labeled APH(3′)-IIIa demonstrate that the apoenzyme is undergoing interconversion between several conformations, as the ^{15}N–^{1}H HSQC spectrum shows all the hallmarks of an unstructured or molten-globule form of a protein. Addition of an antibiotic causes significant changes in the spectrum, indicating that formation of the binary enzyme–AG complex induces a well-defined structure in the enzyme. The nucleotide cannot accomplish this change. In fact, the addition of metal–nucleotide to the apoenzyme or enzyme–AG complexes yields NMR spectra that are consistent with increased flexibility and solvent exposure in the protein.[79] These data in themselves strongly suggest that the protein is very flexible and thus can accommodate a large repertoire of ligands via structural dynamics. A similar observation has been made with the aminoglycoside acetyltransferase(6′)-Ii, which also showed a highly overlapping NMR spectrum that showed much higher resolution after the binding of CoASH.[80] These results are also consistent with this enzyme being highly dynamic in solution.

The dynamic properties of APH(3′)-IIIa were probed by H/D exchange. In the apoenzyme, >90% of all backbone amide groups, even those known to be involved in hydrogen bonds of the secondary structure, can exchange their proton for deuterium within 5 min of exposure to D$_2$O, while the remainder undergo complete exchange within 10 h.[79] MgAMPPCP alone did not provide much protection either; however, binding of antibiotics allows 30–40% of residues to be protected from solvent for >90 h, where both the sites and degree of protection is antibiotic dependent (Fig. 12).

Moreover, protection patterns and antibiotic-dependent chemical shifts are not confined to the antibiotic-binding region but instead reach to diverse regions of the protein. The chemical properties and the location of these residues in this enzyme show a very interesting behavior. In this context, ∼50% of the 25 backbone amide groups that are either completely protected with neomycin and completely exposed with kanamycin (or vice versa) are residues having aliphatic hydrocarbon side chains that are buried in hydrophobic patches of the protein, while the other half comprises residues having charged side chains at the surface of the protein.

The protection afforded by AGs in the binary enzyme–AG complexes becomes less effective when the ternary complexes are formed by the addition of MgAMPPCP. A number of backbone amide groups that were completely protected from H/D exchange in the binary complex become exchangeable. This observation, as much as it was totally unexpected, provided the molecular reasons for the observed very large changes in the binding enthalpy of AGs to the enzyme in the presence of metal–ATP[61] and showed that the enthalpic penalty was the result of weakening or breaking of the H-bonds in which these amides are involved. These data thus provided

EFFECT OF PROTEIN DYNAMICS AND SOLVENT IN LIGAND RECOGNITION 239

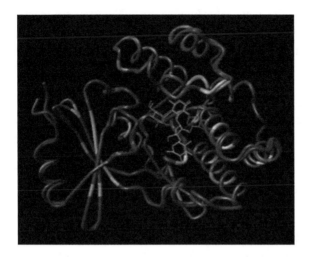

FIG. 12. AG-dependent solvent protection to APH(3′)-IIIa. Amides protected with neomycin only (yellow) and kanamycin only (red) are shown in superimposed structures of the enzyme complexed with both ligands. (See Color Insert.)

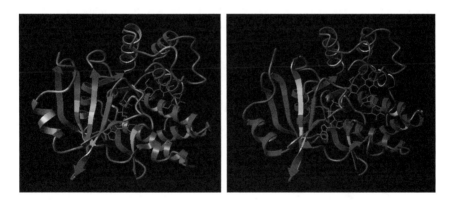

FIG. 13. Residues that become more exposed to solvent (yellow) due to interaction of the nucleotide with APH(3′)-IIIa in complexes of kanamycin (left) and neomycin (right). (See Color Insert.)

a reason for a change in a global thermodynamic parameter and the identity of sites responsible for it. This behavior was, again, AG dependent, and the backbone amide groups that become solvent accessible in the presence of nucleotide are shown in Fig. 13 for the complexes of kanamycin A and neomycin. Overall, these H/D exchange data indicate that the APH(3′)-IIIa "realigns" its conformation from the core to the surface and alters the dynamic properties of various regions, regardless of their distance to the ligand-binding site, to accommodate structurally different AGs. We note that these changes need not be results of large domain movements or structural arrangements that would cause the formation of a different crystal structure in these complexes but clearly highlight the extremely dynamic nature of this enzyme.

For AAC(3)-IIIb, the promiscuity is not due to overall protein flexibility, as with APH(3′)-IIIa, but instead is the responsibility of a large unstructured loop of ~35 amino acids. AAC(3)-IIIb is a member of the large GNAT superfamily of acetyltransferases that share <5% overall sequence- similarity but maintain a core fold. This fold includes a conserved, well-structured binding site for acetyl coenzyme A (motif A), while a flexible loop (motif B) is situated at the adjacent site and prepares for association with a molecule to receive the acetylation. This loop has evolved to produce different proteins, each responsible for a unique acetylating function in the cell. An excellent review on GNAT proteins is available.[81] When coenzyme A is pre-complexed with AAC(3)-IIIb, the affinity of the antibiotic increases several fold, thus suggesting that the coenzyme may be inducing a conformational change at the AG-binding site to make binding of antibiotics more thermodynamically favored (ΔH becomes more negative). Molecular dynamics simulations with AAC(3)-IIIb revealed that the loop of motif B is randomly oriented in the apoenzyme; but upon addition of coenzyme A, it undergoes a concerted movement that opens the antibiotic binding site (Fig. 14).[82] This coincides well with the binding of coenzyme A

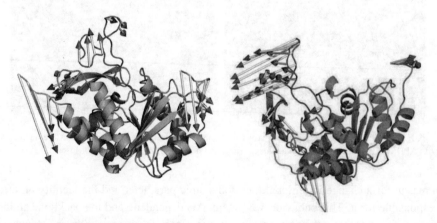

FIG. 14. The flexible, conserved AG-binding loop of AAC(3)-IIIb becomes more flexible upon interaction with CoASH, apoenzyme (left) and enzyme–CoASH complex (right). (Adapted with permission from Ref. 82. Copyright 2011 American Chemical Society). (See Color Insert.)

occurring with a positive entropy and causing shifts of many of the same resonances in the NMR spectrum as antibiotic. The coenzyme A-induced open binding site allows variously structured antibiotics to enter for acetylation, but with an orientation allowing increased favorable interactions between the antibiotic and AAC(3)-IIIb while maintaining 1/1 stoichiometry.

Experimental data supporting the computational studies come from NMR studies. The resonance assignments of AAC(3)-IIIb are not available; however, the ^{15}N–^{1}H HSQC spectrum of the enzyme labeled either with ^{15}N-leucine or ^{15}N-alanine showed that while 28–29 of the expected 29 leucine amide groups are observable, only 30–31 of the expected 39 alanine amide groups could be detected. All leucine residues but one, which is located in the loop, are in well-structured segments of the enzyme. On the other hand, seven alanine residues are located in flexible segments, three are at the N-terminus, and four are in the loop. This information is consistent with the missing number of alanine residues from the spectrum and indicates that these residues are experiencing exchange between two or more conformations over a millisecond range. These properties may be common to GNAT proteins, since the affinity of 2-arylethylamine to arylalkylamine N-acetyltransferase increases with coenzyme A.[80]

VI. Conclusions and Future Considerations

Despite any homology in their amino acid sequence, promiscuous AGMEs that show strong overlap in their ligand spectrum appear to use similar strategies to achieve ligand promiscuity. The shared features of ligand–protein interactions include changes in dynamics of the enzyme and its interaction with solvent, both of which are AG dependent. Additionally, the formation of enzyme–AG complexes is rendered thermodynamically favorable by adjusting the binding enthalpy via changes in low-frequency vibrational modes of the protein and its interactions with the solvent. These common molecular properties are compared in greater detail as follows.

First, among three different AGMEs, with no sequence homology between them and catalyzing three different reactions, there are strong, and in some cases very unusual, antibiotic-dependent solvent effects. These are not predictable from the static structures available but underline a fact that the ability to alter interactions with solvent may be needed to achieve substrate promiscuity.

Next, two highly promiscuous AGMEs, namely APH(3′)-IIIa and AAC(3)-IIIb, both have a large unstructured loop involved in AG binding. In AAC(3)-IIIb, the presence of CoASH causes this loop to be more flexible and achieve more-concerted movement. The opposite trends observed in the dependence of ΔH with temperature between the complexes of this enzyme with neomycin and paromomycin indicate that the binding of paromomycin causes more exposure of hydrophobic surfaces of the protein to solvent, whereas neomycin does the opposite.[83–89] Since ~50% of the residues in the AG-binding loop are those bearing hydrophobic side chains, by adopting different conformations with each AG such changes can easily bring about differential solvent exposure of hydrophobic surfaces. In relation to this change, the loop above the AG-binding site in APH(3′)-IIIa becomes more flexible when the metal–ATP is bound to the enzyme.[78] The H/D exchange patterns of some of the residues on this loop also show AG-dependent differences. For example, Leu-151 and Glu-161 are completely protected in the neomycin complex but are completely exposed in the kanamycin complex. Yet, Glu-160, the residue adjacent to Glu-161, is completely protected with kanamycin A and is exchanged within 41 h with neomycin. The Val-154 residue is protected with both AGs and Leu-147 is exchanged in 12 h with kanamycin A and in 2 h with neomycin. The other residues in this region either exchange quickly or their rates of exchange remain undetermined because of resonance overlap. However, the available data already confirm that this region of the protein adopts different conformations and dynamic properties when complexed with neomycin and kanamycin, causing differential exposure of some amides in this segment. This is also the region of the protein that shows the most significant changes in correlated motions between the kanamycin A- and neomycin-bound APH(3′)-IIIa. It is clear that the increased flexibility in these loops would certainly make it easier to accommodate structurally different AGs. It should also be mentioned that this loop is large in AAC(3)-IIIb and may not need significant changes in other sites, but the smaller loop, as in APH(3′)-IIIa, may need the assistance of changes in other sites as well. Overall, we consider that this is unlikely to be coincidental, and it may represent simple variation in theme rather than a fundamental difference. Thus, flexibility may not only be a common mode of ligand recognition between these two promiscuous AGMEs, but it may also apply to other AGMEs having broad substrate specificity.

It would be interesting to see whether the significant differences observed in the dynamic and thermodynamic properties of enzyme–AG complexes are correlated to their promiscuity. To this end, it might be expected that less-promiscuous enzymes will display less dynamic behavior. Although the accessibility of the active site, or its shape/charge complementarities to AGs, plays a significant role in ligand binding, these factors alone are not sufficient to explain AG-dependent differences in the kinetic, dynamic, and thermodynamic behavior of these enzymes. Therefore, we consider that dynamic properties of these enzymes are the major reason for their substrate promiscuity and have a major impact on their ability to render the binding of structurally diverse AGs thermodynamically favorable. Although the changes discussed are consequences of ligand binding, we emphasize that protein flexibility is necessary to facilitate them.

Finally, we point out that, in order to have any possibility for development of new antimicrobial agents that are effective against pathogenic bacteria, it is very important to understand those molecular determinants of AG recognition that are common among the enzymes. The development of drugs/inhibitors based on the molecular properties of one enzyme has been unsuccessful.[90] In this way, one target AGME can be inhibited, but the other AGMEs, harbored by bacteria, are not affected, nor is the design of inhibitors based on static structures alone likely to show great success. Studies directed toward understanding the dynamic behavior of these enzymes, and an in-depth understanding of the molecular basis of the AG dependence is needed.

Acknowledgments

The work described here was supported by NSF grants (MCB-01110741 and MCB-0842743) and the Alexander von Humboldt Foundation through a stipend to EHS. We also thank Professor David C. Baker of the University of Tennessee for encouraging us to write this article.

References

1. A. Schatz, E. Bugie, and S. A. Waksman, Streptomycin, a substance exhibiting antibiotic activity against gram positive and gram-negative bacteria, *Proc. Soc. Exp. Biol. Med.*, 55 (1944) 66–69.
2. J. E. Davies, Aminoglycoside-aminocyclitol antibiotics and their modifying enzymes, in V. Lorean, (Ed.), *Antibiotics in Laboratory Medicine,* Williams and Wilkins, Baltimore, MD, 1991, pp. 691–713.
3. R. Spotts and R. Y. Stanier, Mechanism of streptomycin action on bacteria: A unitary hypothesis, *Nature*, 192 (1961) 633–637.
4. K. J. Shaw, P. N. Rather, R. S. Hare, and G. H. Miller, Molecular-genetics of aminoglycoside resistance genes and familial relationships of the aminoglycoside-modifying enzymes, *Microbiol. Rev.*, 57 (1993) 138–163.
5. H. Umezawa, Biochemical mechanism of resistance to aminoglycosidic antibiotics, *Adv. Carbohydr. Chem. Biochem.*, 30 (1974) 183–225.
6. J. E. Davies, Inactivation of antibiotics and the dissemination of resistance genes, *Science*, 264 (1994) 375–382.
7. Q. Vicens and E. Westhof, Molecular recognition of aminoglycoside antibiotics by ribosomal RNA and resistance enzymes: An analysis of X-ray crystal structures, *Biopolymers*, 70 (2003) 42–57.
8. D. Fourmy, M. I. Recht, S. C. Blanchard, and J. D. Puglisi, Structure of the A site of *Escherichia coli* 16S ribosomal RNA complexed with an aminoglycoside antibiotic, *Science*, 274 (1996) 1367–1371.
9. M. Recht, D. Fourmy, S. Blanchard, K. Dahlquist, and J. D. Puglisi, RNA sequence determinants for aminoglycoside binding to an A-site rRNA model oligonucleotide, *J. Mol. Biol.*, 262 (1996) 421–436.
10. P. Carter, W. M. Clemons, Jr.,, D. E. Brodersen, R. J. Morgan-Warren, B. T. Wimberly, and V. Ramakrishnan, Functional insights from the structure of the 30s ribosomal subunit and its interactions with antibiotics, *Nature*, 407 (2000) 340–348.

11. S. Yoshizawa, D. Fourmy, and J. D. Puglisi, Structural origins of gentamicin antibiotic action, *EMBO J.*, 17 (1998) 6437–6448.
12. M. Kaul, C. M. Barbieri, and D. S. Pilch, Defining the basis for the specificity of aminoglycoside-rRNA recognition: A comparative study of drug binding to the A sites of *Escherichia coli* and human rRNA, *J. Mol. Biol.*, 346 (2005) 119–134.
13. H. Ryu, A. Litovchick, and R. R. Rando, Stereospecificity of aminoglycoside-ribosomal interactions, *Biochemistry*, 41 (2002) 10499–10509.
14. C. Kim and S. Mobashery, Phosphoryl transfer by aminoglycoside 3′-phosphotransferases and manifestation of antibiotic resistance, *Bioorg. Chem.*, 33 (2005) 149–158.
15. S. Magnet and J. S. Blanchard, Molecular insights into aminoglycoside action and resistance, *Chem. Rev.*, 105 (2005) 477–497.
16. M. H. Perlin, S. A. Brown, and J. N. Dholakia, Developing a snapshot of the ATP binding domain(s) of aminoglycoside phosphotransferases, *Front. Biosci.*, 4 (1999) D63–D71.
17. Y. Tor, Targeting RNA with small molecules, *Chembiochem*, 4 (2003) 998–1007.
18. F. Walter, Q. Vicens, and E. Westhof, Aminoglycoside–RNA interactions, *Curr. Opin. Chem. Biol.*, 3 (1999) 694–704.
19. B. Willis and D. P. Arya, An expanding view of aminoglycoside-nucleic acid recognition, *Adv. Carbohydr. Chem. Biochem.*, 60 (2006) 251–302.
20. D. Wright, Aminoglycoside-modifying enzymes, *Curr. Opin. Microbiol.*, 2 (1999) 499–503.
21. E. H. Serpersu, J. R. Cox, E. L. DiGiammarino, M. L. Mohler, D. R. Ekman, A. Akal-Strader, and M. Owston, Conformations of antibiotics in active sites of aminoglycoside detoxifying enzymes, *Cell Biochem. Biophys.*, 33 (2000) 309–321.
22. E. L. DiGiammarino, K. Draker, G. D. Wright, and E. H. Serpersu, Solution studies of isepamicin and conformational comparisons between isepamicin and butirosin A when bound to an aminoglycoside 6′-N-acetyltransferase determined by NMR spectroscopy, *Biochemistry*, 37 (1998) 3638–3644.
23. D. R. Ekman, E. L. DiGiammarino, E. Wright, E. D. Witter, and E. H. Serpersu, Cloning, overexpression, and purification of aminoglycoside antibiotic nucleotidyltransferase (2″)-Ia: Conformational studies with bound substrates, *Biochemistry*, 40 (2001) 7017–7024.
24. L. M. Mohler, J. R. Cox, and E. H. Serpersu, Aminoglycoside phosphotransferase(3′)-IIIa (APH(3′)-IIIa)-bound conformation of the aminoglycoside lividomycin A characterized by NMR, *Carbohydr. Lett.*, 3 (1997) 17–24.
25. M. A. Owston and E. H. Serpersu, Cloning, overexpression, and purification of aminoglycoside antibiotic 3-acetyltransferase-IIIb: Conformational studies with bound substrates, *Biochemistry*, 41 (2002) 10764–10770.
26. D. H. Fong and A. M. Berghuis, Substrate promiscuity of an aminoglycoside antibiotic resistance enzyme via target mimicry, *EMBO J.*, 21 (2002) 2323–2331.
27. D. Nurizzo, S. C. Shewry, M. H. Perlin, S. A. Brown, J. N. Dholakia, R. L. Fuchs, T. Deva, E. N. Baker, and C. A. Smith, The crystal structure of aminoglycoside-3′-phosphotransferase-IIa, an enzyme responsible for antibiotic resistance, *J. Mol. Biol.*, 327 (2003) 491–506.
28. E. Wright and E. H. Serpersu, Molecular determinants of affinity for aminoglycoside binding to the aminoglycoside nucleotidyltransferase(2″)-Ia, *Biochemistry*, 45 (2006) 10243–10250.
29. L. C. Pedersen, M. M. Benning, and H. M. Holden, Structural investigation of the antibiotic and ATP-binding sites in kanamycin nucleotidyltransferase, *Biochemistry*, 34 (1995) 13305–13311.
30. E. Wolf, A. Vassilev, Y. Makino, A. Sali, Y. Nakatani, and S. K. Burley, Crystal structure of a GCN5-related N-acetyltransferase: *Serratia marcescens* aminoglycoside 3-N-acetyltransferase, *Cell*, 94 (1998) 439–449.
31. W.-C. Hon, G. A. McKay, P. R. Thompson, R. M. Sweet, D. S. C. Yang, G. D. Wright, and A. M. Berghuis, Structure of an enzyme required for aminoglycoside antibiotic resistance reveals homology to eukaryotic protein kinases, *Cell*, 89 (1997) 887–895.

32. L. E. Wybenga-Groot, K. Draker, G. D. Wright, and A. M. Berghuis, Crystal structure of an aminoglycoside 6′-N-acetyltransferase: Defining the GCN5-related N-acetyltransferase superfamily fold, *Struct. Fold. Des.*, 7 (1999) 497–507.
33. M. W. Vetting, S. S. Hegde, F. Javid-Majd, J. S. Blanchard, and S. L. Roderick, Aminoglycoside 2′-N-acetyltransferase from *Mycobacterium tuberculosis* in complex with coenzyme A and aminoglycoside substrates, *Nat. Struct. Biol.*, 9 (2002) 653–658.
34. M. Toth, H. Frase, N. T. Antunes, C. A. Smith, and S. B. Vakulenko, Crystal structure and kinetic mechanism of aminoglycoside phosphotransferase-2″-IVa, *Protein Sci.*, 19 (2010) 1565–1576.
35. P. J. Stogios, T. Shakya, E. Evdokimova, A. Savchenko, and G. D. Wright, Structure and function of APH(4)-Ia, a hygromycin B resistance enzyme, *J. Biol. Chem.*, 286 (2011) 1966–1975.
36. D. L. Burk, W. C. Hon, A. K. Leung, and A. M. Berghuis, Structural analyses of nucleotide binding to an aminoglycoside phosphotransferase, *Biochemistry*, 40 (2001) 8756–8764.
37. P. G. Young, R. Walanj, V. Lakshmi, L. J. Byrnes, P. Metcalf, E. N. Baker, S. B. Vakulenko, and C. A. Smith, The crystal structures of substrate and nucleotide complexes of *Enterococcus faecium* aminoglycoside-2″-phosphotransferase-IIa APH(2″)-IIa provide insights into substrate selectivity in the APH(2″) subfamily, *J. Bacteriol.*, 191 (2009) 4133–4143.
38. K. Shi, D. R. Houston, and A. M. Berghuis, Crystal structures of antibiotic-bound complexes of aminoglycoside 2″-phosphotransferase IVa highlight the diversity in substrate binding modes among aminoglycoside kinases, *Biochemistry*, 50 (2011) 6237–6244.
39. D. D. Boehr, P. R. Thompson, and G. D. Wright, Molecular mechanism of aminoglycoside antibiotic kinase APH(3′)-IIIa—Roles of conserved active site residues, *J. Biol. Chem.*, 276 (2001) 23929–23936.
40. G. A. McKay, P. R. Thompson, and G. D. Wright, Broad spectrum aminoglycoside phosphotransferase type III from Enterococcus: Overexpression, purification and substrate specificity, *Biochemistry*, 33 (1994) 6936–6944.
41. P. R. Thompson, D. W. Hughes, and G. D. Wright, Regiospecificity of aminoglycoside phosphotransferase from Enterococci and Staphylococci (APH(3′)-IIIa), *Biochemistry*, 35 (1996) 8686–8695.
42. G. A. McKay, J. Roestamadji, S. Mobashery, and G. D. Wright, Recognition of aminoglycoside antibiotics by enterococcal-staphylococcal aminoglycoside 3′-phosphotransferase type IIIa: Role of substrate amino groups, *Antimicrob. Agents Chemother.*, 40 (1996) 2648–2650.
43. G. A. McKay and G. D. Wright, Kinetic mechanism of aminoglycoside phosphotransferase type IIIa: Evidence for a Theorell-Chance mechanism, *J. Biol. Chem.*, 270 (1995) 24686–24692.
44. J. Siregar, K. Miroshnikov, and S. Mobashery, Purification, characterization, and investigation of the mechanism of aminoglycoside 3′-phosphotransferase type Ia, *Biochemistry*, 34 (1995) 12681–12688.
45. C. A. Gates and D. B. Northrop, Substrate specificities and structure activity relationships for the nucleotidylation of antibiotics catalyzed by aminoglycoside nucleotidyltransferase-2″-I, *Biochemistry*, 27 (1988) 3820–3825.
46. K. A. Draker, D. B. Northrop, and G. D. Wright, Kinetic mechanism of the GCN5-related chromosomal aminoglycoside acetyltransferase AAC(6′)-Ii from *Enterococcus faecium:* Evidence of dimer subunit cooperativity, *Biochemistry*, 42 (2003) 6565–6574.
47. K. Radika and D. B. Northrop, The kinetic mechanism of kanamycin acetyltransferase derived from the use of alternative antibiotics and coenzymes, *J. Biol. Chem.*, 259 (1984) 2543–2546.
48. K. Radika and D. B. Northrop, Substrate specificities and structure activity relationships for acylation of antibiotics catalyzed by kanamycin acetyltransferase, *Biochemistry*, 23 (1984) 5118–5122.
49. J. E. Van Pelt and D. B. Northrop, Purification and properties of gentamicin nucleotidyltransferase from *Escherichia coli*—Nucleotide specificity, pH optimum, and the separation of 2 electrophoretic variants, *Arch. Biochem. Biophys.*, 230 (1984) 250–263.
50. D. D. Boehr, A. R. Farley, F. J. LaRonde, T. R. Murdock, G. D. Wright, and J. R. Cox, Establishing the principles of recognition in the adenine-binding region of an aminoglycoside antibiotic kinase APH(3′)-IIIa, *Biochemistry*, 44 (2005) 12445–12453.

51. J. Roestamadji, I. Grapsas, and S. Mobashery, Loss of individual electrostatic interactions between aminoglycoside antibiotics and resistance enzymes as an effective means to overcoming bacterial drug-resistance, *J. Am. Chem. Soc.*, 117 (1995) 11060–11069.
52. J. Roestamadji, I. Grapsas, and S. Mobashery, Mechanism-based inactivation of bacterial aminoglycoside 3′-phosphotransferases, *J. Am. Chem. Soc.*, 117 (1995) 80–84.
53. L. B. Magalhaes and J. S. Blanchard, The kinetic mechanism of AAC(3)-IV aminoglycoside acetyltransferase from *Escherichia coli*, *Biochemistry*, 44 (2005) 16275–16283.
54. L. B. Magalhaes, M. W. Vetting, F. Gao, L. Freiburger, K. Auclair, and J. S. Blanchard, Kinetic and, structural analysis of bisubstrate inhibition of the *Salmonella enterica* aminoglycoside 6′-N-acetyltransferase, *Biochemistry*, 47 (2008) 579–584.
55. C. Kim, D. Hesek, J. Zajicek, S. B. Vakulenko, and S. Mobashery, Characterization of the bifunctional aminoglycoside-modifying enzyme ANT(3″)-Ii/AAC(6″)-IId from *Serratia marcescens*, *Biochemistry*, 45 (2006) 8368–8377.
56. T. Vacas, F. Corzana, G. Jimenez-Oses, C. Gonzalez, A. M. Gomez, A. Bastida, J. Revuelta, and J. L. Asensio, Role of aromatic rings in the molecular recognition of aminoglycoside antibiotics: Implications for drug design, *J. Am. Chem. Soc.*, 132 (2010) 12074–12090.
57. M. Hainrichson, O. Yaniv, M. Cherniavsky, I. Nudelman, D. Shallom-Shezifi, S. Yaron, and T. Baasov, Overexpression and initial characterization of the chromosomal aminoglycoside 3′-O-phosphotransferase APH(3′)-IIb from *Pseudomonas aeruginosa*, *Antimicrob. Agents Chemother.*, 51 (2007) 774–776.
58. M. Toth, J. Zajicek, C. Kim, J. W. Chow, C. Smith, S. Mobashery, and S. Vakulenko, Kinetic mechanism of enterococcal aminoglycoside phosphotransferase 2″-Ib, *Biochemistry*, 46 (2007) 5570–5578.
59. C. Ozen, J. M. Malek, and E. H. Serpersu, Dissection of aminoglycoside-enzyme interactions: A calorimetric and NMR study of neomycin B binding to the aminoglycoside phosphotransferase (3′)-IIIa, *J. Am. Chem. Soc.*, 128 (2006) 15248–15254.
60. C. Ozen, A. L. Norris, M. L. Land, E. Tjioe, and E. H. Serpersu, Detection of specific solvent rearrangement regions of an enzyme: NMR and ITC studies with aminoglycoside phosphotransferase (3′)-IIIa, *Biochemistry*, 47 (2008) 40–49.
61. C. Ozen and E. H. Serpersu, Thermodynamics of aminoglycoside binding to aminoglycoside-3′-phosphotransferase IIIa studied by isothermal titration calorimetry, *Biochemistry*, 43 (2004) 14667–14675.
62. K. T. Welch, K. G. Virga, N. A. Whittemore, C. Ozen, E. Wright, C. L. Brown, R. E. Lee, and E. H. Serpersu, Discovery of non-carbohydrate inhibitors of aminoglycoside-modifying enzymes, *Bioorg. Med. Chem.*, 13 (2005) 6252–6263.
63. L. Norris, C. Ozen, and E. H. Serpersu, Thermodynamics and kinetics of association of antibiotics with the aminoglycoside acetyltransferase (3)-IIIb, a resistance-causing enzyme, *Biochemistry*, 49 (2010) 4027–4035.
64. L. Norris and E. H. Serpersu, Antibiotic selection by the promiscuous aminoglycoside acetyltransferase-(3)-IIIb is thermodynamically achieved through the control of solvent rearrangement, *Biochemistry*, 50 (2011) 9309–9317.
65. L. Norris and E. H. Serpersu, Interactions of coenzyme A with the aminoglycoside acetyltransferase (3)-IIIb and thermodynamics of a ternary system, *Biochemistry*, 49 (2010) 4036–4042.
66. E. Wright and E. H. Serpersu, Enzyme-substrate interactions with an antibiotic resistance enzyme: Aminoglycoside nucleotidyltransferase(2″)-Ia characterized by kinetic and thermodynamic methods, *Biochemistry*, 44 (2005) 11581–11591.
67. E. Wright and E. H. Serpersu, Effects of proton linkage on thermodynamic properties of enzyme–antibiotic complexes of the aminoglycoside nucleotidyltransferase (2″)-Ia, *J. Thermodyn. Catal.*, 2 (2011) 105–111.
68. S. S. Hedge, T. K. Dam, C. F. Brewer, and J. S. Blanchard, Thermodynamics of aminoglycoside and acyl-coenzyme A binding to the Salmonella enterica AAC(6′)-Iy aminoglycoside N-acetyltransferase, *Biochemistry*, 41 (2002) 7519–7527.

69. L. Z. Wu and E. H. Serpersu, Deciphering interactions of the aminoglycoside phosphotransferase(3′)-IIIa, *Biopolymers*, 91 (2009) 801–809.
70. D. S. Pilch, M. Kaul, C. M. Barbieri, and J. E. Kerrigan, Thermodynamics of aminoglycoside-rRNA recognition, *Biopolymers*, 70 (2003) 58–79.
71. F. Freire, I. Cuesta, F. Corzana, J. Revuelta, C. Gonzalez, M. Hricovini, A. Bastida, J. Jimenez-Barbero, and J. L. Asensio, A simple NMR analysis of the protonation equilibrium that accompanies aminoglycoside recognition: Dramatic alterations in the neomycin-B protonation state upon binding to a 23-mer RNA aptamer, *Chem. Commun.* (2007) 174–176.
72. M. Baker and K. P. Murphy, Evaluation of linked protonation effects in protein binding reactions using isothermal titration calorimetry, *Biophys. J.*, 71 (1996) 2049–2055.
73. H. Atha and G. K. Ackers, Calorimetric determination of the heat of oxygenation of human hemoglobin as a function of pH and the extent of reaction, *Biochemistry*, 13 (1974) 376–23822.
74. J. M. Sturtevant, Heat-capacity and entropy changes in processes involving proteins, *Proc. Natl. Acad. Sci. U.S.A.*, 74 (1977) 2236–2240.
75. F. Burkhalter, S. M. Dimick, and E. J. Toone, Protein-carbohydrate interaction: Fundamental considerations, in G. W. H. B. Ernst and P. Sinaÿ, (Eds.) *Carbohydrates in Chemistry and Biology*, Vol. 2, Wiley-VCH, New York, 2000, pp. 863–914.
76. T. K. Dam and F. C. Brewer, Thermodynamic studies of lectin-carbohydrate interactions by isothermal titration calorimetry, *Chem. Rev.*, 102 (2002) 387–429.
77. S. Nilapwar, E. Williams, C. Fu, C. Prodromou, L. H. Pearl, M. A. Williams, and J. E. Ladbury, Structural-thermodynamic relationships of interactions in the N-terminal ATP-binding domain of Hsp90, *J. Mol. Biol.*, 392 (2009) 923–936.
78. S. A. Wieninger, E. H. Serpersu, and G. M. Ullmann, ATP binding enables broad antibiotic selectivity of aminoglycoside phosphotransferase(3′)-IIIa: An elastic network analysis, *J. Mol. Biol.*, 409 (2011) 450–465.
79. L. Norris and E. H. Serpersu, NMR detected hydrogen-deuterium exchange reveals differential dynamics of antibiotic- and nucleotide-bound aminoglycosidephosphotransferase 3′-IIIa, *J. Am. Chem. Soc.*, 131 (2009) 8587–8594.
80. L. A. Freiburger, O. M. Baettig, T. Sprules, A. M. Berghuis, K. Auclair, and A. K. Mittermaier, Competing allosteric mechanisms modulate substrate binding in a dimeric enzyme, *Nat. Struct. Mol. Biol.*, 18 (2011) 288–294.
81. F. Dyda, D. C. Klein, and A. B. Hickman, GCN5-related N-acetyltransferases: A structural overview, *Annu. Rev. Biophys. Biomol. Struct.*, 29 (2000) 81–103.
82. X. Hu, A. L. Norris, J. Baudry, and E. H. Serpersu, Coenzyme A binding to the aminoglycoside acetyltransferase (3)-IIIb increases conformational sampling of antibiotic binding site, *Biochemistry*, 50 (2011) 10559–10565.
83. F. Avbelj, P. Z. Luo, and R. L. Baldwin, Energetics of the interaction between water and the helical peptide group and its role in determining helix propensities, *Proc. Natl. Acad. Sci. U.S.A.*, 97 (2000) 10786–10791.
84. R. L. Baldwin, Desolvation penalty for burying hydrogen-bonded peptide groups in protein folding, *J. Phys. Chem. B*, 114 (2010) 16223–16227.
85. A. Cooper, Heat capacity effects in protein folding and ligand binding: A re-evaluation of the role of water in biomolecular thermodynamics, *Biophys. Chem.*, 115 (2005) 89–97.
86. S. Bergqvist, M. A. Williams, R. O'Brien, and J. E. Ladbury, Heat capacity effects of water molecules and ions at a protein-DNA interface, *J. Mol. Biol.*, 336 (2004) 829–842.
87. J. E. Ladbury, J. G. Wright, J. M. Sturtevant, and P. B. Sigler, A thermodynamic study of the Trp repressor-operator interaction, *J. Mol. Biol.*, 238 (1994) 669–681.
88. G. I. Makhatadze and P. L. Privalov, Contribution of hydration to protein folding thermodynamics. 1. The enthalpy of hydration, *J. Mol. Biol.*, 232 (1993) 639–659.

89. W. Kauzmann, Some factors in the interpretation of protein denaturation, *Adv. Protein Chem.*, 14 (1959) 1–63.
90. N. E. Allen, W. E. Alborn, J. N. Hobbs, and H. A. Kirst, 7-Hydroxytropolone—An inhibitor of aminoglycoside-2″-O-adenylyltransferase, *Antimicrob. Agents Chemother.*, 22 (1982) 824–831.

AUTHOR INDEX

Page numbers in roman type indicate that the listed author is cited on that page of an article in this volume; numbers in italic denote the reference number, in the list of references for that article, where the literature citation is given.

A

Aabloo, A., 62, *174*, 62, *177*
Abboud, K.A., 30, *83*
Accorsi, C.A., 42, *106*
Achkar, J., 103, *71*
Ackers, G.K., 229, 73
Adar, R., 6, *19*, 7, *30*, 7, *31*, 7, *32*
Adibekian, A., 98, *38*, 98, *39*, 99, *38*, 99, *39*
Adney, W.S., 61, *170*
Affouard, F., 29, *60*, 29, *61*
Akal-Strader, A., 222, *21*, 223, *21*
Akiba, S., 21, *16*, 63, *16*
Akrigg, D., 29, *55*, 29, *59*
Al-Mafraji, K., 100, *59*, 115, *59*
Al-Maharik, N., 141, *22*
Alais, J., 98, *41*, 98, *44*, 99, *41*, 99, *44*, 100, *41*
Alaranta, S., 127, *110*
Alborn, W.E., 243, *90*
Alderfer, J.L., 30, *68*
Alexander, L.E., 65, *191*
Ali, A., 181, *112*
Allen, A., 6, *24*, 7, *24*
Allen, N.E., 243, *90*
Allinger, N.L., 43, *109*, 43, *110*, 43, *111*, 43, *112*, 45, *109*, 45, *110*, 47, *110*
Almeida, I.C., 140, *19*, 181, *111*
Almond, A., 35, *95*
Alper, P.B., 197, *121*
Altaner, C.M., 68, *194*
Amatayakul, S., 6, *28*
Amster, I.J., 100, *59*, 115, *59*

Andrade, R.B., 208, *143*
Andrianov, V.I., 29, *62*
Angelini, G., 202, *131*
Ångström, J., 7, *31*, 7, *32*
Angulo, J., 109, *84*, 114, *84*, 114, *89*, 114, *91*
Ángyán, J.G., 57, *152*
Anh, N., 57, *152*
Ankri, S., 12, *70*
Antoniou, J., 138, *8*
Antunes, N.T., 225, *34*
Apperley, D.C., 22, *19*, 68, *194*
Arango, R., 7, *30*
Archontis, G., 43, *114*
Ardèvol, A., 27, *34*
Arendt, Y., 30, *91*
Arenzana-Seisdedos, F., 100, *64*, 105, *64*, 114, *64*
Arnott, S., 62, *175*
Aronson, M., 11, *57*
Arungundram, S., 100, *59*, 115, *59*
Arya, D.P., 222, *19*
Asensio, J.L., 225, *56*, 229, *71*
Ashford, D.A., 5, *17*, 6, *28*
Ashkenazi, S., 11, *54*
Ashwell, G., 12, *71*
Asong, J., 100, *59*, 115, *59*
Asztalos, A., 63, *186*
Atalla, R.H., 22, *17*
Atha, H., 229, *73*
Auclair, K., 225, *54*, 238, *80*, 241, *80*
Avbelj, F., 242, *83*

Avci, F.Y., 127, *109*
Avenel, D., 29, *51*
Avizienyte, E., 119, *93*
Ayadi, E., 209, *145*
Azzouz, N., 172, *105*, 207, *142*

B

Baasov, T., 122, *97*, 225, *57*
Bachet, B., 29, *39*
Bader, R.F.W., 22, *21*, 57, *21*
Baeckstroem, G., 97, *18*
Baert, F., 30, *77*
Baeschlin, D.K., 157, *78*, 157, *79*
Baettig, O.M., 238, *80*, 241, *80*
Baker, E.N., 223, *27*, 225, *27*, 225, *37*, 227, *27*
Baker, M., 229, *72*
Baker, R., 144, *50*
Baldwin, R.L., 242, *83*, 242, *84*
Baleux, F., 100, *64*, 105, *64*, 114, *64*
Banerjee, A., 195, *119*
Bannwarth, W., 148, *60*, 152, *60*, 157, *60*, 182, *60*
Bansal, P., 70, *197*
Barath, M., 98, *33*, 99, *33*, 119, *93*
Barbieri, C.M., 222, *12*, 229, *70*
Barkai-Golan, R., 12, *61*
Baron, R., 44, *130*
Bartels, C., 43, *114*
Bartolozzi, A., 102, *67*, 102, *68*, 114, *68*, 119, *68*
Bascou, A., 98, *41*, 99, *41*, 100, *41*
Bastida, A., 225, *56*, 229, *71*
Baudry, J., 240, *82*
Baum, L.G., 9, *46*
Bayer, E.A., 63, *187*
Becker, C., 141, *28*, 181, *29*
Becker, P., 57, *155*
Beckham, G.T., 63, *184*, 63, *185*, 63, *187*, 63, *188*, 63, *189*
Becquart, J., 29, *51*
Bédué, O., 22, *20*
Beeler, D.L., 127, *107*
Beetz, T., 98, *31*, 100, *31*
Beitsch, D.D., 12, *72*

Bellesia, G., 63, *186*
Bender, S.L., 174, *106*, 179, *106*
Benning, M.M., 225, *29*
Bergenstråhle, M., 63, *182*, 63, *184*, 63, *185*, 63, *189*
Berghuis, A.M., 223, *26*, 224, *26*, 225, *26*, 225, *31*, 225, *32*, 225, *36*, 225, *38*, 227, *26*, 232, *26*, 232, *31*, 232, *36*, 233, *26*, 238, *80*, 241, *80*
Berglund, L.A., 63, *182*
Berglund, M., 11, *56*
Bergqvist, S., 242, *86*
Bernhard, W., 11, *58*
Bertolasi, V., 42, *106*
Bertozzi, C.R., 138, *4*, 140, *18*, 141, *18*, 141, *32*, 203, *133*, 203, *134*, 206, *141*
Besemer, A.C., 100, *53*
Betzel, C., 30, *90*
Beyeh, K., 29, *54*
Bhandary, K.K., 29, *50*, 30, *50*
Biarnés, X., 27, *34*
Billington, D.C., 144, *50*
Bilodeau, M.T., 124, *103*
Bindschaedler, P., 98, *38*, 99, *38*
Bishop, L., 141, *21*
Bjork, I., 96, *3*, 97, *3*
Blackwell, J., 21, *7*
Blanchard, J.S., 222, *15*, 225, *33*, 225, *53*, 225, *54*, 226, *68*, 227, *33*, 229, *68*, 234, *33*
Blanchard, S.C., 222, *8*, 222, *9*, 223, *8*, 224, *8*
Blatchford, C.G., 30, *72*
Bock, K., 42, *103*
Boehr, D.D., 225, *39*, 225, *50*
Bolanos, J.G.F., 122, *95*
Bols, M., 122, *95*, 122, *96*
Boltje, T.J., 100, *56*
Bomble, Y.J., 63, *187*, 63, *188*
Bommarius, A.S., 70, *197*
Bonnaffe, D., 98, *34*, 98, *41*, 98, *44*, 99, *34*, 99, *41*, 99, *44*, 100, *41*, 100, *63*, 100, *64*, 105, *64*, 113, *86*, 114, *64*, 114, *86*
Bono, F., 100, *50*, 100, *51*

Boons, G.-J., 100, *52*, 100, *59*, 107, *52*, 113, *85*, 115, *59*, 145, *55*, 157, *81*, 157, *85*, 158, *81*, 210, *146*
Boresch, S., 43, *114*
Bork, P., 138, *6*
Borodkin, V.S., 188, *115*, 193, *115*, 193, *117*
Bouckaert, J., 11, *56*
Boyer, P.D., 4, *1*
Bracha, R., 12, *66*, 12, *67*, 12, *69*, 12, *70*
Brady, J.W., 21, *15*, 45, *134*, 56, *149*, 61, *170*, 63, *15*, 63, *184*, 63, *185*, 63, *189*
Brecker, L., 30, *89*, 35, *89*, 37, *89*
Brewer, C.F., 6, *23*, 226, *68*, 229, *68*
Brewer, F.C., 232, *76*
Bridges, A.S., 127, *115*, 130, *115*, 130, *116*
Brimacombe, J.S., 144, *52*
Brisse, F., 29, *40*
Brodersen, D.E. Jr., 222, *10*
Brooks, B.R., 43, *114*
Brooks, C.L. III., 43, *114*
Brown, C.L., 225, *62*
Brown, G.M., 40, *97*, 40, *98*
Brown, S.A., 222, *16*, 223, *27*, 225, *27*, 227, *27*
Bruhn, C., 30, *91*
Bruno, I.J., 24, *31*, 25, *31*, 57, *31*
Budhu, R.J., 174, *106*, 179, *106*
Bugie, E., 222, *1*
Bulach, V., 29, *66*
Bünsch, A., 153, *75*
Burger, M., 8, *41*
Burgula, S., 200, *128*
Burk, D.L., 225, *36*, 232, *36*
Burkhalter, F., 232, *75*
Burley, S.K., 225, *30*
Byrnes, L.J., 225, *37*

C

Caflisch, A., 43, *114*
Cai, F., 205, *137*
Caldwell, J.W., 43, *124*
Callow, P., 68, *194*
Camargo, M.M., 181, *111*

Campbell, A.S., 151, *66*, 151, *70*, 152, *66*, 156, *66*, 162, *70*
Campen, R.K., 44, *131*
Cañada, F.J., 58, *158*
Canivenc-Gansel, E., 199, *124*
Car, R., 43, *108*
Carmichael, I., 35, *96*
Carroll, F.I., 202, *130*
Carter, P., 222, *10*
Cartoni, A., 202, *131*
Carver, J.P., 51, *143*
Case, D.A., 43, *113*, 43, *124*
Castelli, R., 141, *28*, 181, *29*
Casu, B., 96, *11*, 96, *13*, 97, *20*, 105, *13*, 127, *110*
Caulfield, T.J., 160, *90*
Caves, L., 43, *114*
Cha, J.K., 197, *120*, 201, *120*, 205, *120*
Chandrasekhar, J., 43, *123*
Chang, S.-W., 98, *35*, 99, *35*, 117, *35*
Chang, W., 100, *55*, 102, *55*, 119, *55*
Chanzy, H.D., 21, *10*, 21, *11*, 21, *12*, 21, *13*, 23, *22*, 27, *12*, 29, *41*, 30, *22*, 35, *10*, 35, *11*, 40, *10*, 40, *11*, 62, *10*, 62, *11*, 62, *13*, 67, *10*
Chaperon, A.R., 157, *78*, 157, *79*
Charbonneau, V., 157, *79*
Cheatham, T.E. III., 43, *113*, 43, *124*
Cheinska, L., 59, *163*
Chen, C., 100, *60*, 115, *60*
Chen, J., 100, *60*, 115, *60*, 127, *108*, 127, *109*
Chen, J.P., 29, *58*
Chen, J.Y.-J., 58, *156*
Chen, K.-H., 43, *109*, 43, *110*, 45, *109*, 45, *110*, 47, *110*
Chen, L., 30, *82*
Chen, M., 127, *109*, 127, *115*, 130, *115*
Cherniavsky, M., 225, *57*
Chi, F.-C., 98, *35*, 98, *36*, 99, *35*, 99, *36*, 117, *35*, 117, *36*
Chiara, J.-L., 160, *91*
Chipman, D.M., 4, *8*
Choay, J., 97, *22*, 98, *30*, 100, *30*
Choudhury, D., 11, *56*

Chow, J.W., 225, *58*
Christensen, C., 206, *140*
Chu, J.-W., 43, *120*
Chu, S.S.C., 36, *94*
Chui, D., 10, *47*
Cifonelli, J.A., 96, *2*
Clausen, H., 167, *96*
Clayette, P., 100, *64*, 105, *64*, 114, *64*
Clemons, W.M., 222, *10*
Cmoch, P., 208, *144*
Cocinero, E.J., 47, *136*
Codée, J.D.C., 98, *27*, 100, *56*, 100, *61*, 101, *66*, 121, *66*
Cole, C.L., 119, *93*
Cole, J.C., 24, *31*, 25, *31*, 57, *31*
Coles, S.J., 30, *73*
Condon, B., 61, *173*
Conradt, H.S., 167, *99*
Conzelmann, A., 199, *124*
Cook, N.J., 127, *112*, 127, *113*
Cools, M., 11, *56*
Cooper, A., 242, *85*
Coppens, P., 57, *154*
Coquerel, G., 29, *52*
Corzana, F., 225, *56*, 229, *71*
Cottaz, S., 144, *52*
Cox, E.G., 23, *26*
Cox, J.R., 222, *21*, 223, *21*, 223, *24*, 225, *50*
Cramer, C.J., 42, *107*, 45, *132*, 49, *107*, 50, *107*, 51, *144*, 51, *147*, 52, *132*, 52, *147*, 52, *148*, 54, *132*
Cremer, D., 24, *30*
Crich, D., 195, *119*, 205, *137*
Cross, G.A.M., 138, *10*
Crowley, M.F., 63, *184*, 63, *185*, 63, *187*, 63, *188*, 63, *189*
Császár, P., 58, *159*
Csizmadia, I.G., 57, *152*, 57, *153*, 58, *159*
Csonka, G.I., 43, *122*, 52, *148*, 56, *150*, 57, *152*, 57, *153*, 58, *150*, 58, *159*
Cuesta, I., 229, *71*
Cuevas, G., 58, *158*
Cui, Q., 43, *114*
Cui, Y.-X., 29, *64*

Czechura, P., 109, *82*
Czernecki, S., 209, *145*

D

Dahlquist, K., 222, *9*
Dale, S.H., 30, *72*
Dam, T.K., 226, *68*, 229, *68*, 232, *76*
Damus, P.S., 96, *4*
Daniels, C.R., 43, *121*, 45, *121*
Daniels, M.A., 10, *48*
Danishefsky, S.J., 124, *103*
Dann, S.E., 30, *72*
Danon, D., 6, *25*
Daragics, K., 119, *92*
Darden, T., 43, *113*
Date, V., 152, *72*
Davies, J.E., 222, *2*, 222, *6*
Davis, B.G., 47, *136*
Davis, N.J., 100, *54*
De, A.P.L., 127, *115*, 130, *115*
De Genst, L., 11, *56*
De Greve, H., 11, *56*
de Jong, A.J.M., 98, *31*, 100, *31*
De Koning, J.H., 198, *122*, 210, *122*
de Nooy, A.E.J., 100, *53*
de Paz, J.-L., 105, *73*, 109, *83*, 109, *84*, 111, *83*, 112, *83*, 114, *73*, 114, *84*, 114, *88*, 114, *90*, 119, *90*
de Witte, W., 100, *61*
DeBolt, S., 43, *124*
Deguchi, H., 98, *31*, 98, *37*, 99, *37*, 100, *31*
del Carmen Fernández-Alonso, M., 58, *158*
Delbaere, L.T.J., 42, *103*
Demchenko, A.V., 122, *99*, 145, *55*
Derollez, P., 29, *60*, 29, *61*
Derr, P., 167, *99*
DeShong, P., 174, *107*
DeSolms, S.J., 154, *76*
Dessen, A., 6, *23*
Deva, T., 223, *27*, 225, *27*, 227, *27*
Dholakia, J.N., 222, *16*, 223, *27*, 225, *27*, 227, *27*
Díaz, M.D., 58, *158*

DiGiammarino, E.L., 222, *21*, 223, *21*, 223, 22, 223, *23*
Dilhas, A., 98, *34*, 98, *41*, 99, *34*, 99, *41*, 100, *41*, 100, *63*
Dimick, S.M., 232, *75*
Ding, L.-S., 29, *57*, 30, *81*
Ding, S.Y., 63, *190*, 61, *170*
Dinkelaar, J., 100, *61*
Dinner, A.R., 43, *114*
Dobler, M., 29, *49*
Doering, T.L., 199, *123*
Dordick, J.S., 127, *111*
Dorfman, A., 96, *2*
Dorland, L., 5, *16*
Douwes, M., 144, *51*
Dowd, M.K., 27, *33*, 30, *76*, 37, *76*, 42, *107*, 44, *127*, 45, *132*, 48, *127*, 49, *107*, 50, *107*, 51, *144*, 52, *132*, 54, *132*, 59, *164*, 59, *165*
Draker, K.A., 223, *22*, 225, *32*, 225, *46*
Dreef, C.E., 142, *40*, 144, *51*
Dubois, P., 10, *49*
Duchaussoy, P., 97, *22*, 98, *30*, 100, *30*, 100, *51*
Duchet, D., 29, *44*
Duguid, J.P., 10, *52*
Dulaney, S., 100, *58*, 101, *58*, 124, *58*
Dumarcay-Charbonnier, F., 29, *66*
Duverger, V., 98, *41*, 99, *41*, 100, *41*
Dwek, R.A., 5, *17*, 139, *13*, 6, *28*
Dyda, F., 240, *81*

E

Eardley, D.D., 138, *9*
Edgington, P.R., 24, *31*, 25, *31*, 57, *31*
Eguchi, T., 30, *86*
Eisenhaber, B., 138, *6*
Eisenhaber, F., 138, *6*
Ek, M., 144, *48*
Ekman, D.R., 222, *21*, 223, *21*, 223, *23*
El-Dakdouki, M., 100, *57*, 105, *57*
Eliás, K., 58, *159*
Elie, C.J.J., 142, *40*, 144, *51*
Eller, S., 108, *80*

Elsegood, M.R.J., 30, *72*
Emeis, J., 127, *110*
Englund, P.T., 138, *2*, 167, *2*, 199, *123*
Ernst, A., 30, *85*
Esko, J.D., 97, *14*
Espinosa, E., 58, *161*, 58, *162*
Evans, K., 140, *16*, 177, *16*
Evdokimova, E., 225, *35*

F

Fan, R.-H., 103, *71*
Farley, A.R., 225, *50*
Favor, D.A., 174, *107*
Feig, M., 43, *114*
Feld, R., 57, *155*
Ferguson, A.J., 181, *111*
Ferguson, D., 43, *124*
Ferguson, M.A.J., 138, *1*, 138, *5*, 139, *1*, 139, *13*, 144, *52*, 167, *1*, 167, *98*, 188, *115*, 193, *115*, 193, *117*, 200, *98*
Fernandes, A.N., 68, *194*
Fernandez-Megia, E., 203, *135*
Fernandez, R., 98, *45*, 99, *45*
Fernig, D.G., 96, *8*
Ferretti, V., 42, *106*
Ferrières, V., 162, *92*
Finazzi-Agrò, A., 202, *131*
Finne, J., 138, *7*
Firgang, S.I., 29, *62*, 29, *63*
Firon, N., 11, *54*, 11, *55*
Fischer, S., 43, *114*
Fleischer, M., 113, *87*
Fleming, D.N., 202, *130*
Flitsch, S.L., 100, *54*
Flowers, H.M., 4, *10*
Fokin, V.V., 206, *139*
Foley, B.L., 43, *118*, 43, *121*, 45, *121*
Fong, D.H., 223, *26*, 224, *26*, 225, *26*, 227, *26*, 232, *26*, 233, *26*
Ford, Z.M., 62, *178*
Foriers, A., 7, *38*
Forstner, M.B., 140, *18*, 141, *18*, 141, *32*
Forsyth, V.T., 63, *183*, 67, *183*, 68, *183*, 68, *194*

Fouret, R., 30, *77*
Fourmy, D., 222, *8*, 222, *9*, 222, *11*, 223, *8*, 223, *11*, 224, *8*, 224, *11*
Fournet, B., 30, *77*
Foust, T.D., 61, *170*
Fransson, L.-A., 97, *18*
Frase, H., 225, *34*
Fraser-Reid, B., 142, *38*, 151, *38*, 151, *65*, 151, *66*, 151, *67*, 151, *68*, 151, *69*, 151, *70*, 152, *66*, 152, *68*, 152, *72*, 152, *73*, 153, *74*, 155, *69*, 156, *66*, 157, *83*, 158, *83*, 162, *70*, 172, *103*, 172, *104*, 173, *68*, 173, *73*, 179, *109*, 182, *73*, 211, *109*
Freiburger, L.A., 225, *54*, 238, *80*, 241, *80*
Freire, F., 229, *71*
French, A.D., 21, *5*, 21, *9*, 22, *18*, 23, *25*, 27, *33*, 30, *76*, 30, *89*, 34, *25*, 35, *25*, 35, *89*, 37, *76*, 37, *89*, 42, *107*, 43, *110*, 43, *122*, 44, *125*, 44, *127*, 44, *128*, 45, *110*, 45, *132*, 45, *133*, 47, *110*, 47, *138*, 48, *127*, 48, *138*, 48, *140*, 48, *141*, 49, *107*, 50, *107*, 50, *140*, 51, *133*, 51, *144*, 51, *145*, 51, *146*, 52, *132*, 52, *133*, 52, *146*, 52, *148*, 54, *132*, 55, *138*, 56, *150*, 58, *150*, 59, *164*, 59, *165*, 62, *174*, 62, *177*, 62, *178*, 63, *183*, 65, *193*, 66, *193*, 67, *183*, 67, *193*, 68, *183*, 68, *193*
Frey, W., 30, *74*
Fu, C., 234, *77*
Fuchs, R.L., 223, *27*, 225, *27*, 227, *27*
Fügedi, P., 98, *32*, 99, *32*, 107, *32*, 107, *79*, 119, *92*
Fujita, M., 140, *14*, 199, *14*, 200, *127*
Furukawa, J., 98, *40*, 99, *40*

G

Gachelin, G., 10, *49*
Gallagher, J.T., 97, *15*
Galun, E., 12, *62*
Gamblin, D.P., 47, *136*
Gandhi, N.S., 96, *5*, 97, *5*
Gangadharmath, U.B., 122, *99*
Ganguly, A.K., 30, *80*
Gao, F., 225, *54*

Gao, J., 43, *114*
García, S., 141, *23*
Gardiner, J.M., 98, *33*, 99, *33*, 119, *93*
Gardner, K.H., 21, *7*
Garegg, P.J., 141, *25*, 144, *48*, 144, *49*, 146, *56*, 148, *56*, 149, *49*, 154, *49*, 168, *56*, 196, *56*
Garnier, S., 29, *52*
Gates C.A., 225, *45*
Gavard, O., 98, *41*, 98, *44*, 99, *41*, 99, *44*, 100, *41*, 113, *86*, 114, *86*
Gazzinelli, R.T., 181, *111*
Gbarah, A., 11, *58*
Gessler, K., 30, *90*
Gigg, J., 141, *20*
Gigg, R., 141, *20*
Gilboa-Garber, N., 7, *29*
Gill, M., 98, *42*, 99, *42*
Gillespie, W., 9, *46*
Gilli, G., 42, *106*
Gillier-Pandraud, H., 29, *44*, 29, *51*, 57, *155*
Gimenez-Gallego, G., 107, *78*, 109, *84*, 114, *84*
Gin, D.Y., 120, *94*, 124, *94*
Gnanakaran, S., 47, *138*, 48, *138*, 55, *138*, 63, *186*
Go, K., 29, *50*, 29, *58*, 29, *65*, 30, *50*
Godbout, L., 21, *14*, 63, *14*
Goebel, M., 162, *94*
Gohlke, H., 43, *113*
Goldberg, A.R., 8, *41*
Gomez, A.M., 225, *56*
Gonzalez, C., 225, *56*, 229, *71*
González-Nilo, F., 43, *117*
González-Outeiriño, J., 43, *121*, 45, *121*
Goridis, C., 138, *7*
Gottfridson, E., 127, *112*, 127, *113*
Götze, S., 207, *142*
Gowda, D.C., 181, *112*
Grabowski, S.J., 58, *157*
Grapsas, I., 225, *51*, 225, *52*
Gray, D.G., 21, *14*, 63, *14*
Green, L.G., 157, *78*, 157, *79*, 206, *139*
Greenberg, W.A., 203, *132*

Gress, M.E., 40, *99*
Gribben, J.G., 167, *97*
Grice, P., 157, *81,* 157, *87,* 158, *81*
Griffiths, G., 127, *112,* 127, *113*
Gross, A.S., 43, *120*
Groves, J.T., 140, *18,* 141, *18,* 141, *32*
Grundler, G., 149, *63,* 152, *71,* 163, *71,* 177, *63,* 179, *63,* 182, *63,* 201, *71,* 208, *71*
Guedes, N., 109, *82*
Guerrant, R.L., 12, *63,* 12, *64*
Guerrini, M., 100, *49,* 113, *87*
Guiry, K.P., 30, *73*
Guo, X., 141, *33,* 141, *34,* 141, *35,* 172, *33,* 172, *34,* 172, *35*
Guo, Z., 141, *21,* 141, *30,* 141, *31,* 141, *33,* 141, *34,* 141, *35,* 166, *30,* 166, *31,* 167, *30,* 167, *31,* 167, *95,* 167, *101,* 170, *102,* 172, *30,* 172, *31,* 172, *33,* 172, *34,* 172, *35,* 185, *113,* 185, *114,* 187, *114,* 195, *118,* 200, *128,* 203, *136*
Gupta, D., 6, *23*
Guvench, O., 43, *119,* 49, *119*

H

Ha, M.-A., 22, *19*
Ha, S.N., 56, *149*
Habibi, Y., 61, *169*
Hagihara, S., 205, *138*
Hahm, H.S., 207, *142*
Hahn, M.G., 157, *78*
Hainrichson, M., 225, *57*
Hakomori, S.-I., 167, *96*
Hale, G., 167, *99*
Hall, M., 70, *197*
Hallek, M., 167, *97*
Haller, M., 100, *52,* 107, *52*
Ham, J.T., 30, *75,* 37, *75,* 60, *75*
Han, L., 202, *129*
Hanley, S.J., 21, *14,* 63, *14*
Hans, L., 159, *88*
Hansen, H.S., 44, *130*
Hansen, S.U., 98, *33,* 99, *33,* 119, *93*
Hanson, S.R., 203, *132*
Hara, O., 124, *104*

Hara, S., 98, *24*
Hare, R.S., 222, *4,* 225, *5*
Hart, G.W., 199, *123*
Hatcher, E., 43, *119,* 48, *139,* 49, *119,* 55, *139*
Hattori, M., 29, *64*
Hawkins, G.D., 202, *130*
Hayashi, K., 98, *24*
Hayashi, S., 21, *16,* 63, *16*
Hazell, R., 122, *95*
Hazen, K.C., 152, *73,* 173, *73,* 182, *73*
He, H.T., a138, *7*
Hederos, M., 141, *26*
Hedge, S.S., 226, *68,* 229, *68*
Hegde, S.S., 225, *33,* 227, *33,* 234, *33*
Hendrickson, T., 140, *15*
Hendriks, K.B., 143, *42,* 160, *42,* 185, *42*
Henrissat, B., 23, *22,* 23, *23,* 30, *22*
Herbert, J.-M., 100, *50,* 100, *51*
Hernandez-Torres, J.M., 103, *71*
Hersant, Y., 98, *41,* 99, *41,* 100, *41,* 100, *64,* 105, *64,* 114, *64*
Hertz, B., 35, *96*
Hesek, D., 225, *55*
Hess, B., 43, *115*
Heux, L., 27, *36,* 61, *36,* 63, *36*
Hewitt, M.C., 140, *16,* 140, *17,* 141, *17,* 177, *16,* 177, *17*
Hewlett, E.L., 12, *64*
Hickey, A.M., 127, *111*
Hickman, A.B., 240, *81*
Hicks, M., 96, *4*
Himmel, M.E., 21, *15,* 61, *170,* 63, *15,* 63, *184,* 63, *185,* 63, *187,* 63, *188,* 63, *189*
Hirschmann, R., 174, *107*
Hoagland, M.B., 4, *1*
Hobbs, J.N., 243, *90*
Hodoscek, M., 43, *114*
Hodson, N., 127, *113*
Hoeoek, M., 97, *18*
Hogquist, K.A., 10, *48*
Holden, H.M., 225, *29*
Hollinshead, D.M., 157, *85*
Homans, S.W., 6, *28,* 139, *13*
Hon, W.-C., 225, *31,* 225, *36,* 232, *31,* 232, *36*

Hook, M., 96, *3*, 97, *3*
Hopwood, J., 96, *3*, 97, *3*
Horenstein, B.A., 30, *83*
Hori, H., 155, *77*
Houjou, T., 200, *127*
Houston, D.R., 225, *38*
Howley, P.S., 21, *9*
Hricovini, M., 229, *71*
Hsieh, Y.-L., 30, *84*
Hu, M., 30, *84*
Hu, S., 124, *103*
Hu, X., 30, *71*, 123, *102*, 240, *82*
Hu, Y.-P., 100, *55*, 102, *55*, 119, *55*
Huang, C.-Y., 100, *55*, 102, *55*, 119, *55*
Huang, L., 100, *46*, 100, *47*, 122, *98*, 123, *102*, 124, *106*
Huang, X., 100, *46*, 100, *47*, 100, *57*, 100, *58*, 100, *62*, 101, *58*, 105, *57*, 122, *98*, 123, *102*, 124, *58*, 124, *106*, 130, *116*
Huff, J.R., 154, *76*
Hughes, D.W., 225, *41*
Hughes, R.C., 4, *3*
Huibers, M., 105, *74*
Hultberg, H., 144, *48*
Hultgren, S.J., 11, *56*
Humeres, E., 43, *117*
Hünenberger, P.H., 44, *130*
Hung, C.S., 11, *56*
Hung, S.-C., 29, *46*, 98, *35*, 98, *36*, 99, *35*, 99, *36*, 100, *55*, 102, *55*, 102, *70*, 105, *72*, 117, *35*, 117, *36*, 117, *72*, 119, *55*, 197, *121*
Hwa, K., 138, *12*

I

Iglesias, J.L., 6, *27*
Ikeda, T., 98, *40*, 99, *40*
Ikezawa, H., 138, *3*
Im, W., 43, *114*
Imberty, A., 29, *39*, 30, *69*
Impey, R.W., 43, *123*
Inbar, M., 8, *42*
Ince, S.J., 157, *78*
Ingram, L.J., 105, *77*
Inoue, N., 138, *11*, 167, *100*

Ishihara, M., 98, *23*, 98, *24*
Islam, T.F., 105, *76*
Isojima, S., 167, *96*
Ito, H., 124, *104*
Ito, Y., 143, *45*, 145, *54*, 205, *138*
Itzicovitch, L., 10, *50*
Iverson, T., 144, *49*, 149, *49*, 154, *49*

J

Jacob, F., 10, *49*
Jacquinet, J.C., 29, *44*, 98, *30*, 100, *30*
Jaeger, C., 30, *89*, 35, *89*, 37, *89*
Jager, C., 30, *69*, 30, *70*
Jager, V., 30, *74*
Jain, R.K., 30, *68*
James, K., 143, *42*, 160, *42*, 185, *42*
Jameson, C.S., 10, *48*
Jaradat, D.M.M., 59, *163*
Jarvis, M.C., 22, *19*, 68, *194*
Javid-Majd, F., 225, *33*, 227, *33*, 234, *33*
Jayaprakash, K.N., 172, *103*, 172, *104*
Jayson, G.C., 98, *33*, 99, *33*, 119, *93*
Jeanloz, R.W., 4, *3*, 4, *4*, 4, *10*
Jeffrey, G.A., 23, *26*, 24, *29*, 36, *94*, 40, *99*, 42, *105*, 51, *142*
Jensen, H.H., 122, *96*
Jensen, L.H., 29, *43*
Jewett, J.C., 203, *134*, 206, *141*
Jia, Z.J., 179, *109*, 211, *109*
Jigami, Y., 140, *14*, 199, *14*, 200, *127*
Jiménez-Barbero, J., 58, *158*, 229, *71*
Jimenez-Oses, G., 225, *56*
Jing, J., 127, *114*, 129, *114*
Johansson, R., 144, *49*, 149, *49*, 154, *49*
Johnson, D.K., 61, *170*
Johnson, G.P., 23, *25*, 30, *76*, 34, *25*, 35, *25*, 37, *76*, 42, *107*, 43, *110*, 43, *122*, 44, *125*, 45, *110*, 45, *132*, 47, *110*, 47, *138*, 48, *138*, 48, *140*, 48, *141*, 49, *107*, 50, *107*, 50, *140*, 51, *144*, 51, *145*, 51, *146*, 52, *132*, 52, *146*, 52, *148*, 54, *132*, 55, *138*, 59, *164*, 59, *165*, 62, *178*, 63, *183*, 65, *193*, 66, *193*, 67, *183*, 67, *193*, 68, *183*, 68, *193*

Joly, J.-P., 29, *66*
Jones, C.L., 127, *108*
Jorgensen, W.L., 43, *123*

K

Kabat, E.A., 6, *26*
Kaizu, T., 167, *96*
Kajtar-Peredy, M., 98, *32*, 99, *32*, 107, *32*
Kakinuma, K., 30, *86*
Kalenius, E., 29, *54*
Kamena, F., 172, *105*, 180, *110*
Kanda, T., 98, *24*
Kanemitsu, T., 141, *27*
Kapoor, N., 10, *51*
Kariya, Y., 98, *23*
Karlsson, K.-A., 7, *31*, 7, *32*
Karplus, M., 43, *114*
Karst, N.A., 98, *28*, 105, *76*
Kartha, G., 29, *58*, 29, *65*
Kataoka, H., 160, *90*
Katchalski, E., 5, *13*, 5, *14*
Kaul, M., 222, *12*, 229, *70*
Kauzmann, W., 242, *89*
Kawada, T., 30, *70*
Kawatkar, S., 210, *146*
Ke, W., 98, *42*, 99, *42*
Keglevic, D., 29, *45*
Keith, T., 60, *166*
Kekalainen, T., 29, *54*
Keller-Schierlein, W., 29, *49*
Kelm, S., 9, *46*
Kelterer, A.-M., 42, *107*, 45, *132*, 49, *107*, 50, *107*, 51, *144*, 52, *132*, 54, *132*
Kennedy, C.J., 68, *194*
Kerrigan, J.E., 229, *70*
Kessler, M., 24, *31*, 25, *31*, 57, *31*
Khiar, N., 141, *23*
Kikuchi, H., 98, *23*, 98, *24*
Kim, C., 222, *14*, 225, *55*, 225, *58*
Kim, I.J., 124, *103*
Kim, J.-H., 210, *146*
Kinoshita, T., 138, *11*, 167, *100*, 200, *127*
Kinzy, W., 143, *47*
Kirchhoff, C., 167, *99*

Kirkpatrick, D., 10, *51*
Kirschner, K.N., 43, *121*, 45, *121*
Kirst, H.A., 243, *90*
Kjellberg, A., 7, *35*
Kjellen, L., 97, *16*
Klein, D.C., 240, *81*
Klein, M.L., 43, *123*
Klein, R.A., 58, *160*
Klemm, P., 11, *56*
Klepach, T., 35, *96*
Knight, S.D., 11, *56*
Knorr, E., 143, *43*, 152, *43*
Kobiler, D., 12, *67*
Koch, U., 57, *151*, 60, *151*
Koenigs, W., 143, *43*, 152, *43*
Koessler, J.-L., 141, *23*
Kohn, J., 5, *18*
Koike, K., 143, *45*
Koizumi, K., 29, *42*
Kojic-Prodic, B., 29, *45*
Koll, P., 30, *87*
Kollman, P.A., 43, *124*
Kolossvary, I., 58, *159*
Kondo, K., 30, *86*
Konradsson, P., 141, *25*, 141, *26*, 142, *38*, 151, *38*, 151, *69*, 155, *69*, 157, *83*, 158, *83*
Koole, L.A., 12, *70*
Kopf, J., 30, *87*
Kopitzki, S., 109, *82*
Koritsanszky, T.S., 57, *154*
Koshland, D.E., 4, *5*, 4, *6*
Koshland, M.E., 138, *9*
Kosma, P., 30, *69*, 30, *70*, 30, *89*, 35, *89*, 37, *89*
Koto, S., 42, *103*
Kovensky, J., 100, *51*
Králová, B., 27, *35*
Krässig, H.A., 70, *196*
Kratzer, B., 147, *58*, 149, *62*, 162, *58*, 190, *58*
Krauss, N., 30, *90*
Krengel, U., 7, *35*
Kreuger, J., 107, *78*
Krochta, J.M., 30, *84*

Kroon-Batenburg, L.M.J., 62, *181*
Kroon, J., 62, *181*
Kuberan, B., 127, *107*
Kubicki, J.D., 44, *131*
Kuczera, K., 43, *114*
Kuksis, A., 200, *126*
Kulagowski, J.J., 144, *50*
Kulkarni, K.A., 7, *33*
Kulkarni, S.S., 29, *46*, 102, *70*, 105, *72*, 117, *72*
Kupfer, E., 29, *49*
Kurth, M.J., 30, *84*
Kusche-Gullberg, M., 127, *110*
Kuttel, M., 45, *134*, 45, *135*
Kutzner, C., 43, *115*
Kvick, Å., 23, *22*, 30, *22*
Kwon, O., 124, *103*
Kwon, Y.-U., 141, *27*, 141, *28*, 177, *108*, 180, *110*

L

La Ferla, B., 100, *49*
Lachekar, H., 58, *161*
Ladbury, J.E., 234, *77*, 242, *86*, 242, *87*
Lahmann, M., 122, *100*
Laine, R.A., 48, *141*
Laio, A., 27, *34*
Lakin, M., 62, *179*
Lakshmanan, T., 30, *79*
Lakshmi, V., 225, *37*
Lallana, E., 203, *135*
Lam, L.H., 97, *17*, 97, *19*
Land, M.L., 225, *60*, 231, *60*, 232, *60*, 234, *60*
Lang, S., 185, *114*, 187, *114*
Langan, P.J., 21, *10*, 21, *11*, 21, *12*, 21, *13*, 27, *12*, 35, *10*, 35, *11*, 40, *10*, 40, *11*, 47, *138*, 48, *138*, 55, *138*, 62, *10*, 62, *11*, 62, *13*, 63, *183*, 63, *186*, 67, *10*, 67, *183*, 68, *183*
Langermann, S., 11, *56*
Larocque, S., 98, *42*, 99, *42*
LaRonde, F.J., 225, *50*
Lassaletta, J.-M., 98, *45*, 99, *45*, 109, *84*, 114, *84*
Lawrence, S.E., 30, *73*

Lay, L., 98, *29*, 100, *29*, 100, *49*, 113, *87*
Lazaridis, T., 43, *114*
Leach, F.E. III, 100, *59*, 115, *59*
Lech, M.Z., 127, *107*
Lecomte, C., 58, *161*, 58, *162*
Lederman, I., 97, *22*, 98, *30*, 100, *30*
Lee, J., 197, *120*, 201, *120*, 205, *120*
Lee, J.-C., 29, *46*, 98, *35*, 99, *35*, 105, *72*, 117, *35*, 117, *72*
Lee, J.H., 70, *197*
Lee, R.E., 225, *62*
Lefebvre, J., 29, *60*
Lehmann, M., 57, *155*
Lemieux, R., 7, *36*
Lemieux, R.U., 42, *103*, 143, *42*, 160, *42*, 185, *42*
Leslie, R., 157, *80*, 157, *81*, 157, *85*, 158, *81*
Leung, A.K., 225, *36*, 232, *36*
Leung, F., 29, *41*, 30, *88*
Levy, H.A., 40, *97*, 40, *98*
Levy, H.M., 4, *5*, 4, *6*
Ley, S.V., 157, *78*, 157, *79*, 157, *80*, 157, *81*, 157, *82*, 157, *85*, 157, *87*, 158, *81*
Li, J.-P., 127, *110*
Li, Q.-F., 29, *57*, 30, *81*
Li, X., 123, *102*
Liang, F.-F., 30, *82*
Lidholt, K., 127, *112*, 127, *113*
Lifely, M.R., 167, *98*, 200, *98*
Lii, J.-H., 43, *109*, 43, *110*, 43, *112*, 45, *109*, 45, *110*, 47, *110*
Lin, S.-Y., 100, *55*, 102, *55*, 119, *55*
Lind, T., 127, *112*, 127, *113*
Lindahl, E., 43, *115*
Lindahl, U., 96, *3*, 96, *11*, 97, *3*, 97, *14*, 97, *16*, 97, *18*, 97, *20*, 107, *78*, 127, *110*
Lindberg, B., 144, *49*, 149, *49*, 154, *49*
Lindberg, J., 141, *25*
Lindemann, E., 4, *6*
Lindh, I., 143, *44*, 146, *44*, 177, *44*, 182, *44*, 208, *44*
Lindner, B., 43, *116*
Ling, C.C., 7, *36*

Linhardt, R.J., 96, *1*, 98, *28*, 105, *76*, 127, *109*, 127, *111*, 127, *114*, 127, *115*, 129, *114*, 130, *115*
Linker, A., 97, *18*
Linker-Israeli, M., 9, *44*
Lipmann, F., 4, *1*, 4, *2*
Lis, H., 5, *12*, 5, *13*, 5, *14*, 5, *16*, 5, *17*, 6, *20*, 6, *21*, 6, *22*, 6, *27*, 6, *28*, 7, *34*, 7, *37*, 8, *21*, 8, *43*, 13, *37*, 13, *76*, 13, *77*, 13, *78*
Litjens, R.E.J.N., 102, *67*
Litovchick, A., 222, *13*
Liu, D., 96, *10*
Liu, J., 100, *58*, 101, *58*, 124, *58*, 127, *108*, 127, *109*, 127, *114*, 127, *115*, 129, *114*, 130, *115*, 130, *116*
Liu, J.-Y., 100, *55*, 102, *55*, 119, *55*
Liu, R., 100, *58*, 101, *58*, 124, *58*, 127, *114*, 127, *115*, 129, *114*, 130, *115*, 130, *116*
Liu, X., 141, *28*, 177, *108*, 180, *110*, 181, *29*
Liu, X.-H., 29, *64*
Lodder, G., 100, *61*
Loganathan, D., 30, *79*
Lohman, G.J.S., 102, *68*, 117, *68*
Longchambon, F., 29, *44*, 57, *155*
Lopez, O., 122, *95*
Lortat-Jacob, H., 100, *64*, 105, *64*, 113, *86*, 114, *64*, 114, *86*
Lotan, R., 6, *20*, 6, *21*, 6, *25*, 6, *26*, 8, *21*, 8, *40*, 12, *62*, 12, *73*
Loureiro-Morais, L., 100, *64*, 105, *64*, 114, *64*
Low, M.G., 200, *126*
Lozano, R.M., 109, *84*, 114, *84*
Lu, J., 172, *103*, 172, *104*
Lu, L.-D., 102, *70*
Lu, X.-A., 29, *46*, 102, *70*, 105, *72*, 117, *72*
Lubineau, A., 98, *44*, 99, *44*, 113, *86*, 114, *86*
Lucas, R., 114, *89*
Lucia, L., 61, *169*
Lücking, U., 157, *79*
Luger, P., 59, *163*
Luo, P.Z., 242, *83*
Luo, R., 43, *113*
Ly, M., 127, *111*

M

Ma, C.-M., 29, *64*
Ma, J., 43, *114*
Maccarrone, M., 202, *131*
Mach, M., 152, *73*, 173, *73*, 182, *73*
MacKerell, A.D. Jr., 43, *114*, 43, *119*, 48, *139*, 49, *119*, 55, *139*
Mackie, W., 29, *55*, 29, *59*
Macrae, C.F., 24, *31*, 25, *31*, 57, *31*
Madsen, L.J., 56, *149*
Madsen, R., 151, *65*, 151, *68*, 151, *69*, 152, *68*, 153, *74*, 155, *69*, 173, *68*
Madura, J.D., 43, *123*
Maeda, Y., 200, *127*
Maeta, H., 143, *41*, 145, *41*, 170, *41*
Magalhaes, L.B., 225, *53*, 225, *54*
Magnet S., 222, *15*
Maier, T., 151, *64*
Makhatadze, G.I., 242, *88*
Makino, Y., 225, *30*
Malek, J.M., 225, *59*, 229, *59*, 231, *59*, 233, *59*
Mallet, A., 157, *86*
Mallet, F., 29, *61*
Mallet, J.M., 100, *50*, 157, *86*
Malz, F., 30, *70*
Mancera, R.L., 96, *5*, 97, *5*
Manoso, A.S., 174, *107*
Manuzi, A., 105, *74*
Marchessault, R.H., 28, *37*, 28, *38*, 29, *41*, 30, *88*, 36, *37*, 36, *38*, 61, *168*
Marenich, A.V., 51, *147*, 52, *147*
Margonelli, A., 202, *131*
Mario Pinto, S.B., 157, *84*, 158, *84*
Mark, H., 21, *3*, 31, *3*
Maron, L., 146, *56*, 148, *56*, 168, *56*, 196, *56*
Marsura, A., 29, *66*
Marth, J.D., 10, *47*
Martín-Lomas, M., 109, *82*, 109, *83*, 109, *84*, 111, *83*, 112, *83*, 114, *84*, 114, *88*, 114, *89*, 114, *90*, 114, *91*, 119, *90*, 141, *23*, 160, *91*
Martin-Zamora, E., 98, *45*, 99, *45*
Maruyama, T., 124, *105*

Mascayano, C., 43, *117*
Mason, P.E., 21, *15*, 63, *15*
Masterson, W.J., 199, *123*
Masuko, S., 127, *114*, 129, *114*
Mathew, F., 152, *73*, 173, *73*, 182, *73*
Mathieson, A.Mcl., 23, *24*
Matsui, M., 143, *46*
Matsumoto, T., 143, *41*, 145, *41*, 170, *41*
Matsuo, I., 205, *138*
Matta, K.L., 30, *68*
Matthews, J.F., 21, *15*, 63, *15*, 63, *184*, 63, *185*, 63, *188*, 63, *189*
Mawatari, A., 98, *37*, 99, *37*, 114, *37*, 115, *37*
Mawer, I.M., 144, *50*
May, C.L., 151, *67*
Mayer, T.G., 147, *58*, 147, *59*, 149, *62*, 162, *58*, 190, *58*, 190, *116*
Mazeau, K., 27, *36*, 61, *36*, 62, *179*, 63, *36*, 63, *182*
McCabe, P., 24, *31*, 25, *31*, 57, *31*
McConville, M.J., 199, *125*
McDonnell, C., 122, *95*
McDowell, L.M., 127, *109*
McGivern, P.G., 193, *117*
McKay, G.A., 225, *31*, 225, *40*, 225, *42*, 225, *43*, 232, *31*
McPhail, A.T., 30, *80*
Mebs, S., 59, *163*
Medalia, O., 11, *57*
Meguro, H., 155, *77*
Mehlert, A., 181, *111*
Mehta, S., 157, *84*, 158, *84*
Meldal, M., 206, *140*
Mellor, A.L., 138, *8*
Mendonca, S., 48, *141*
Menuel, S., 29, *66*
Mercier, J.P., 22, *20*
Mereiter, K., 30, *70*, 30, *89*, 35, *89*, 37, *89*
Merritt, J.R., 153, *74*
Mertens, J.M.R., 98, *31*, 100, *31*
Merz, K.M. Jr., 43, *113*
Meseguer, B., 157, *82*
Meshorer, A., 10, *50*
Metcalf, P., 225, *37*

Meyer, K.H., 21, *3*, 21, *4*, 31, *3*, 38, *4*, 39, *4*
Michel, J., 148, *61*, 189, *61*, 196, *61*
Miller, D.P., 62, *174*
Miller, G.H., 222, *4*, 225, *5*
Miller, R.W., 30, *80*
Millrain, M., 138, *8*
Minamisawa, T., 98, *23*
Mino, Y., 124, *104*
Mio, S., 157, *82*
Mirelman, D., 4, *9*, 10, *53*, 11, *54*, 11, *57*, 11, *58*, 12, *61*, 12, *62*, 12, *65*, 12, *66*, 12, *67*, 12, *68*, 12, *69*, 12, *70*
Miron, T., 5, *18*
Miroshnikov, K., 225, *44*
Misch, L., 21, *4*, 38, *4*, 39, *4*
Mischnick, P., 20, *2*, 22, *2*
Mitani, M., 160, *89*
Mitra, N., 7, *33*
Mitschler, A., 57, *155*
Mittermaier, A.K., 238, *80*, 241, *80*
Miyazaki, A., 205, *138*
Mizoue, K., 30, *86*
Mizrachi, L., 7, *29*
Mizushima, S.-I., 33, *93*, 40, *93*
Mo, F., 29, *43*, 29, *48*
Mobashery, S., 222, *14*, 225, *42*, 225, *44*, 225, *51*, 225, *52*, 225, *55*, 225, *58*
Mohler, L.M., 223, *24*
Mohler, M.L., 222, *21*, 223, *21*
Molins, E., 58, *162*
Momany, F.A., 44, *126*, 47, *137*, 51, *126*, 51, *137*, 52, *126*
Momcilovic, D., 20, *2*, 22, *2*
Montreuil, J., 30, *77*
Moody, A.M., 10, *47*
Mootoo, D.R., 142, *38*, 151, *38*, 151, *69*, 152, *72*, 155, *69*, 157, *83*, 158, *83*
Moran, R.A., 29, *67*
Morell, A.G., 12, *71*
Moreno, E., 7, *31*, 7, *32*
Morf, M., 30, *87*
Morgan-Warren, R.J., 222, *10*
Mori, M., 145, *54*

Morita, T., 160, *89*
Mouhous-Riou, N., 29, *39*
Mousa, S.A., 127, *114*, 129, *114*
Moynihan, H.A., 30, *73*
Muggli, R., 21, *6*
Munoz, E.M., 127, *109*
Murakata, C., 142, *36*, 142, *37*, 142, *39*, 157, *36*
Murdock, T.R., 225, *50*
Murphy, C.F., 12, *64*
Murphy, K.P., 229, *72*
Murphy, P., 122, *95*
Myasnikova, R.M., 29, *62*, 29, *63*
Myher, J.J., 200, *126*

N

Naggi, A., 127, *110*
Naidoo, K.J., 45, *134*, 58, *156*
Nakahara, Y., 143, *45*
Nakatani, Y., 225, *30*
Nakayasu, E.S., 140, *19*
Narayanasami, U., 96, *9*
Neitola, R., 29, *54*
Neuberger, A., 6, *24*, 7, *24*
Neuman, A., 29, *51*
Neupert-Laves, K., 29, *49*
Newman, R.H., 22, *19*
Nguyen, H.M., 120, *94*, 124, *94*
Nicolaou, K.C., 160, *90*
Nicolas, J.F., 10, *49*
Nieto, P.M., 109, *84*, 114, *84*, 114, *89*, 114, *91*, 141, *23*
Nifant'ev, N.E., 29, *39*
Nikitin, A.V., 29, *62*, 29, *63*
Nikolaev, A.V., 140, *19*, 141, *22*, 188, *115*, 193, *115*, 193, *117*
Nilapwar, S., 234, *77*
Nilsson, C.L., 7, *35*
Nilsson, L., 43, *114*
Nimlos, M.R., 61, *170*, 63, *184*, 63, *185*
Nimtz, M., 167, *99*
Nishida, Y., 155, *77*
Nishimura, S., 21, *16*, 63, *16*

Nishiyama, Y., 21, *10*, 21, *11*, 21, *12*, 21, *13*, 27, *12*, 35, *10*, 35, *11*, 40, *10*, 40, *11*, 62, *10*, 62, *11*, 62, *13*, 63, *183*, 65, *193*, 66, *193*, 67, *10*, 67, *183*, 67, *193*, 68, *183*, 68, *193*
Nohara, L.L., 140, *19*
Noll, B.C., 29, *56*, 30, *71*
Nordstrom, L.U., 122, *96*
Norris, A.L., 225, *60*, 231, *60*, 232, *60*, 234, *60*, 240, *82*
Norris, L., 225, *63*, 225, *64*, 225, *65*, 229, *63*, 230, *63*, 231, *63*, 234, *64*, 236, *64*, 238, *79*
Nørskov-Lauritsen, L., 43, *111*
Northrop, D.B., 225, *45*, 225, *46*, 225, *47*, 225, *48*, 225, *49*
Nothofer, H.-G., 162, *94*
Noti, C., 98, *25*, 98, *26*, 98, *38*, 99, *38*, 105, *73*, 114, *73*
Novogrodsky, A., 6, *21*, 8, *21*, 8, *40*
Nowell, P., 7, *39*, 8, *39*
Nudelman, E., 167, *96*
Nudelman, I., 225, *57*
Nurizzo, D., 223, *27*, 225, *27*, 227, *27*

O

O'Brien, R., 242, *86*
Ofek, I., 10, *53*, 11, *54*, 11, *55*, 11, *57*, 11, *59*
Ogawa, T., 142, *36*, 142, *37*, 142, *39*, 143, *45*, 143, *46*, 144, *53*, 145, *54*, 146, *57*, 157, *36*
Ohanessian, J., 29, *44*
Ohashi, Y., 30, *86*
Ohrui, H., 155, *77*
Ojeda, R., 109, *83*, 111, *83*, 114, *90*, 114, *91*, 119, *90*
Old, D.C., 10, *52*
Oliver, A.G., 30, *71*
Ollmann, I.R., 122, *97*
Olmstead, M.M., 30, *84*
Olschewski, D., 141, *28*, 181, *29*
Olsson, L., 179, *109*, 211, *109*
Oltvoort, J.J., 198, *122*, 210, *122*
Onufriev, A., 43, *113*

Oreste, P., 127, *110*
Orgueira, H.A., 98, *43,* 99, *43,* 102, *67,* 102, *68,* 114, *68,* 119, *68*
Osaki, K., 29, *42*
Osawa, T., 4, *10*
Osborn, H.M.I., 157, *87*
Oscarson, S., 11, *56,* 122, *100,* 141, *25,* 144, *48*
Osztrovszky, G., 98, *32,* 99, *32,* 107, *32*
Ottosson, H., 153, *74*
Ovchinnikov, V., 43, *114*
Overkleeft, H.S., 98, *27,* 100, *56,* 100, *61,* 101, *66,* 121, *66*
Owston, M.A., 222, *21,* 223, *21,* 223, *25*
Ozaki, S., 160, *89*
Ozen, C., 225, *59,* 225, *60,* 225, *61,* 225, *62,* 225, *63,* 227, *61,* 229, *59,* 229, *61,* 229, *63,* 230, *61,* 230, *63,* 231, *59,* 231, *60,* 231, *63,* 232, *60,* 233, *59,* 234, *60,* 238, *61*

P

Paci, E., 43, *114*
Padilla-Vaca, F., 12, *70*
Pakulski, Z., 208, *144*
Palmacci, E.R., 100, *48,* 102, *67,* 108, *81,* 208, *143*
Palmer, M., 185, *114,* 187, *114*
Pan, G.-R., 102, *70*
Pan, Q., 29, *56,* 30, *71*
Pang, M., 9, *46*
Panza, L., 100, *49*
Pareja, C., 98, *45,* 99, *45*
Parikh, D.V., 61, *173*
Parish, C.R., 96, *7*
Park, J., 210, *146*
Park, T.K., 124, *103*
Parrinello, M., 27, *34,* 43, *108*
Pastor, R.W., 43, *114,* 43, *119,* 49, *119*
Patterson, A., 68, *195*
Patterson, M.C., 174, *107*
Pauli, J., 30, *69*
Paulick, M.G., 138, *4,* 140, *18,* 141, *18,* 141, *32*

Paulsen, H., 100, *65,* 153, *75*
Paulson, J.C., 9, *46*
Pearl, L.H., 234, *77*
Pearlman, D.A., 43, *124*
Pearson, J., 24, *31,* 25, *31,* 57, *31*
Pedersen, L.C., 127, *109,* 225, *29*
Pekari, K., 141, *24,* 151, *24,* 162, *24,* 162, *92,* 162, *93*
Peng, S.-L., 29, *57,* 30, *81*
Peralta-Inga, Z., 30, *76,* 37, *76*
Pereira, M.E.A., 6, *26*
Pérez, S., 20, *1,* 22, *1,* 23, *23,* 29, *39,* 29, *40,* 29, *41,* 29, *55,* 30, *69,* 42, *104,* 62, *179*
Peric-Hassler, L., 44, *130*
Perlin, M.H., 222, *16,* 223, *27,* 225, *27,* 227, *27*
Pertsin, A.I., 29, *62*
Peters, B., 63, *188*
Peterson, L., 48, *140,* 50, *140*
Petit, S., 29, *52*
Petitou, M., 97, *20,* 97, *21,* 97, *22,* 98, *21,* 98, *30,* 100, *30,* 100, *50,* 100, *51*
Petrella, R.J., 43, *114*
Petridis, L., 43, *116*
Petrus, L., 30, *87*
Petrusova, M., 30, *87*
Pietruszka, J., 157, *87*
Pilch, D.S., 222, *12,* 229, *70*
Pinkner, J., 11, *56*
Piskorz, C.F., 30, *68*
Planas, A., 27, *34*
Plante, O.J., 108, *81,* 208, *143*
Platteau, C., 29, *60,* 29, *61*
Pluschke, G., 180, *110*
Polat, T., 122, *101*
Poletti, L., 98, *29,* 100, *29,* 100, *49,* 113, *87*
Polito, L., 105, *73,* 114, *73*
Pollack, M.S., 10, *51*
Poole, J.L., 120, *94,* 124, *94*
Popelier, P.L.A., 57, *151,* 60, *151*
Pople, J.A., 51, *142*
Poppleton, B.J., 23, *24*
Pornsuriyasak, P., 122, *99*
Porwanski, S., 29, *66*

Post, C.B., 43, *114*
Potthast, A., 30, *69*
Pourhossein, M., 127, *113*
Powell, A.K., 96, *8*
Prabhu, A., 113, *85*
Priatel, J.J., 10, *47*
Priepke, H.W.M., 157, *87*
Pritchard, R.G., 98, *33*, 99, *33*
Privalov, P.L., 242, *88*
Procópio, D.O., 181, *111*
Prodromou, C., 234, *77*
Pu, J.Z., 43, *114*
Puglisi, J.D., 222, *8*, 222, *9*, 222, *11*, 223, *8*, 223, *11*, 224, *8*, 224, *11*
Puranik, R., 98, *36*, 99, *36*, 117, *36*

Q
Qi, Y., 96, *10*
Qiao, Y.-F., 30, *86*
Qui, D.T., 23, *22*, 30, *22*

R
Rademacher, T.W., 5, *17*, 6, *28*, 139, *13*, 141, *23*
Radika, K., 225, *47*, 225, *48*
Radom, L., 51, *142*
Rahman, M., 43, *112*
Ralton, J.E., 199, *125*
Ramachandran, G.N., 41, *100*, 44, *100*
Ramakrishnan, C., 41, *100*, 44, *100*
Ramakrishnan, V., 222, *10*
Rånby, B.G., 61, *167*, 62, *180*
Rando, R.R., 222, *13*
Randolph, J.T., 124, *103*
Rao, C.S., 153, *74*
Rao, V.S., 42, *103*
Raper, J., 199, *123*
Rath, N.P., 122, *99*
Rather, P.N., 222, *4*, 225, *5*
Ravdin, J.I., 12, *63*, 12, *64*, 12, *65*
Ravid, A., 8, *40*
Raymond, S., 23, *22*, 30, *22*
Raz, A., 12, *73*
Razdan, R.K., 202, *129*

Realff, M.J., 70, *197*
Reche, P.A., 10, *47*
Recht, M.I., 222, *8*, 222, *9*, 223, *8*, 224, *8*
Redondo-Horcajo, M., 109, *84*, 114, *84*
Redondoe, A., 63, *186*
Rees, B., 57, *155*
Rees, D.A., 42, *102*
Reeves, R.E., 23, *28*
Reggiori, F., 199, *124*
Reichardt, N.-C., 109, *82*, 114, *90*, 119, *90*
Reilly, P., 48, *140*, 50, *140*
Reilly, P.J., 27, *33*, 44, *127*, 48, *127*
Reinherz, E.L., 10, *47*
Reisner, Y., 9, *44*, 9, *45*, 10, *49*, 10, *50*, 10, *51*
Rencurosi, A., 30, *69*
Rendleman, J.A., 30, *76*, 37, *76*
Revol, J.-F., 21, *14*, 63, *14*
Revuelta, J., 225, *56*, 229, *71*
Riadi, G., 43, *117*
Richards, G.F., 29, *67*
Riguera, R., 203, *135*
Rihs, G., 29, *47*
Rissanen, K., 29, *54*
Roberts, C., 151, *65*, 151, *67*, 151, *68*, 151, *69*, 152, *68*, 153, *74*, 155, *69*, 173, *68*
Roberts, I.S., 127, *110*, 127, *112*, 127, *113*
Roberts, W.L., 200, *126*
Robinson, P.J., 138, *8*
Roderick, S.L., 225, *33*, 227, *33*, 234, *33*
Roehrig, S., 98, *43*, 99, *43*
Roestamadji, J., 225, *42*, 225, *51*, 225, *52*
Rohrling, J., 30, *69*
Rojas, O.J., 61, *169*
Roß, G., 162, *94*
Rosenau, T., 30, *70*, 30, *89*, 35, *89*, 37, *89*
Rosenberg, R.D., 96, *4*, 97, *17*, 97, *19*, 127, *107*
Rosenberry, T.L., 200, *126*
Ross, W.S., 43, *124*
Rostovtsev, V.V., 206, *139*
Roux, B., 43, *114*
Rovira, C., 27, *34*
Rozenblatt, S., 6, *19*, 7, *30*
Ruda, K., 141, *25*

Rushton, G., 119, *93*
Russo, G., 100, *49*
Rutjes, F.P.J.T., 105, *74*
Ryu, H., 222, *13*

S

Sabesan, S., 6, *23*
Sacchettini, J.C., 6, *23*
Sachs, L., 8, *42*, 8, *43*
Saenger, W., 30, *90*
Saito, A., 98, *37*, 99, *37*, 114, *37*, 115, *37*
Salameh, B.A.B., 98, *33*, 99, *33*
Salata, R.A., 12, *64*
Sali, A., 225, *30*
Salmivirta, M., 100, *51*, 107, *78*, 127, *110*
Samain, D., 20, *1*, 22, *1*
Sanderson, R.D., 96, *12*
Sansonetti, P.J., 11, *58*, 12, *68*
Sardzik, R., 30, *74*
Sarko, A., 21, *6*, 21, *8*, 28, *37*, 30, *90*, 36, *37*, 62, *176*
Sarrazin, S., 113, *86*, 114, *86*
Sarre, O.Z., 30, *80*
Sasajima, K., 146, *57*
Sasisekaharan, V., 41, *100*, 44, *100*
Sasisekharan, R., 96, *9*, 96, *10*
Sato, S., 143, *45*, 145, *54*
Savchenko, A., 225, *35*
Säwén, E., 48, *139*, 55, *139*
Schaefer, M., 43, *114*
Schatz, A., 222, *1*
Schechter, B., 6, *21*, 8, *21*
Schell, P., 98, *43*, 99, *43*, 102, *67*, 102, *68*, 114, *68*, 119, *68*
Schembri, E., 11, *56*
Schiattarella, M., 101, *66*, 121, *66*
Schlueter, U., 152, *73*, 172, *104*, 173, *73*, 182, *73*
Schmidt, R.R., 141, *24*, 143, *47*, 147, *58*, 147, *59*, 148, *61*, 149, *62*, 149, *63*, 151, *24*, 151, *64*, 152, *71*, 162, *24*, 162, *58*, 162, *92*, 162, *93*, 163, *71*, 177, *63*, 179, *63*, 182, *63*, 189, *61*, 190, *58*, 190, *116*, 196, *61*, 201, *71*, 208, *71*

Schneider, P., 167, *98*, 200, *98*
Schnupf, U., 44, *126*, 51, *126*, 52, *126*
Schofield, L., 140, *16*, 177, *16*
Schollmeyer, D., 30, *78*
Schori, L., 11, *57*
Schröter, S., 167, *99*
Schuetzenmeister, N., 98, *38*, 99, *38*
Schulz, R., 43, *116*
Schwarz, F.P., 7, *35*
Seeberger, P.H., 96, *6*, 98, *25*, 98, *26*, 98, *38*, 98, *39*, 98, *43*, 99, *38*, 99, *39*, 99, *43*, 100, 48, 102, *67*, 102, *68*, 102, *68*, 105, *73*, 108, *80*, 108, *81*, 114, *68*, 114, *73*, 117, 68, 119, *68*, 140, *16*, 140, *17*, 141, *17*, 141, *27*, 141, *28*, 172, *105*, 177, *16*, 177, *17*, 177, *108*, 180, *110*, 181, *29*, 207, *142*, 208, *143*
Seible, G., 43, *124*
Seidel, R., 141, *28*, 181, *29*
Sela, B.A., 8, *43*
Seltzman, H.H., 202, *130*
Serianni, A.S., 29, *53*, 29, *56*, 30, *71*, 35, *96*
Serpersu, E.H., 222, *21*, 223, *21*, 223, *22*, 223, *23*, 223, *24*, 223, *25*, 224, *28*, 225, *28*, 225, *59*, 225, *60*, 225, *61*, 225, *62*, 225, *63*, 225, *64*, 225, *65*, 226, *28*, 226, *66*, 226, *67*, 227, *61*, 227, *66*, 227, *69*, 229, *28*, 229, *59*, 229, *61*, 229, *63*, 230, *61*, 230, *63*, 231, *59*, 231, *60*, 231, *63*, 232, *60*, 233, *59*, 234, *28*, 234, *60*, 234, *64*, 235, *78*, 236, *64*, 238, *79*, 238, *61*, 240, *82*
Shaanan, B., 6, *22*, 7, *34*
Shakya, T., 225, *35*
Shallom-Shezifi, D., 225, *57*
Shang, M., 29, *53*
Shao, N., 141, *30*, 141, *31*, 166, *30*, 166, *31*, 167, *30*, 167, *31*, 172, *30*, 172, *31*
Sharon, N., 4, *1*, 4, *2*, 4, *3*, 4, *4*, 4, *5*, 4, *6*, 4, *7*, 4, *8*, 4, *9*, 4, *10*, 4, *11*, 5, *12*, 5, *13*, 5, *14*, 5, *15*, 5, *16*, 5, *17*, 6, *19*, 6, *20*, 6, *21*, 6, *22*, 6, *24*, 6, *25*, 6, *26*, 6, *27*, 6, *28*, 7, *24*, 7, *30*, 7, *31*, 7, *32*, 7, *33*, 7, *34*, 7, *35*, 7,

36, 7, *37*, 7, *38*, 8, *21*, 8, *40*, 8, *43*, 9, *44*, 10, *49*, 10, *50*, 10, *53*, 11, *54*, 11, *55*, 11, *57*, 11, *58*, 11, *59*, 11, *60*, 12, *61*, 12, *62*, 13, *37*, 13, *74*, 13, *75*, 13, *76*, 13, *77*, 13, *78*
Sharpless, K.B., 206, *139*
Shaw, K.J., 222, *4*, 225, *5*
Sheehan, J.K., 35, *95*
Sheldrick, B., 29, *55*, 29, *59*
Shen, T., 47, *138*, 48, *138*, 55, *138*, 63, *186*
Shen, Z., 185, *114*, 187, *114*
Shewry, S.C., 223, *27*, 225, *27*, 227, *27*
Shi, K., 225, *38*
Shibanova, T.A., 29, *63*
Shie, C.-R., 102, *70*
Shimanouchi, T., 33, *93*, 40, *93*
Shoham, M., 6, *22*
Shriver, Z., 96, *9*, 96, *10*
Siegelman, H.W., 6, *20*
Sigler, P.B., 242, *87*
Silbert, J.E., 97, *17*
Silman, I., 12, *72*
Silva, L.S., 181, *111*
Simmerling, C., 43, *113*
Simons, J.P., 47, *136*
Simpson, E., 138, *8*
Simpson, L.S., 105, *75*
Sinaÿ , P., 29, *44*
Sinaÿ, P., 29, *51*, 97, *22*, 98, *30*, 100, *30*, 100, *50*, 100, *51*, 157, *86*
Siomos, M.-A., 140, *16*, 177, *16*
Siregar, J., 225, *44*
Sizun, P., 100, *51*
Skerrett, R.J., 42, *102*
Skopec, C.E., 21, *15*, 63, *15*
Skutelsky, E., 6, *25*
Slättegård, R., 11, *56*
Sletten, E.M., 203, *133*, 206, *141*
Smith, A.B., 174, *107*
Smith, C.A., 223, *27*, 225, *27*, 225, *34*, 225, *37*, 225, *58*, 227, *27*
Smith, J.C., 43, *116*
Smith, J.H., 30, *72*
Smith, P.J.C., 62, *175*

Snyder, D.A., 140, *17*, 141, *17*, 141, *27*, 141, *28*, 177, *17*
Sobel, M., 98, *37*, 99, *37*, 114, *37*, 115, *37*
Soli, E.D., 174, *107*
Sopin, V.F., 29, *62*, 29, *63*
Soucy, R.L., 141, *27*, 141, *28*
Souhassou, M., 58, *161*
Spek, A.L., 25, *32*
Spik, G., 30, *77*
Spiwok, V., 27, *35*, 44, *129*
Spotts, R., 222, *3*
Sprules, T., 238, *80*, 241, *80*
Srikrishnan, T., 30, *68*
Sriram, D., 30, *79*
Srivastava, A., 7, *33*
Stadler, R., 30, *78*
Stanier, R.Y., 222, *3*
Stauch, T., 145, *55*
Stawinski, J., 143, *44*, 146, *44*, 177, *44*, 182, *44*, 208, *44*
Steinborn, D., 30, *91*
Steiner, T., 30, *90*
Steinmeier, H., 61, *172*
Stenutz, R., 29, *53*, 35, *96*
Stevens, E.D., 30, *76*, 37, *76*, 59, *164*, 59, *165*, 62, *178*
Stick, R.V., 143, *42*, 160, *42*, 185, *42*
Stimpson, W.T., 98, *33*, 99, *33*
Stogios, P.J., 225, *35*
Stortz, C.A., 43, *122*, 44, *128*
Strati, G.L., 47, *137*, 51, *137*
Strecker, G., 30, *77*
Streicher, H., 6, *19*, 7, *31*, 7, *36*
Strosberg, A.D., 7, *38*
Stubba, B., 101, *66*, 121, *66*
Stulberg, M.P., 4, *1*
Sturiale, L., 107, *78*
Sturtevant, J.M., 231, *74*, 242, *87*
Stylianides, N.A., 160, *90*
Suda, Y., 98, *37*, 99, *37*, 114, *37*, 115, *37*
Suen, A., 127, *111*
Sugimoto, M., 143, *45*
Sugiyama, J., 21, *11*, 21, *15*, 35, *11*, 40, *11*, 62, *11*, 63, *15*

Suguna, K., 7, *33*
Sumiya, S., 29, *42*
Sun, B., 100, *58*, 101, *58*, 124, *58*
Sundararajan, R., 28, *38*, 36, *38*
Surolia, A., 7, *33*
Suzuki, K., 143, *41*, 145, *41*, 170, *41*
Suzuki, T., 205, *138*
Svensson, C., 7, *35*
Swarts, B.M., 141, *33*, 141, *34*, 172, *33*, 172, *34*, 195, *118*, 200, *128*, 203, *136*
Sweet, R.M., 225, *31*, 232, *31*
Szabó, K.J., 24, *30*

T

Tabeur, C., 100, *50*
Taga, T., 29, *42*
Taguchi, R., 200, *127*
Tailler, D., 141, *24*, 151, *24*, 162, *24*, 162, *92*
Takano, R., 98, *24*
Takeda, J., 138, *11*
Takieddin, M., 127, *114*, 129, *114*
Tamborrini, M., 180, *110*
Tashima, Y., 200, *127*
Tatai, J., 98, *32*, 99, *32*, 107, *32*, 107, *79*
Tatami, A., 205, *138*
Taylor, C., 127, *110*
Taylor, G.N., 6, *28*
Taylor, R., 24, *31*, 25, *31*, 42, *105*, 57, *31*
Taylor, S.D., 105, *77*
Teichberg, V.I., 4, *11*, 12, *72*
Teneberg, S., 7, *32*, 7, *35*
Terenti, O., 109, *83*, 111, *83*, 112, *83*
Tessier, M.B., 43, *118*
Teumelsan, N., 100, *47*
Thibaudeau, C., 35, *96*
Thibodeaux, D.P., 61, *173*
Thomas, L.H., 68, *194*
Thompson, F., 180, *110*
Thompson, P.R., 225, *31*, 225, *39*, 225, *40*, 225, *41*, 232, *31*
Thopate, S.R., 98, *35*, 99, *35*, 117, *35*
Thunberg, L., 97, *18*
Tidor, B., 43, *114*
Tiffin, P.D., 157, *80*

Timmer, M.S.M., 98, *38*, 98, *39*, 99, *38*, 99, *39*
Tiruchinapally, G., 100, *57*, 100, *58*, 101, *58*, 105, *57*, 124, *58*
Tjioe, E., 225, *60*, 231, *60*, 232, *60*, 234, *60*
Tomisic, Z. B., 29, *45*
Toone, E.J., 232, *75*
Toporek, S.S., 30, *83*
Toporowicz, A., 9, *45*
Tor, Y., 222, *17*
Torget, R.W., 21, *15*, 63, *15*
Tornøe, C.W., 206, *140*
Torrecilhas, A.C.T., 140, *19*
Torri, G., 98, *30*, 100, *30*, 113, *87*, 127, *110*
Toth, M., 225, *34*, 225, *58*
Travassos, L.R., 181, *111*
Traxler, P., 29, *47*
Treumann, A., 200, *98*, 167, *98*
Truhlar, D.G., 51, *147*, 52, *147*
Trzeciak, A., 148, *60*, 152, *60*, 157, *60*, 182, *60*
Tsai, Y.-H., 207, *142*
Tschampel, S.M., 43, *121*, 45, *121*
Tsuchihashi, G.-i., 143, *41*, 145, *41*, 170, *41*
Tsuji, Y., 167, *96*
Tsvetkov, Y.E., 29, *39*
Turnbull, J.E., 96, *8*, 100, *59*, 115, *59*
Tvaroška, I., 23, *23*, 27, *35*, 42, *104*, 44, *129*, 51, *143*

U

Udodong, U.E., 151, *65*, 151, *69*, 153, *74*, 155, *69*, 157, *83*, 158, *83*
Uekusa, H., 30, *86*
Ugi, I., 162, *94*
Ullmann, G.M., 235, *78*
Umezawa, H., 222, *5*
Usov, A.I., 29, *62*, 29, *63*
Utamura, T., 29, *42*

V

Vacas, T., 225, *56*
Vacca, J.P., 154, *76*
Vainiotalo, P., 29, *54*

AUTHOR INDEX

Vakulenko, S., 225, *58*
Vakulenko, S.B., 225, *34*, 225, *37*, 225, *55*
van Aelst, S.F., 98, *31*, 100, *31*
van Bekkum, H., 100, *53*
van Boeckel, C.A.A., 97, *21*, 98, *21*, 98, *27*, 98, *31*, 100, *31*, 101, *66*, 121, *66*, 198, *122*, 210, *122*
van Boom, J.H., 100, *56*, 101, *66*, 121, *66*, 142, *40*, 144, *51*, 198, *122*, 210, *122*
van Delft, F.L., 105, *74*
van den Bos, L.J., 100, *56*
van den Bosch, R.H., 98, *31*, 100, *31*
van der Marel, G.A., 98, *27*, 100, *56*, 100, *61*, 101, *66*, 121, *66*, 142, *40*, 144, *51*
van der Spoel, D., 43, *115*
van der Toorn, J.C., 100, *56*
van der Vlugt, F.A., 98, *31*, 100, *31*
van Dijk, J., 100, *61*
van Halbeek, H., 5, *16*
Van Pelt, J.E., 225, *49*
Varma, Y., 140, *15*
Varon Silva, D., 207, *142*
Vasella, A., 30, *85*, 160, *91*
Vassilev, A., 225, *30*
Vazquez, N., 109, *82*
Venable, R.M., 43, *114*, 43, *119*, 49, *119*
Venkataraman, G., 96, *9*, 96, *10*
Venot, A., 100, *59*, 113, *85*, 115, *59*
Verde, A.V., 44, *131*
Verdon, R., 11, *58*, 12, *68*
Verduyn, R., 142, *40*
Veromaa, T., 127, *110*
Vetting, M.W., 225, *33*, 225, *54*, 227, *33*, 234, *33*
Vicens, Q., 222, *7*, 222, *18*
Viëtor, R.J., 22, *19*, 62, *179*
Vijayan, M., 7, *33*
Virga, K.G., 225, *62*
Vishwakarma, R.A., 181, *112*
Vliegenthart, J.F.G., 5, *16*
Vogel, C., 113, *87*
von Braunmuhl, V., 30, *78*
Vos, J.N., 98, *31*, 100, *31*

W

Wakao, M., 98, *37*, 99, *37*, 114, *37*, 115, *37*
Waksman, S.A., 222, *1*
Walanj, R., 225, *37*
Walker, A., 97, *15*
Walter, F., 222, *18*
Walther, E., 157, *79*
Walvoort, M.T.C., 100, *61*
Wang, B., 43, *113*
Wang, C.-C., 98, *35*, 99, *35*, 117, *35*
Wang, H., 124, *106*
Wang, Q., 141, *33*, 141, *34*, 172, *33*, 172, *34*
Wang, X.-L., 29, *57*, 30, *81*
Wang, Z., 100, *57*, 100, *58*, 100, *62*, 101, *58*, 105, *57*, 122, *98*, 124, *58*, 127, *111*, 130, *116*
Ward, K. Jr., 61, *171*
Warin, V., 30, *77*
Warriner, S.L., 157, *87*
Watanabe, Y., 160, *89*
Wei, A., 103, *71*
Weingart, R., 141, *24*, 151, *24*, 162, *24*
Weishaupt, M., 108, *80*
Weiwer, M., 127, *115*, 130, *115*
Welch, K.T., 225, *62*
Welply, J.K., 6, *28*
Wen, Y.-S., 29, *46*, 98, *35*, 99, *35*, 105, *72*, 117, *35*, 117, *72*
Wertz, J.-L., 22, *20*
Werz, D.B., 96, *6*
Westhof, E., 222, *7*, 222, *18*
Whitfield, D.M., 98, *42*, 99, *42*, 100, *62*
Whittemore, N.A., 225, *62*
Widlanski, T.S., 105, *75*
Widmalm, G., 48, *139*, 55, *139*
Wieninger, S.A., 235, *78*
Wiest, R., 57, *155*
Wilchek, M., 5, *18*
Willart, J.-F., 29, *61*
Willett, J.L., 47, *137*, 51, *137*
Williams, A.F., 138, *1*, 139, *1*, 167, *1*
Williams, D.G., 30, *75*, 37, *75*, 60, *75*
Williams, E., 234, *77*
Williams, M.A., 234, *77*, 242, *86*

Willis, B., 222, *19*
Wimberly, B.T., 222, *10*
Winter, W.T., 23, *23*
Wischnat, R., 122, *97*
Wise, A.R., 141, *32*
Witter, E.D., 223, *23*
Witzig, C., 160, *91*
Wolf, E., 225, *30*
Won, Y., 43, *114*
Wong, C.-H., 98, *40*, 99, *40*, 122, *97*, 122, *101*, 197, *121*, 203, *132*
Woodcock, C., 21, *8*
Woodcock, H.L., 43, *114*
Woods, M., 157, *85*, 157, *80*
Woods, R.J., 43, *113*, 43, *118*, 43, *121*, 45, *121*
Wright, D., 222, *20*
Wright, E., 223, *23*, 224, *28*, 225, *28*, 225, *62*, 226, *28*, 226, *66*, 226, *67*, 227, *66*, 229, *28*, 234, *28*
Wright, G.D., 223, *22*, 225, *31*, 225, *32*, 225, *35*, 225, *39*, 225, *40*, 225, *41*, 225, *42*, 225, *43*, 225, *46*, 225, *50*, 232, *31*
Wright, J.G., 242, *87*
Wu, L.Z., 227, *69*
Wu, Q., 35, *96*
Wu, X., 43, *114*, 185, *113*, 185, *114*, 187, *114*
Wu, Z., 141, *34*, 141, *35*, 153, *74*, 172, *34*, 172, *35*
Wu, Z.L., 127, *107*
Wuhrer, M., 11, *56*
Wuilmart, C., 7, *38*
Wybenga-Groot, L.E., 225, *32*
Wyns, L., 11, *56*

X

Xia, J., 30, *68*
Xing, G.-W., 30, *82*
Xu, H., 127, *114*, 129, *114*
Xu, W., 100, *60*, 115, *60*
Xu, Y., 100, *58*, 101, *58*, 124, *58*, 127, *114*, 127, *115*, 129, *114*, 130, *115*, 130, *116*
Xue, J., 141, *30*, 141, *31*, 166, *30*, 166, *31*, 167, *30*, 167, *31*, 167, *95*, 167, *101*, 170, *102*, 172, *30*, 172, *31*

Y

Yamada, T., 124, *104*, 124, *105*
Yamago, S., 124, *104*, 124, *105*
Yang, B., 100, *58*, 101, *58*, 124, *58*
Yang, D.S.C., 225, *31*, 232, *31*
Yang, F., 205, *137*
Yang, W., 43, *114*
Yang, X.-W., 29, *64*
Yaniv, O., 225, *57*
Yaron, S., 225, *57*
Yashunsky, D.V., 140, *19*, 188, *115*, 193, *115*, 193, *117*
Yates, E.A., 96, *8*
Yates, J.H., 24, *29*
Ye, X.-S., 122, *97*, 124, *106*
Yeung, L.L., 157, *81*, 158, *81*
Yin, Z., 100, *57*, 105, *57*
Yoko-o, T., 200, *127*
Yonath, A., 6, *22*
Yoneda, Y., 30, *70*, 30, *89*, 35, *89*, 37, *89*
Yongye, A.B., 43, *121*, 45, *121*
York, D.M., 43, *114*
Yoshida, J.-I., 124, *104*, 124, *105*
Yoshida, K., 98, *23*, 98, *24*
Yoshizawa, S., 222, *11*, 223, *11*, 224, *11*
Young, P.G., 225, *37*
Yu, B., 100, *60*, 115, *60*
Yu, H.N., 98, *40*, 99, *40*
Yu, K.-B., 29, *57*, 30, *81*
Yu, S.-H., 98, *42*, 99, *42*
Yui, T., 21, *16*, 63, *16*

Z

Zajicek, J., 225, *55*, 225, *58*
Zamecnik, P.C., 4, *1*
Zavialov, A., 11, *56*
Zehavi, U., 4, *7*
Zeng, X., 185, *114*, 187, *114*
Zeng, Y., 100, *62*
Zhang, F., 127, *111*
Zhang, L., 127, *109*
Zhang, L.-H., 29, *64*
Zhang, Q., 127, *115*, 130, *115*
Zhang, Z., 122, *97*

Zhao, J., 29, *64*
Zhao, S., 35, *96*
Zhong, W., 127, *111*
Zhou, X., 127, *115,* 130, *115*
Zhou, Y., 29, *57,* 100, *60,* 115, *60*

Zimmer, B., 30, *87*
Zoppetti, G., 127, *110*
Zuccato, P., 21, *15,* 63, *15*
Zugenmaier, P., 41, *101,* 61, *172,* 62, *176*
Zulueta, M.M.L., 100, *55,* 102, *55,* 119, *55*

SUBJECT INDEX

Note: Page numbers followed by "*f*" indicate figures, "*t*" indicate tables and "*s*" indicate schemes.

A

N-Acetylglucosamine–asparagine, 5
Aminoglycoside (AG)
  APH(3')-IIIa, 223–224
  enzyme/RNA-bound conformations, 223–224, 224*f*
  kanamycins and neomycin, 222–223, 223*f*
  modifying enzymes
    AAC(3)-IIIb and APH(3')-IIIa, 224–225
    crystal structures, AGME, 225
    enzyme–AG complexes, 225
    superimposability, primed and 2-DOS rings, 225
  NMR and X-ray crystallography, 223–224
  primed and unprimed rings, 222–223
Aminoglycoside-modifying enzyme (AGMEs). *See* Protein dynamics
Atoms-in-Molecules (AIM) theory
  application, 59
  bond path and BCP, 57–58, 58*f*
  β-cellobiose, 58, 59*f*
  description, 57
  hydrogen bond, 59–61

B

BCP. *See* Bond critical point (BCP)
Bond critical point (BCP), 57–58, 58*f*
Branched GPI anchors, synthesis
  galactosyl phosphate and Man-I acceptor, 209–210
  pseudoheptasaccharide, orthogonal protection, 210, 210*s*

  retrosynthesis, 207–208, 208*s*
Toxoplasma gondii
  completion of synthesis, 210, 211*s*
  disaccharide, Man-I, 210–211
  retrosynthesis, 208, 209*s*
  trisaccharide, 208, 209*s*

C

Cellulose
  β-cellobiose, 23
  Cellulose Science and Technology, 22
  crystallinity, 22
  crystal packing and intermolecular interactions
    α-cellobiose complex, 37, 38*f*
    cellotetraose hemihydrate, 39
    IFOVOO01 refcode and ZILTUJ refcode structures, 38, 39*f*
    MOVGIN refcode, 37
    YONVUT01 refcode, 39–40, 40*f*
  crystals, 61–68
  description, 20
  energy calculation, 41–57
  fiber crystallography, 21
  D-glucopyranose ring (*see* D-glucopyranose ring)
  hypothesis herein, 22
  linkage geometry
    β-cellobiose, 28, 28*s*
    glycosidic conformation, 31, 32*f*
    hydrogen bonds, 31–33
    oligosaccharides, 28, 29*t*

Cellulose (cont.)
    tetrahydropyran rings, 28
    torsion angle, 31
    trioside crystal, 33
    molecular structure drawings, 70–83
    O-6, orientation
        β-glucose, 37
        distribution, 35, 36f
    parallel/antiparallel packing, 20–21
    polymer shape, 33–35
    stabilizing interactions, detection, 57–61
    twofold screw axis, 21–22
Cellulose crystals
    biomass recalcitrance, 61
    cellulose III$_I$ and Iα, 62
    Debye scattering equation, 65
    description, 61
    fiber-diffraction patterns, 62, 65
    fibrils, 66–67, 66f
    juxtaposed half-diffraction patterns, 67, 67f
    "mini-crystal" approach, 62–63
    models, 63, 64f
    periodic structures, 65
    Scherrer equation, 68
    small-angle scattering, 65
Chemoenzymatic synthesis
    approach, 127–130
    N-deacetylase/N-sulfotransferase (NDST), 125–127
    pentasaccharide, 127–130
Clickable GPI anchors, synthesis
    Alkynyl-GPI and Azido-GPI, reactions, 206, 207s
    click tags, 203, 203f
    completion of synthesis
        Alkynyl-GPI, 205, 205s
        Azido-GPI, 205–206, 206s
    pseudodisaccharides, 204, 204s
    retrosynthetic analysis, 203–204, 204s

E

Energy calculation, cellulose
    AMBER and CHARMM programs, 43
    Boltzmann equation, 41
    disaccharide modeling, 44
    ϕ/ψ mapping
        dielectric constant, 54–55
        intramolecular hydrogen bonds, 54
    higher-energy conformations, 44
    hydroxyl groups orientations, 56–57
    isolated molecules
        approach, 47
        CHARMM MD, 48
    dielectric constant, 47
    ϕ/ψpoint, 45
    HF/6-31G(d) and MM3(96), 49–50, 50f
    methyl α and β-cellobioside, 47–48
    4-thiocellobiose, 48–49
    MM3 and MM4, 43
    models, 41–42
    molecular mechanics, 42
    QM methods, 51–53
    rigid-residue programs, 42

G

D-Glucopyranose ring
    CSD software, 24, 25f
    fiber-diffraction, 27
    PLATON software, 25
    powder-diffraction methods, 25
    puckering parameters, 24, 26f
    pyranosyl rings, 27
    tangled-chain model, 27
D-glucosamine (GlcN), 96
Glycosylphosphatidylinositol (GPI) anchors, synthesis
    branched (*see* Branched GPI anchors, synthesis)
    clickable (*see* Clickable GPI anchors, synthesis)
    description, 138
    development, 140–141
    human CD52 antigen
        lymphocyte (*see* Human lymphocyte CD52 antigen GPI anchor, synthesis)
        sperm (*see* Human sperm CD52 antigen GPI anchor, synthesis)

SUBJECT INDEX

P. falciparum (see Plasmodium falciparum GPI anchor, synthesis)
rat brain Thy-1 (*see* Rat brain Thy-1 GPI anchor, synthesis)
Saccharomyces cerevisiae (*see* Yeast GPI anchor, synthesis)
structure and function
  description, 139–140, 139*f*
  "GPIomics", 140
  retrosynthesis, 140, 141*s*
  tetrasaccharide core, 139–140
T. brucei (*see Trypanosoma brucei* GPI anchor, synthesis)
T. cruzi (*see Trypanosoma cruzi* GPI anchor, synthesis)
unsaturated lipid chains
  azido glucosyl donor and inositol derivative, coupling, 198
  completion of synthesis, 198–199, 199*s*
  description, 195, 195*s*
  inositol derivative, 197, 197*s*
  preparation, trimannoside, 195–196, 196*s*
  pseudodisaccharide, 197–198, 198*s*
  Schmidt glycosylation method, 196–197

H

Heparin/heparan sulfate oligosaccharides
  active–latent strategy
    1,6-anhydro sugars, 117–118, 118*s*
    description, 113
    disaccharide, 115–117, 117*s*
    dodecamer, 113–114, 115*s*
    silyl protecting groups, 114, 116*s*
  chemoenzymatic synthesis, 125–130
  chemoselective glycosylation
    armed–disarmed strategy, 121, 121*s*
    monosaccharide, 122–123, 122*s*
    relative reactivity values (RRVs), 122, 123
  description, 96
  differentiation, 96–97
  GlcN, 96
  polymer-supported synthesis
    carboxylate group, 111–112, 111*s*
    heparin oligomers, 109, 110*s*
    monosaccharide building blocks, 112–113, 112*s*
    MPEG, 109–111, 110*s*
    oligosaccharide, 107–108
  pre-activation-based strategy
    divergent synthesis, 124, 125*s*
    glycosylation, 123–124, 123*s*
  selective-activation method, 118–121
  solution phase
    late-stage oxidation, 107, 109*s*
    linear approach, 105–106
    oligosaccharides, 106–107, 107*s*
    trisaccharides, 106–107, 108*s*
  synthesis
    factors, 98
    L-IdoA, 98–99
    protecting-group, 102–105
    stereochemical control, 100–102, 103*s*
    uronic acid *vs.* pyranoside, 100, 101*s*
Human lymphocyte CD52 antigen GPI anchor, synthesis
  completion of synthesis, 202, 202*s*
  construction, 202–203
  description, 199–200
  preparation, tetramannosyl donor, 201–202, 201*s*
  retrosynthetic analysis, 200*s*, 200–201
  tetramannosyl and pseudodisaccharide, 200*s*, 201
Human sperm CD52 antigen GPI anchor, synthesis
  acylated inositol, Guo group (2003)
    completion of synthesis, 171, 171*s*
    cyclophosphitamidation reaction, 167, 169*s*
    description, 167
    GPI–peptide/glycopeptide conjugates, 171–172, 172*s*
    inositol derivative, 168–170, 170*s*
    Koenigs–Knorr and Suzuki glycosylations, 171–172

Human sperm CD52 antigen GPI anchor, synthesis (cont.)
  phospholipidated pseudodisaccharide, 170–171, 170s
  retrosynthesis, 167, 168s
  trimannose thioglycoside, 168, 169s
  phosphorylation and lipidation, Guo group (2007)
    CAMP factor, 187
    completion of synthesis, 186–187, 187s
    pseudodisaccharide, 185–186, 186s
    retrosynthesis, 185, 186s

I
L-Iduronic acid (L-IdoA)
  monosaccharide precursors, 98–99, 99s
  preparation, 98
Isothermal titration calorimetry (ITC)
  isotherms, AGs-AGME, 227, 228f
  kanamycin A and neomycin, 234
ITC. See Isothermal titration calorimetry (ITC)

L
Lectins
  animal lectins in cell recognition, 12–13
  biological effects, plant lectins, 7–8
  host–microbial recognitions
    Entamoeba histolytica and *Shigella flexneri*, 12
    Escherichia coli, 10–11
    FimH–ligand complexes, 10–11
    innate immunity, 11
    mannose-binding FimH, 11
    methyl α-D-mannoside, 10–11
    soil microorganism, 11–12
  membranes and separations, cells
    diverse biological systems, 8–9
    graft-versus-host disease (GVHD), 10
    growth and differentiation pathological processes, 8
    immune system, 9
    Lotus tetragonolobus, 9
    MHC-peptide complex, 9–10
    PNA receptor, 9
    single glycosyltransferase, 9–10
    transplantation, bone marrow, 10
Ligand recognition. *See* Protein dynamics

M
Monomethyl polyethylene glycol (MPEG), 109–111, 110s

N
N-Deacetylase/N-sulfotransferase (NDST), 125–127

P
Peanut agglutinin (PNA)
  carbohydrate-recognition, 6
  terminal sialic acid residues, 7–8
  thymic cells maturation, 9
PHA. *See* Phytohemagglutinin (PHA)
Phytohemagglutinin (PHA)
  immunocompetence, 7–8
  N-acetylgalactosamine residues, 7–8
  wheat-germ agglutinin (WGA), 7
Plasmodium falciparum GPI anchor, synthesis
  Fraser-Reid group (2005)
    Bender's *de novo* inositol synthesis, 176–177
    completion of synthesis, 175–176, 176s
    description, 172
    mannosyl NPOEs, 173, 174s
    pseudodisaccharide, 174–175, 174s
    pseudopentasaccharide, 175, 175s
    retrosynthesis, 172–173, 173s
  Seeberger group (2005)
    completion of synthesis, 179, 180s
    description, 177
    development, 180–181
    GPI–prion conjugate, 180–181, 181s
    H-phosphonate method, 177
    lipidation and phosphorylation, 179–180
    pseudodisaccharide, 179, 179s
    retrosynthesis, 177, 178s
    tetramannosyl donor, 177–179, 178s

PNA. *See* Peanut agglutinin (PNA)
Polymer shape, cellulose
  characterization, 33–34
  DIGOXN10 and ACCELL10 refcodes, 34–35, 34f
  dixylose component, 34–35
  helical parameters, 35, 36f
  hydrogen bonding, 35
Promiscuity
  AG-dependent solvent protection, 237–238, 239f
  antibiotic binding site, 240–241, 240f
  APH(3')-IIIa, 237–238
  experimental data, 241
  fluid-breathing motions, 237–238
  GNAT superfamily, acetyltransferases, 240–241
  kanamycin A and neomycin, 238–240, 239f
  MgAMPPCP, 238–240
  N-H HSQC spectrum, 237–238
Protein dynamics
  AGMEs and AGs, 222
  AG recognition, 243
  aminocyclitols, 222
  aminoglycoside antibiotics, 222–224
  aminoglycoside-modifying enzymes (AGMEs), 224–225
  APH(3')IIIa and AAC(3)-IIIb, 241–242
  enzyme–AG complexes, 241
  promiscuity, AGMEs, 237–241
  thermodynamics, enzyme–AG complexes, 225–237

Q

Quantum (QM) methods
  COSMO solvation, 51
  HF and B3LYP calculation, 51
  hydroxyl-group orientations, 52
  starting geometries, 52
  vacuum-state, 52

R

Rat brain Thy-1 GPI anchor, synthesis
  Fraser-Reid group (1995)
    completion of synthesis, 155, 156s
    description, 151
    β-GalNAc-Man disaccharide, 153–154, 154s
    hydroxyl groups, differentiation, 154
    inositol acceptor and azido glucosyl bromide, 155
    Man-II acceptor and Man-III donor, 153
    n-pentenyl glycoside (NPG), 155, 156–157
    n-pentenyl trimannoside, 152–153, 153s
    phosphoramidite method, 151–152
    pseudodisaccharide, 154–155, 154s
    pseudoheptasaccharide, 156
    retrosynthesis, 151–152, 152s
  Schmidt group (1999, 2003)
    completion of synthesis, 165–166, 166s
    description, 162
    disaccharide, 163–164
    inositol acceptor, 164–165
    orthogonal amino-group protection, 166
    pentasaccharide, 162–163, 164s
    pseudodisaccharide, 164, 165s
    retrosynthesis, 162, 163s
Relative reactivity values (RRVs), 122

S

SBA. *See* Soybean agglutinin protein (SBA)
Selective-activation method
  glycosylation strategy, 118–119
  hemiacetals and thioglycosides, 120–121, 120s
  trichloroacetimidate donors, 119–120, 119s
Sharon, Nathan
  interest in glycoproteins (*see* Soybean agglutinin protein (SBA))
  interest in lectins (*see* Lectins)
Soybean agglutinin protein (SBA)
  N-acetylglucosamine–asparagine, 5
  affinity chromatography, 6
  antibodies, 5–6
  ECorL lectin and ConA, 6–7
  Erythrina cristagalli, 6–7
  glucosamine, 5

Soybean agglutinin protein (SBA) (cont.)
  and PNA, 6
  wheat-germ agglutinin (WGA), 5–6

T

Thermodynamics
  Gibbs energy changes, AG binding
    aminoglycoside nucleotidyltransferase, 225–227
    binding enthalpies, 227
    binding enthalpy, AGs-AGMEs, 227
    CoASH, 229
    entropic compensation, 227–228
    isotherms binding, 227, 228f
    OH vs. NH_2, 227–228
  heat-capacity changes
    binding, neomycin, 233–234
    crystal structure, enzyme, 232–233
    Cys-156, 232–233
    dynamic properties, 231
    ligand-binding event, 231
    neomycin and kanamycin, 232, 232f
    neomycin and paromomycin, 233–234, 234f
    substrate recognition and discrimination, 231
    superimposed structures, APH, 232–233, 233f
  proton linkage
    AAC(3)-IIIb, 229, 230f
    AAC(3)-IIIb–CoASH–kanamycin A complex, 230
    AG–nucleic acid interactions, 229
    amine groups, neomycin, 230–231
    heats, ionization, 229
    protonation and deprotonations, 230–231
  solvent effects
    AG-recognition process, 235–237
    aminoglycoside-modifying enzymes (AGMEs), 235–237
    enzyme–AG complexes, 235–237
    ITC, 234
    kanamycins and neomycins, 234–235
    nonsuperimposable rings, neomycin and kanamycin, 235
    superimposed structures, 235, 236f
    temperature and antibiotic dependent, 235–237
    temperature dependency, 235–237, 237f
Trypanosoma brucei GPI anchor, synthesis
  Ley group (1998)
    bis(dihydropyrans), 161–162
    completion of synthesis, 160–161, 161s
    pentasaccharide, 158–160, 159s
    pseudodisaccharide, 160, 160s
    retrosynthesis, 157, 158s
    seleno- and thio-glycosides, 157–158, 159s
  Ogawa group (1991)
    azidoglucose acceptor, 143–144
    description, 142
    digalactosyl fluoride, 143, 143s
    digalactosyl group, 145–146
    Garegg's inositol derivative, 144
    glycosylation, 145
    H-phosphonate method, 146
    pseudotetrasaccharide and anhydro sugar, 145–146, 146s
    pseudotrisaccharide, 143–144, 144s
    retrosynthesis, 142–143, 142s
Trypanosoma cruzi GPI anchor, synthesis
  unsaturated lipids, Nikolaev group (2006)
    acetal- and silyl-based protection, 193, 194s
    completion of synthesis, 191–192, 192s
    description, 188–189
    intermediates, reactions, 192
    pseudodisaccharide products, 190, 191, 191s
    retrosynthesis, 189, 189s
    tetramannosyl donor, 189–190, 190s
  Vishwakarma group (2005)
    completion of synthesis, 184–185, 184s
    description, 181
    pseudodisaccharide, 183–184, 183s

retrosynthesis, 181–182, 182s
tetramannosyl donor, 182–183, 183s

Y

Yeast GPI anchor, synthesis
  ceramide phospholipid, 151
  completion of synthesis, 150, 150s
  formation, phosphotriester, 150–151
  Man-I acceptor and Man-II trichloroacetimidate donor, 149
  phosphoramidite method, 147–148
  pseudodisaccharide, 149–150, 149s
  retrosynthesis, 147–148, 147s
  tetramannosyl donor, 148–149, 148s

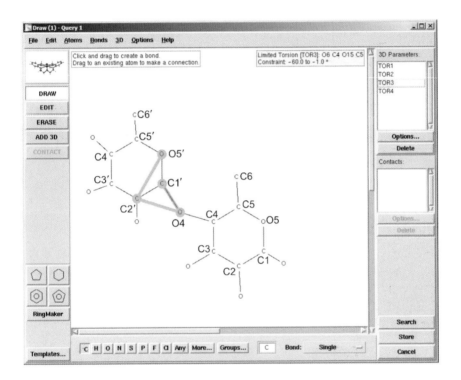

PLATE 1 Screen shot from the query builder in the Conquest module of the CSD software.[31] Atom labels were added. The heavy lines that connect C-1′, O-4, C-2′, and O-5′ define an improper torsion angle used to assure the configuration at C-1′. In the search about to be initiated, the selected structures must have values between −60° and -1° for this improper torsion angle.

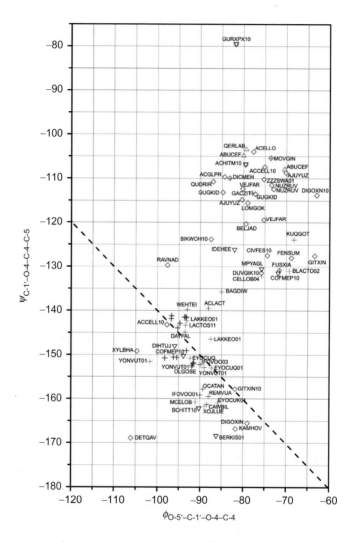

PLATE 2 ϕ/ψ Plot of glycosidic linkage conformations from experimental crystal structures that have β-1→4 linkages and two $^4C_1$ rings. All structures are labeled with their CSD "REFCODE" except for the structures of methyl cellotrioside (TAQYAL) and cellotetraose (ZILTUJ), which are indicated by red-+ symbols. Black + symbols indicate structures having both O-3—H and O-2′—H in equatorial disposition. Inverted blue triangles indicate structures with O-3—H but no O-2′—H, green triangles indicate structures with O-2′—H but no O-3—H group, and magenta diamond symbols indicate structures with either substituted or missing O-2′ and O-3—H groups. Orange diamonds indicate structures with O-3—H but which do not form the O-3—H...O-5′ hydrogen bond. The dashed diagonal line indicates ϕ/ψ combinations that would correspond to twofold screw-axis symmetry if applied to cellulose.

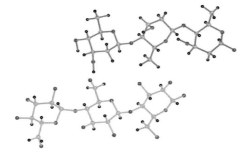

PLATE 3 Similarities between the linkage geometries in DIGOXN10 (upper) and ACCELL10 (lower). The molecular fragments are shifted so that the two linkages giving helical geometries with $n=2.88$ and 2.91 residues per turn are placed over and under each other, while the residues with nearly twofold linkages are on the left (lower) and right (upper) ends. The acetate groups for ACCELL10 are not shown, nor is the remaining part of the digitalis structure on the reducing end of DIGOXN10. In the twofold linkages, the C-1'—H and C-4—H bonds are parallel to each other.

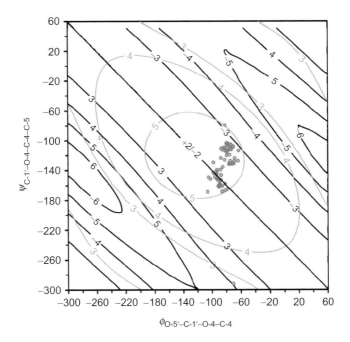

PLATE 4 Conversion of ϕ and ψ values to helical parameters, n and h, based on the geometry of the nonreducing ring in β-cellobiose (CELLOB02)[94] and a glycosidic angle of 115°. The $h=0$ line is not shown, but the contours for $n=5$ and 6 change sign when they cross it. Orange circles indicate the linkage geometries observed experimentally, which are also indicated in Fig. 3. Adapted from Ref. 25.

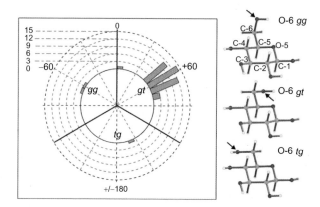

PLATE 5 Distribution of O-6 orientations on rings at the nonreducing ends of cellobiose moieties that have O-3—H groups on the reducing-end ring of the moiety, as found in examples from the CSD. The *gt* orientations correspond to $+60°$, *gg* to $-60°$, and *tg* to $180°$ (the O-6—C-6—C-5—O-5 torsion angle).[37,38] The plot was created with the Vista module of the CCDC software. The O-6 that is oriented to 160.7° (*tg*) is on a galactose residue in the KUQGOT structure (methyl 4-*O*-β-D-galactopyranosyl α-D-mannopyranoside methanol solvate). Also shown are glucose rings with the O-6 atoms in the three orientations as indicated by the arrows. Their secondary hydroxyl groups are oriented counterclockwise.

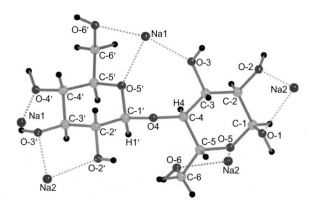

PLATE 6 Structure of the α-cellobiose complex with NaI and H$_2$O (MOVGIN, iodine, and water not shown). The most visible indication of twisting from the twofold screw structure of native cellulose is the difference in orientation of the C-1'—H and C-4—H bonds. In twofold structures, these bonds are parallel to each other.

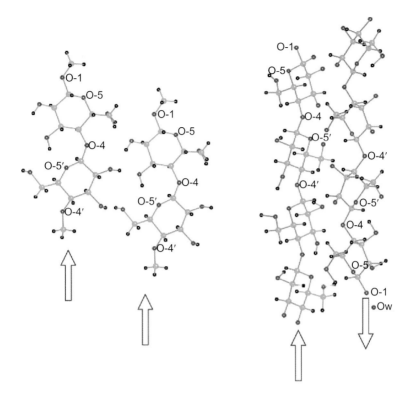

PLATE 7 Molecules from the crystal structures of methyl 4-*O*-methyl-β-D-glucopyranosyl-(1→4)-β-D-glucopyranoside (methyl 4′-*O*-methyl-β-cellobioside, IFOVOO01, left) and cellotetraose hemihydrate (ZILTUJ, right). Arrows indicate the direction of the chains in the respective unit cells (up molecules have O-1 higher than O-4). Some hydrogen atoms were not located during the ZILTUJ structure determination.

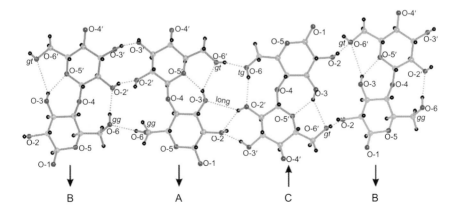

PLATE 8 The carbohydrate component of the crystal structure of cyclohexyl 4′-*O*-cyclohexyl-β-cellobioside cyclohexane solvate (YONVUT01). The three unique molecules in the unit cell (A, B, and C) are shown, with a copy of B to allow visualization of all hydrogen bonds in the particular crystal plane. The O-6 orientations are labeled *gt*, *gg*, and *tg*, and arrows show the molecular orientations in the unit cell. The cyclohexyl components attached at O-1 and O-4′ are not shown. The figure is from Ref. 89.

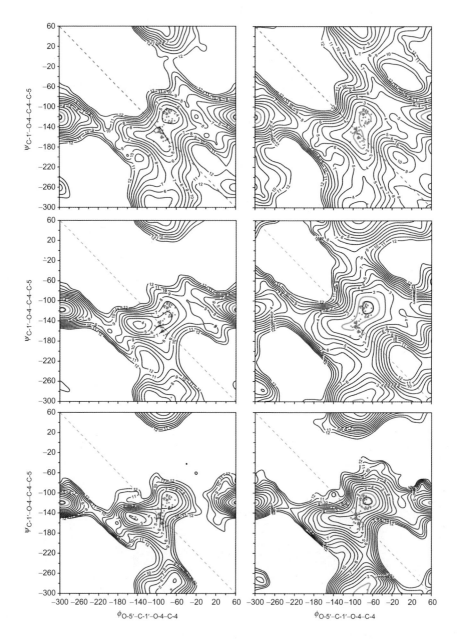

PLATE 9 Energy maps for MM4 (top left: $\varepsilon = 1.5$, top right $\varepsilon = 7.5$), AMBER/Glycam (middle left $\varepsilon = 1.0$, middle right $\varepsilon = 8.0$), and CHARMM/CSFF (bottom left 1.0, bottom right 4.0). Magenta contours are for 0.25 kcal/mol and the blue for 1.0 kcal/mol. Orange dots represent the crystal structures described in Table I and Fig. 3. Calculations by G. P. Johnson and A. D. French.

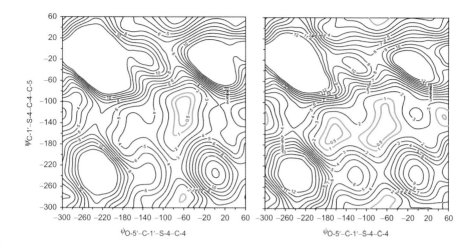

PLATE 10 Energy maps using HF/6-31G(d) (left) and MM3(96) (right) for the analogue of 4-thiocellobiose (tetrahydropyran rings, linked through a sulfur atom). The MM3 calculation utilized parameters generated automatically for the ϕ and ψ torsional energies because parameters for C—O—C—S and O—C—S—C angles had not been part of the official development of MM3. Green contour lines indicate a 0.5-kcal/mol contour. The substantial differences in energies led to the use of a hybrid method for mapping the disaccharide.[140] In that simple, nonintegral hybrid method,[107] the difference between these two maps at each ϕ/ψ point was added to the MM3 map for the disaccharide. Calculations by G. P. Johnson, L. Peterson, A. D. French, and P. Reilly.

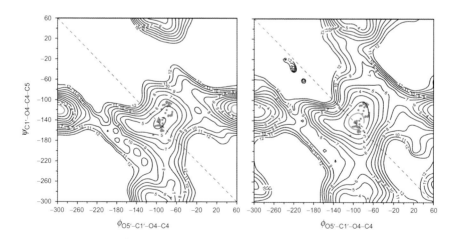

PLATE 11 Energy surfaces for β-cellobiose calculated with geometries from mixed-basis-set energy minimization (see text). The left map is for a vacuum calculation, and the right incorporates SMD solvation. Used with permission from Ref. 148.

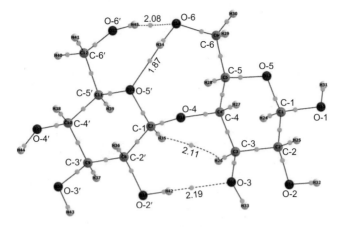

PLATE 12 Molecular graph for β-cellobiose, created with the AIMStudio module of the AIMAll package. The structure having $\phi = -100°$ and $\psi = -300°$ corresponds to the third-lowest minimum, at the bottom of the vacuum map (Fig. 13). With O-6 on the reducing residue in the *tg* orientation and O-6 on the nonreducing residue in the *gt* orientation, it has a relative energy of 4.56 kcal/mol. Distances (Å) for atom–atom noncovalent interactions have been added. The short hydrogen bond (1.87 Å) between O-6 and O-5' merits a solid bond path according to the defaults in the software, whereas weaker interactions are shown with dotted lines. Green dots represent the bond critical points.

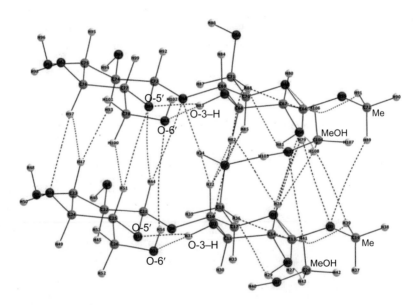

PLATE 13 Atoms-In-Molecules molecular graphs for the methyl β-cellobioside–methanol adduct, showing bond paths (BPs) and the atoms, based on the MCELOB crystal structure.[75] Two complete cellobioside molecules are shown, one above the other, with the reducing-end methyl groups at the right side (Me). The methanol molecules (MeOH) are also shown. The white spheres are hydrogen atoms, the large red spheres are oxygen atoms, and the gray spheres are carbon. Covalent BPs are solid black lines, and weak interactions, including O—H...O and C—H...O hydrogen bonds, are shown as dashed lines. A strong conventional hydrogen bond from the upper MeOH to a lower O-6 is shown as a solid line, and the slightly weaker bifurcated O-3-H...O-5 and ...O-6 bonds are shown as dashed lines. This drawing was produced by AIMStudio from the AIMAll[166] software, based on B3LYP/6-311++G(3df,3pd) quantum calculations.

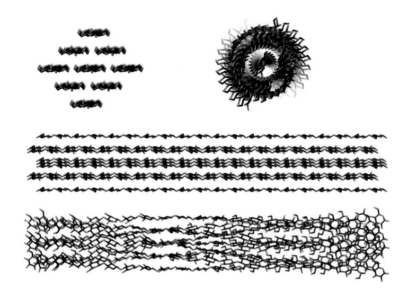

PLATE 14 Models of cellulose crystals with nine chains (3 × 3) before (upper left and center) and after (upper right and bottom) energy minimization with AMBER/Glycam06. Compare the extent of twisting with the lesser amount in the 100-chain model in Fig. 19.

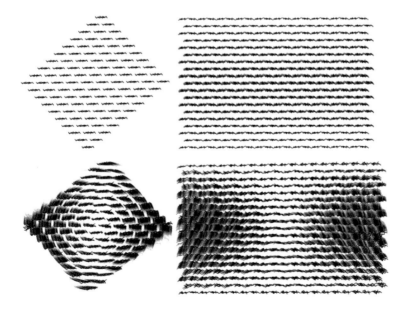

PLATE 15 Models of cellulose crystals with 100 chains (10 × 10) before (upper) and after (lower) energy minimization with AMBER/Glycam06.

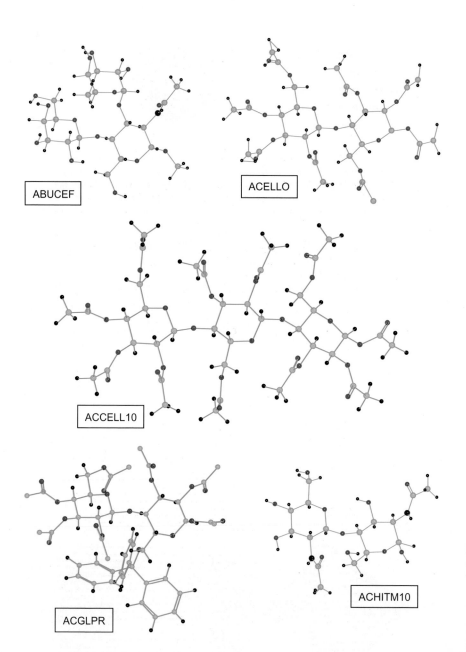

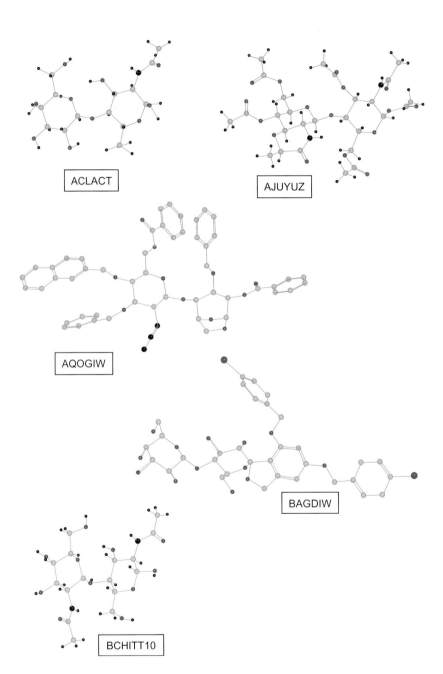

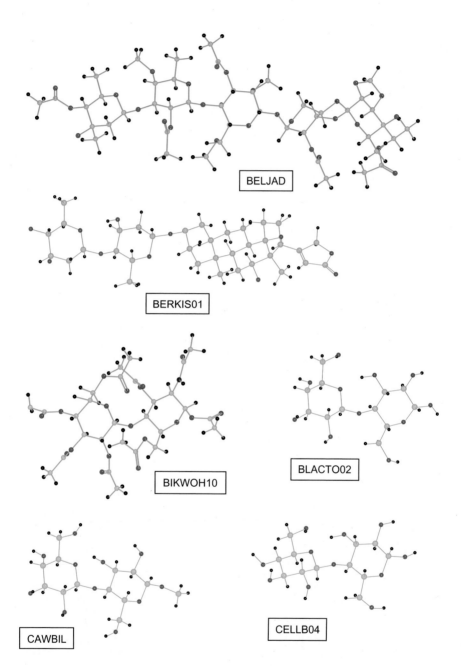

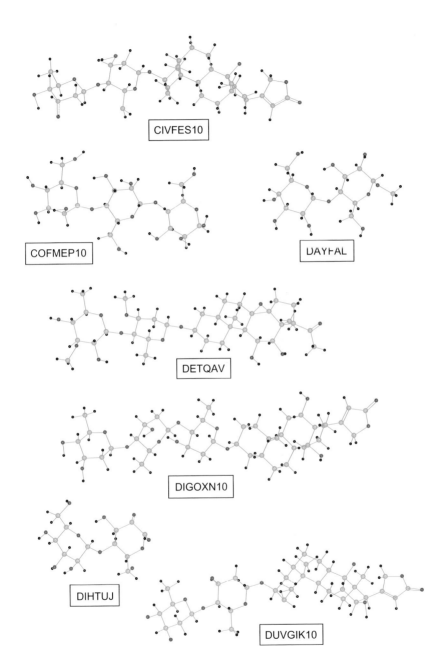

EYOCUQ and EYOCUQ01

FENSUM

FUSXIA

GACZIT

GITXIN10

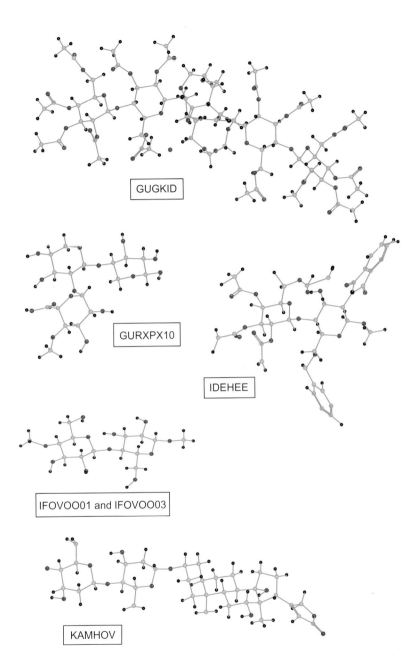

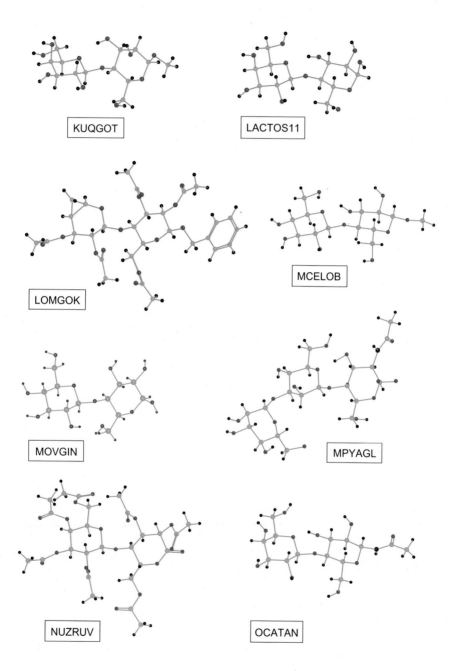

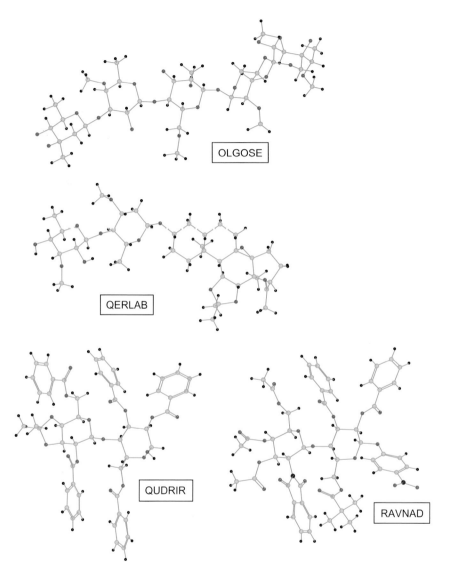

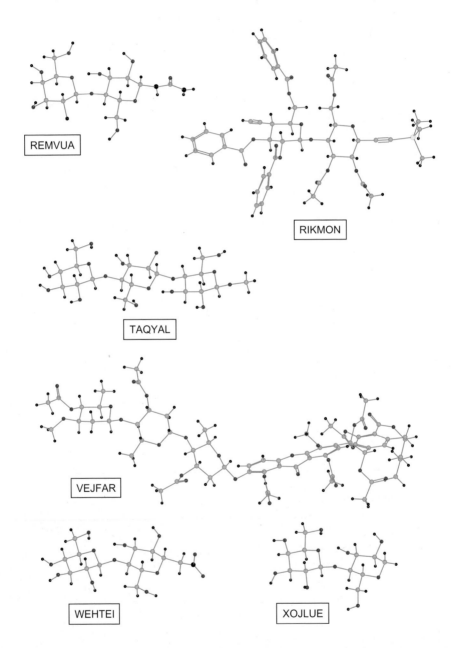

XYLBHA

YONVUT01

ZILTUJ

ZZZSWA01

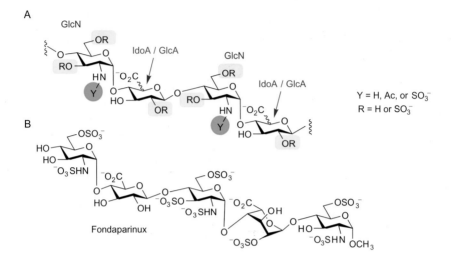

PLATE 16 (A) Structures of heparin/HS; (B) structure of fondaparinux (Arixtra®). (Note idose and iduronic acid are arbitrarily presented in the $^1C_4$ conformation following common usage in the field. This does not necessarily represent the conformations in solution of the various heparin derivatives depicted throughout the article.)

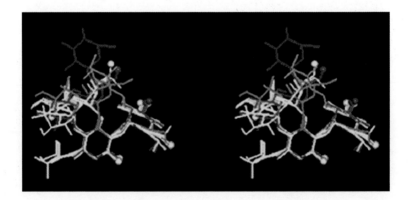

PLATE 17 Stereoview of enzyme/RNA-bound conformations of aminoglycoside antibiotics. In yellow are enzyme-bound AG structures determined with four different enzymes, while blue structures are RNA-bound gentamicin[11] and paromomycin[8] (for simplicity only three rings are shown), all derived via NMR. Red structures are APH(3′)-IIIa-bound kanamycin A and neomycin as determined by X-ray crystallography.[26] All structures are overlaid at the A and B rings. (Reprinted with permission from Ref. 25. Copyright 2002, American Chemical Society.)

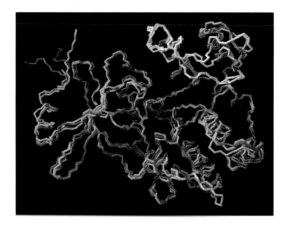

PLATE 17 Superimposed structures of APH(3′)-IIIa in apo (yellow), nucleotide (blue), and nucleotide–AG complexes with neomycin (green) and kanamycin (red).

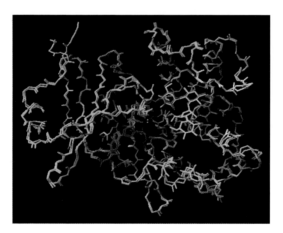

PLATE 18 Superimposed structures of APH(3′)-IIIa–MgADP–neomycin and APH(3′)-IIIa–MgADP–kanamycin A. Neomycin (green) and kanamycin (purple) are on the right side. The purple molecule on the left is MgADP. Yellow highlights those residues showing AG-dependent differences in their correlated motions.

PLATE 19 AG-dependent solvent protection to APH(3′)-IIIa. Amides protected with neomycin only (yellow) and kanamycin only (red) are shown in superimposed structures of the enzyme complexed with both ligands.

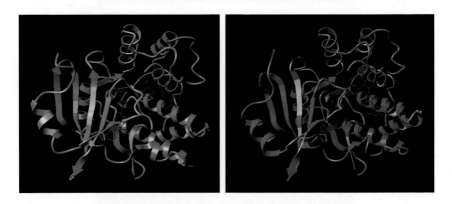

PLATE 20 Residues that become more exposed to solvent (yellow) due to interaction of the nucleotide with APH(3′)-IIIa in complexes of kanamycin (left) and neomycin (right).

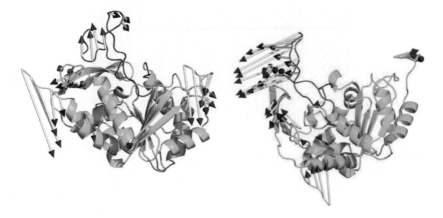

PLATE 21 The flexible, conserved AG-binding loop of AAC(3)-IIIb becomes more flexible upon interaction with CoASH, apoenzyme (left) and enzyme–CoASH complex (right). (Adapted with permission from Ref. 82. Copyright 2011 American Chemical Society.)